Essential Manufacturing

Essential Manufacturing

Gordon Mair
University of Strathclyde
UK

Registered Offices
John Wiley & Sons, Inc., 111 River Street, Hoboken, NJ 07030, USA
John Wiley & Sons Ltd, The Atrium, Southern Gate, Chichester, West Sussex, PO19 8SQ, UK

Editorial Office
The Atrium, Southern Gate, Chichester, West Sussex, PO19 8SQ, UK

For details of our global editorial offices, customer services, and more information about Wiley products visit us at www.wiley.com.

Wiley also publishes its books in a variety of electronic formats and by print-on-demand. Some content that appears in standard print versions of this book may not be available in other formats.

Library of Congress Cataloging-in-Publication Data

Names: Mair, Gordon M., 1949- author.
Title: Essential manufacturing / Gordon Mair, University of Strathclyde.
Description: Hoboken, NJ : John Wiley & Sons, Inc., 2019. | Include
    index. |
Identifiers: LCCN 2018039779 (print) | LCCN 2018041977 (ebook) | ISBN
    9781119061687 (Adobe PDF) | ISBN 9781119061670 (ePub) | ISBN 9781119061663
    (pbk.)
Subjects: LCSH: Manufacturing processes.
Classification: LCC TS183 (ebook) | LCC TS183 .M233 2019 (print) | DDC
    670–dc23
LC record available at https://lccn.loc.gov/2018039779

Cover Design: Wiley
Cover Images: © Monty Rakusen/Getty Images, © Digital Vision/Getty Images

Set in 10/12pt WarnockPro by SPi Global, Chennai, India

Printed and bound by CPI Group (UK) Ltd, Croydon, CR0 4YY

10 9 8 7 6 5 4 3 2 1

Contents

Preface

Essential Manufacturing is the title of this book for two reasons. First, the book's main purpose is to provide the reader with the essential technological, managerial and economic ingredients of manufacturing industry. The second reason is that I believe manufacturing is essential in that it is of crucial importance to the wellbeing of a large nation and its population. Our society benefits from manufacturing industry as it creates wealth. It does this by *adding value* to raw materials through work. This wealth is used to provide services such as health and education and to generally improve the quality of life. The products created by manufacturing industry form the fabric of our civilisation, whether they are used in transport, entertainment, telecommunications, medicine, education or even within the manufacturing industry itself. Unless a country has unique assets such as an abundance of natural resources that are in high demand, then without an efficient manufacturing capability most large developed countries will become relatively poor and experience economic and social decline. It is also significant that the manufacture of a complex product such as a car or aircraft is now a global activity. Thus many countries, by utilising their human resources and depending on their own particular skills, wage rates and quality of work, will all benefit from the production of manufactured products.

A nation therefore needs individuals with the ability to understand how manufacturing industry works, and how to put that knowledge to profitable use. The spectrum of manufacturing today requires knowledge of markets, product and system design, innovation, materials and technologies, logistics and management, manufacturing finance, human factors, environmental concerns and a grasp of the concept of the whole life cycle of a product.

Although much has changed in the manufacturing environment since my earlier book on this subject, the basic manufacturing processes and the importance of manufacturing still exist. However, the Internet has now revolutionised the organisational aspects of manufacturing with companies able to access information rapidly from a wide range of networks and the World Wide Web from anywhere on the globe. This has had an extremely beneficial effect on aspects such as supply chain management and with the emergence of the Industrial Internet of Things (IIoT) a wide range of opportunities are available; for example, the monitoring and control of assembly machines in a UK factory can be carried out in real time at the company headquarters in China or the USA or vice versa. Processes such as casting and metal cutting are unchanged at a basic level but new processes have become mainstream. In particular, the additive manufacturing processes, only made possible by the confluence of digital design, digital communication and digital control, have opened up the possibility of mass personalisation of products. In the product design arena immersive virtual reality and augmented reality have begun to have real practical applications as the initial hyperbole has receded and they have become useful tools, and on the assembly line there has been the introduction of safe humanoid robots that can work alongside human workers. It should also be recognised that the pace of change is such that there are continual improvements in manufacturing processes. Thus although the process specifications included in the book are as correct as could be ascertained at the time of writing the reader should check for current capabilities with equipment suppliers when appropriate.

The book has been written as a result of a conviction that there is a need for an introductory text covering the basic elements found throughout manufacturing. There are many excellent books, some of them quite large that cover in detail the product design process, manufacturing technology or the management of manufacturing. This book is different in that I have attempted to show most facets of the subject in one concise volume thus allowing their interrelationships to be understood. Despite the broad range of the book, each topic is covered in sufficient detail to provide a basic working knowledge. Engineering and business students should understand the importance and content of manufacturing industry. Engineering accreditation bodies expect courses in all disciplines to provide students with a broad background in order to see their topic in context. This often involves an introduction to manufacturing industry. Usually engineering students only acquire knowledge of manufacturing in discrete and isolated topics making it difficult for them to grasp how these topics fit together. This book will therefore provide a holistic appreciation of the topic but with enough detail to be of practical use. As the book is relatively concise it should be capable of being read through from beginning to end providing a complete overview of the significance of manufacturing and its essential elements. However, the specific topics are covered in sufficient depth to have practical applicability in order to satisfy the requirements of college and first and second year university engineering courses. The book should therefore be useful as an introductory text to those beginning courses in manufacturing or other engineering disciplines, and to others who need background knowledge of the subject, for example, those in business studies such as finance, human resource management and marketing. There is no prerequisite knowledge needed for the book as it is at an introductory level.

Part 1 is an introduction. It contains chapters on the significance of manufacturing, its history, an overview of typical manufacturing industries, how to design for manufacture and basic manufacturing principles and elements. These chapters put the subject into economic, social and technological perspective. Part 2 explains the materials used in manufacturing, how they are obtained and their various uses. Part 3 looks in more detail at the basic manufacturing processes to be found throughout industry. This should allow an understanding of how the processes work, their capabilities and relative advantages and disadvantages. This should produce the ability to decide on the fitness of a process for a particular task. Part 4 examines manufacturing automation much of which is microprocessor-based and is applicable in all areas of the manufacturing process. Part 5 introduces manufacturing management by considering financial aspects and the general organisation of manufacturing. Manufacturing companies need to be world class. This stimulates them to rigorously examine their own operations and constantly compare themselves with other companies to ensure that their products stay ahead and their organisation remains dynamic. Since the competition is doing the same a state of constant change arises, hence the need for the management of change, a task recognised as necessary for company survival. Thus concepts such as good supply chain management are seen as essential and phrases such as 'continuous improvement', 'towards excellence' and 'zero defects' have been coined. Techniques have been formalised to assist achievement of these concepts, for example, total quality management, just in time manufacture, lean manufacturing, the focused factory, and simultaneous and concurrent engineering. Part 6 is a combination of management and technology as the nature of quality and quality measurement is presented. Part 7 presents the human considerations needed for a safe and healthy working environment for people that work in manufacturing industry. The book therefore includes the five 'Ms' of manufacture – machines, materials, money, management and the almost archaic term for human resources – manpower. Part 8 concludes by looking at some current trends and having a speculative look at some relevant aspects of the future.

I close this Preface with a personal note. When I started my career in manufacturing as a young apprentice press-toolmaker and designer over half a century ago the world and manufacturing industry was vastly different from today. After my university education and then while working in a number of manufacturing engineering and management roles in a variety of companies, I could see how the rate of change in the industry was accelerating. At the age of 30, I joined academia and for the next 36 years, I observed, and was thrilled to be part of, what seems to be an exponential growth in manufacturing innovation and knowledge. It is an industry that is important, constantly changing and financially and emotionally rewarding. I have enjoyed continuously learning and applying new knowledge, working with interesting and sometimes challenging people and imparting my knowledge and experience to others. In my academic work I have taught, carried out research and formed my own research group and a spin-out company, carried out consultancy work for a wide variety of manufacturing companies, contributed to international committees and participated in conferences around the world. Therefore, based on my own experience, I have found a career in manufacturing exciting, challenging and satisfying and I hope, for you, that this book proves a stimulating introduction to essential manufacturing.

Part I

Introduction

1

Introduction

Most countries need to use the work of their population to create wealth. If they have a significant size of population then for them, a healthy manufacturing industry is an essential ingredient for prosperity. In a general sense, modern manufacturing now covers a broad range of economic activity. As well as including the technology and management of the manufacturing process it also now spans the spectrum from understanding markets through to distribution and services. However, this book will focus on the core aspects of manufacturing and this chapter describes what manufacturing is, considers what is meant by the term 'prosperity', shows why manufacturing is an indispensable creator of wealth and concludes by considering the complex global environment in which it exists.

Around 10 000 BC the world population was about 10 million, by 5000 BC it was 100 million, by 1650 AD 500 million and today, in the early decades of the twenty-first century, it is approximately 7500 million and increasing. Early humans could survive off the land by hunting and gathering but as the population increased so the need for planned agriculture arose to ensure an adequate food supply. However, since the population of just one city such as Shanghai or Karachi is double that of the early world, manufacturing industry has now become an essential factor in creating the wealth necessary to support the world's 7.5 billion people.

Deriving from two Latin words, '*manus*' meaning 'hand', and '*facere*' meaning 'to make', 'manufacturing' is the process whereby materials are changed from one state into another by work. The new state is now worth more than the old, thus *value has been added*. Today the manufacturing process often involves the use of sophisticated machinery and complex organisation. For example, the Airbus A380, see Figure 1.1, is a manufactured product comprised of approximately four million manufactured components. It is the result of nearly 20 years of research, design and development and has its components manufactured by around 1500 companies distributed throughout 30 different countries.

Unless you are reading this somewhere in the countryside or beach, everything around you and on you is manufactured, such as the clothes you are wearing and the seat or floor you are sitting on. In fact, whether or not you are using hardcopy or electronic media it is manufacturing that has allowed you to read this book. Without modern manufacturing techniques many products would not be as affordable as they are today. Domestic labour saving devices like washing machines, fridges and vacuum cleaners are only possible at today's prices because of mass production techniques; the same applies

Essential Manufacturing, First Edition. Gordon Mair.
© 2019 John Wiley & Sons Ltd. Published 2019 by John Wiley & Sons Ltd.

Figure 1.1 The Airbus A380. *Source*: Reproduced with permission of Pixabay.

to home entertainment equipment such televisions and music systems. The motor car, which may contain around 15 000 individual components, is now regarded in developed countries as an essential possession, but unless produced using carefully designed and selected manufacturing methods and materials it would be an unattainable luxury for most people.

As well as the fabric of our way of life being held together by the use of manufactured goods, so are the economics of our society also dependent on manufacturing. The wealth created by manufacturing industry gives employment to individuals and enables a country to pay for services at a national level, such as health and education.

In order to put the rest of the book in perspective some aspects of these statements are now looked at in more detail.

1.1 Wealth and Prosperity

Consider first the economic environment within which we work. The economies we are concerned with are 'market economies', in that they operate on the basis that things, whether they be items or services, are bought and sold to bring benefit to the individual. It works on the basis of *supply and demand*. The price of material, food, products and labour, is determined by a combination of the demand for them and their availability or supply. Another type of economy, one that has not proved successful, is the 'centrally planned economy', in which the state owns the means of production and the means of distribution. In this type of economy the state, rather than market forces, would determine the level of wages and the price of goods.

The standard of living available in a country is determined by that nation's economic wealth. In our type of economy, 'wealth' consists of all things that satisfy human wants. These things can be transferred or exchanged and because they are limited in supply they have value in exchange. By this definition not everything that we need and enjoy is 'wealth' in the economic sense. For example, without fresh air to breathe we would all die but as air is unlimited in supply to all it is not economic wealth. Sunlight is also essential for life but as it is not limited in supply it would not be regarded as part of economic wealth. Thus, by definition if something cannot be bought or sold it does not count in the measure of a nation's economic wealth. However, it is now realised that the essentials of life such as fresh air and sunlight may no longer be regarded as being in unlimited supply forever. This is due to the manner of creation of the economic wealth and the pursuit of increasing standards of living at the expense of damage to our natural environment. Those involved in manufacturing industry therefore have their part to play in ensuring that industry does not contribute to environmental pollution. It also means that the full life cycle of the product has to be considered, including disposal, or re-manufacture of the product or recycling of its materials.

The creation of economic wealth on its own is no longer an adequate gauge of a country's true prosperity and *quality* of life. It is significant that the word 'wealth' comes from the Anglo-Saxon word *'wela'* meaning 'wellbeing', which is the condition of being contented, healthy or successful. It is therefore the resulting quality of life, rather than the material standard of living, that is meaningful. The manner in which the wealth is created, and the equity with which it is distributed throughout the population, contribute to that quality of life. These concepts are not new. In 1862 the scholar John Ruskin wrote the following: 'There is no wealth but life, life including all its powers of love, of joy, and of admiration. That country is richest which nourishes the greatest number of noble and happy human beings'. It is therefore interesting that 128 years later in 1990, the United Nations Human Development Report was launched in which a country's progress towards true wealth was measured by using a Human Development Index (HDI). The HDI index, revised in 2010, is produced annually and is derived from the factors of life expectancy, education and standard of living based on Gross National Income (GNI) per capita, rather than, for example, the overall Gross Domestic Product (GDP). The GDP is one of a number of traditional but important and useful statistics that measures economic wealth and is the total value of all the final goods and services produced annually within the country.

The human perspective of the concept of wealth as well as the purely economic view has been highlighted because throughout the rest of this book we will be focusing on how to increase wealth by making the manufacturing process *financially profitable* but we should also remember that the purpose of financial success is ultimately to improve the *quality, and standard, of life for all*. This includes the manufacturing company employees, the company shareholders and the public in general.

1.2 Manufacturing Industry

In an economy, industry is generally regarded as existing at three levels, *Primary*, *Secondary* and *Tertiary*. Sometimes an additional term *Quaternary* is used to include activities such as consultancy and research and development, however, for our purposes these may simply be considered under tertiary. It is important to realise that the primary and tertiary industries are very much dependent on the secondary manufacturing sector. Estimates suggest that about half of those employed in the primary and tertiary industries depend indirectly on manufacturing industry for their jobs.

Primary industry is the first level and includes agriculture, fishing, forestry and mining; it provides the essentials of food and fuel, plus the raw materials for the secondary industries. Efficient operation of the primary industries is dependent on the availability of manufactured goods. For example, the efficiency of farming owes much to the use of tractors, combine harvesters and other mechanised equipment. Fishing fleets require well-made vessels with reliable navigation and communication equipment. Forestry workers use machines to cultivate the ground for tree planting and to assist with felling and handling mature trees. Cost effective mining relies on the use of modern automated digging, cutting and tunnelling machines. The primary industries rely on manufactured goods to keep themselves competitive in the world market, and to maintain safe and tolerable conditions for their employees. Food is now often a manufactured item when the amount of processing and packaging that is done on it is considered.

Secondary industry is basically the manufacture of goods from the raw materials produced in the first level. Manufacture is the process of changing materials from one state into another; this causes value to be added to the material or product. For example, if iron ore is taken, combined with certain other materials and melted, then steel ingots can be produced by pouring the melt into moulds. This steel is much more valuable than the raw ore because it is easier to put to use. If these ingots are now squeezed in a forging press to produce wheels for railway carriages then once more value is added. If these wheels are now assembled together with other components to form a complete carriage then further value is added. An example of a product with very high value added would be a communications satellite; this requires many expensive labour-years of work in the form of research, development and high-skill manufacturing in relation to the cost of the original raw materials used. Manufacturing companies today also have research and development, design, marketing and finance as part of their structure; these would seem to be tertiary industry activities but this highlights the merging of these industry levels in the modern economy.

Tertiary industry involves services rather than extraction or manufacture. Examples of this type of industry are transport, entertainment, banking, police, health and education. Tertiary industry also relies heavily on manufactured products. Transport services need reliable trains, buses and lorries or trucks to operate satisfactorily. The entertainment industry uses many artefacts for lighting and sound and video recording and storage, and of course there are communication satellites and the television and radio receivers in our living rooms. Banking requires the use of extensive computing facilities and customer terminals. The police use computers, communication equipment and transport. The health service uses much of what has been mentioned already plus pharmaceuticals manufactured to stringent specifications. Education also uses many manufactured items such as computers, books and workshop and laboratory equipment. Section 1.3 uses this book you are reading to illustrate the way in which manufacturing stimulates economic activity.

1.3   Manufacturing as a Stimulant

This may be understood by considering the following. Initially this book was written on a computer. Thus there was a need to purchase the computer, which would first have had to be manufactured. In the computer manufacturing company people would be required to carry out a variety of jobs and a whole range of other manufactured products such as industrial robots for assembly, special purpose equipment for adding integrated circuits to printed circuit boards and so on. In turn, these products would require to have been made by companies each specialising in their own product; for example, industrial robots or integrated circuits. At each of these stages value was being added to a raw material then to the manufactured components through work.

As the book was in progress, communication between the author and the publisher had to take place, hence the need for communication systems that require manufacture and operation. The process of producing the book involves the work of many people including the editor, the reviewers and the people involved in production and printing of the book. If you are reading this in hardcopy rather than electronically, the printing machines also had to be made and the paper and ink produced. Transport facilities had to be used to distribute the book to shops or a direct order distribution warehouse, thus providing further employment. Computers were required to monitor stock levels and raise orders. Throughout the whole process financial personnel would be involved in ensuring that money was available to finance the project and monitor the costs incurred. Sales and marketing personnel would also be involved in promoting the book.

It can therefore be seen that stimulus to economic activity arises from even a relatively simple product like this book. Consider the vastly greater implications of manufacturing more complex products such as motor cars and aircraft; for example, the Airbus 380 noted earlier. The wages earned by all of those involved in the production process, and the profits made by the shareholders, are used to purchase more goods and services, thus further stimulating the economy. Unfortunately, this is all too apparent when working in reverse. For example, if a large shipyard or steel making plant closes down a few thousand people may lose their jobs directly. But the numbers of unemployed do not stop there as there will be the companies that relied on their business for sub-contract work. These companies may have supplied materials and services. Trade in the surrounding community will also be affected as there will no longer be spare money available for the ex-employees to spend in the retail shops and the local entertainment and leisure industries – and so it goes on. These multiplier or 'knock on' effects are well known and verify the validity of the statement that a healthy manufacturing industry is essential for prosperity.

In the twentieth century, the economist Nicholas Kaldor in his short book *Strategic Factors in Economic Development* noted that the growth of GDP is positively related to the growth of the manufacturing sector and other stimulating effects of manufacturing. One of the most important characteristics of manufacturing industry is its inherent ability to generate economic activity and growth without recourse to other sectors of the economy. In fact, manufacturing determines the prosperity of the other sectors. For example, the advent of microprocessor-based automation, in the form of computer controlled machines and industrial robots, made it more economic to produce smaller batches of products and to change quickly from making one type of product to another. This flexibility made it possible to do such things as offering a wide variety of car options to the customer such as colour, trim, engine size, accessories and so on. This perpetuates itself as customers want more and more options, greater freedom of choice and subsequently more frequent releases of new models. This puts more pressure on the car industry to respond quickly to new styles and demands and so it needs to invest more heavily in capital equipment – and so the cycle repeats itself. Also, to ensure that the manufactured products are always competitive there has to be research carried out. This again provides employment and advances the scientific and technical knowledge base of the countries involved.

Thus manufacturing is an extremely effective means of creating wealth and is more efficient than many of the service industries that also add value but rely on funds already generated elsewhere. Thus it is important that nations give priority to work in which wealth is created, rather than simply redistributed or dissipated, otherwise the prosperity of local communities and whole countries decline.

All countries have to trade with each other as no country is entirely self-sufficient in everything its population needs. Therefore, to be able to purchase goods from other countries (imports), without going into debt, it is necessary to sell goods to other countries (exports). Naturally, if the goods sold have a high value added then this makes it possible to purchase many more goods of lower value added without going into debt. Manufactured products provide this high value-added factor, hence the absolute necessity for a country to have a strong manufacturing base to prevent increasing balance of payments problems.

1.4 The Supply Chain

Today, manufacturing is a global business and this is particularly true for complex products such as are found in the automotive and aerospace industries. Manufacturers have to consider the whole process from where and when to purchase their materials and components through to when they provide the finished product to a satisfied customer. This requires 'Supply Chain Management' that takes into account all of the participants in the process of satisfying customer needs. These participants may be located in various parts of the world and so transport and logistics will also be important parts of the equation.

In order to explain this consider a car manufacturer that will usually have as its core competency the design and assembly of its models. The manufacturers of contributing elements such as the tyres, windscreen, electric window motors and the electronic engine control module (ECM) will all be done by other manufacturers who have these products as their core competencies. These companies will require further companies to supply them with the raw materials such as rubber, glass, copper, steel and, in the case of the ECM manufacturer, electronic components. Also, the car manufacturer has to consider the market demand and individual customer requirements and in order to get the product to the customer will use intermediaries, for example, a distributor to take the product to the retailer who will then sell directly to the customer. This all has to be orchestrated to ensure that the *supply of the product matches the demand* for the product. This means that the product has to be of the right quality, price and quantity, and be

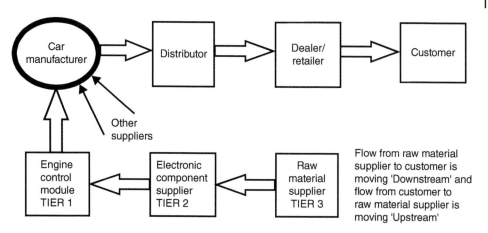

Figure 1.2 A simplified supply chain for one component contributing to a finished car assembly.

available at the correct time and place in order to have a satisfied customer. A simplified example for one component in a car is shown in Figure 1.2. However, it is evident that there will be thousands of individual components contributing to the finished car all of which will have to be carefully managed.

As the supply chain may span the globe with suppliers, retailers and customers widespread so the need for real-time data collection and information is essential as is the need for good communication between components of the supply chain. With complex products, the chain may also resemble a network with various suppliers needing to coordinate supplies to the final manufacturer and also integrate with logistics companies for on time deliveries. Also anticipating changes in market demand, costs of materials, environmental and legal issues, technological developments and political developments are important. All of this real-time activity is facilitated by the Internet. As well as the electronic transfer of information through Internet-based software, emails and telephones, the Internet also provides the platform for the World Wide Web, which is an essential element for data gathering and distributing information. Additionally, 'cloud computing' facilities allow storage of, and access to, large amounts of data that can be held securely yet also be made available to all necessary elements of the supply chain. The physical structures for this that include fibre-optic and copper cables, short- and long-range wireless systems, communication satellites and, of course computers, are constantly being improved and extended.

1.5 Conclusion

This chapter has attempted to explain what wealth and prosperity are, and how a healthy manufacturing industry is essential for creation of that wealth and prosperity. It has also tried to indicate the breadth of modern manufacturing and how a modern manufacturing company relies on close relationships with suppliers and customers all of who may be spread around the world. Chapter 2 now considers how manufacturing industry has developed over the years to reach its present level of importance.

Review Questions

1 What is a Market Economy?

2 How can the wealth of a country be measured?

3 Describe the three levels of industry present in an economy and discuss their composition.

4 Explain what you understand by the term 'value added'.

5 In your own words discuss why a healthy manufacturing industry is essential for a successful economy.

6 Why is it important for a modern manufacturing company to fully understand its own supply chain?

2

Manufacturing History

2.1 Toolmaking Humans

In a classic science fiction film from half a century ago, '2001 A Space Odyssey', there is a short scene that captures the essence of the following discussion. It occurs when what may be an early hominid, specifically a hominim, throws a bone into the air in triumph after discovering its usefulness as a weapon. As it spins upwards, it is transformed by the camera into a rotating toroidal space station 'waltzing' to the strains of 'The Blue Danube'. Whether or not the ape-like creature is a valid representation of early man the imagery is powerful. From the first use of simple primitive tools for weapons, humans have developed the manufacturing technology and culture of today. In fact, some sources suggest the word 'tool' comes from the Old Norse 'tol' meaning 'weapon'.

Although this scene is poetically inspiring, it also highlights the fact that humans are unique in the animal kingdom in their ability to *design and make tools*. Humans progressed from the selective phase of simply picking a natural object from the environment to use as a tool, just as a chimpanzee selects a twig to poke into anthills to catch ants or a fish uses a stone to crack a snail shell. Humans also passed through the adaptive phase where a stone was chipped to create a sharp cutting edge. Now the inventive phase has been reached where an object not found in nature is conceived and manufactured; for example, a bow and arrow or a supersonic aircraft.

An early manlike creature, *Australopithecus*, is thought to have made a simple cutting and chipping tool by striking the edge of one stone against another to produce a sharp edge. Since these creatures lacked dexterity the tool would have been held crudely, with the thick blunt end pressed against the palm while being used to cut meat, see Figure 2.1. Australopithecus lived from about four million years ago in the open in Africa and they probably cooperated in hunting. However, stone tools are not likely to have been generally used until around two or three million years ago.

Peking man, a form of *Homo erectus*, possibly lived as long as 750 000 years ago. They were hunters and travellers but built camps and primitive-huts for shelter. Many stone 'chopper' type tools of varying size and shape have been found from this period, though they do not differ much from those used during the previous two million years. There is evidence, however, that fire was used to harden the tips of sharpened wooden tools and weapons.

Between 70 000 and 40 000 years ago there lived Neanderthal man, an early member of *Homo sapiens*. The average size of his brain was probably slightly larger than

Essential Manufacturing, First Edition. Gordon Mair.
© 2019 John Wiley & Sons Ltd. Published 2019 by John Wiley & Sons Ltd.

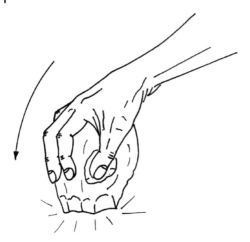

Figure 2.1 How one of the first stone tools may have been used.

modern humans and there is disputed evidence that he had a primitive religion. He was probably a well-organised hunter working in groups to ambush animals or drive them over cliffs. The tools and weapons he used were still of fire hardened wood and shaped stone, although they were now more elegant in design.

2.2 The New Stone Age

Modern man, Homo sapiens, appears just as Neanderthal man becomes extinct though they may have coexisted for a period. The use of stone tools continued into the New Stone Age. This was the Neolithic period that lasted in south west Asia from about 9000 to 6000 BC, and in Europe from about 4000 to 2400 BC. Europe developed later as the last ice age finished there only 10 000 years ago. During the Neolithic period, man made the change from a hunter gatherer to a farmer. Crop growing and cattle rearing were started. Tools had now become quite sophisticated although still made of wood, bone and types of stone such as flint. The flint tools were often the master tools from which others could be made. Stone headed axes with bone handles, and knives and arrow heads of flint were common. Wooden hoes and sickles, originally made from setting flints into straight or curved bones, were used for cultivation. Clay pots were made and clothing was woven using a primitive loom. The sling and the bow and arrow were early inventions and, by the end of this period, man had learned to make as well as use fire and had produced the lever, the wedge and the wheel. Since Neolithic society was relatively peaceful most weapons were probably used for hunting.

At this stage we can see the beginning of the primary industries. With the advent of agriculture man would be able to produce more food than his immediate community needed. This would create the opportunity of *trading* some of his food for other items, perhaps dried fish from a tribe that lived by the river or sea. The use of appropriate tools would allow him to increase the amount of food grown and so this would stimulate him to make improvements in these tools. Some tribes may have lived near sources of flint. They would become adept at fashioning tools and a 'manufacturing' industry would arise. These tools would be traded with the fishers and farmers. The toolmakers were adding value to the flint by working it into shape, the fishermen adding value by drying the fish to prevent it from spoiling and the farmers would be adding value to their produce by harvesting and transporting it to where it could be traded. These principles are unchanged today, except of course money is now used to facilitate the sale and purchase of labour and goods.

In the following sections, it should be remembered that the dates stated for both the Bronze and Iron Ages are generalisations since their start points depended on geographical location.

2.3   The Bronze Age

Technological progress does not really begin to accelerate until about 5000 BC when we begin to move out of prehistory. The first urban civilisations existed along fertile riverbanks. These were the Sumerians in the valleys of the Tigris and Euphrates rivers in Mesopotamia, the Egyptians along the Nile, the Indus Valley civilisation in India and the Chinese along the banks of the Yellow river. It is around this time that metal first began to be worked and it was metal that allowed manufacturing industry to have such an influence on society.

Originally copper was used; it was probably obtained in its pure form and beaten into shape. Relatively soft in this state, it has limited use for tools and was used mostly for ornaments. However, it was soon found that when melted it could be alloyed with tin to produce a much stronger and harder material – bronze. The Bronze Age lasted from its beginnings with the Egyptians and Sumerians around 3500 BC to the start of the Iron Age around 1200 BC. Europe was about 1000 years behind Asia Minor in using bronze, with the Bronze Age in Britain lasting roughly from 2000 to 500 BC.

Pure copper became scarce in supply, therefore it was necessary to obtain or 'win' the metal by smelting copper rich ores. Excavations have been made of copper smelting furnaces built into the ground in beds of sand. Stones of varying size were used to support the internal walls of the structure and burning charcoal was used as a source of heat. The furnace pit was lined with mortar and had clay tubes set into the side for bellows' access. The bellows provided a forced draught to increase the furnace temperature. About 10 cm above the base of the furnace there was a tapping hole that led downwards into a tapping pit. As the ore was smelted the impurities would rise to the top of the melt and eventually be drawn off through the tapping hole. The pure copper would be left in the bottom of the furnace. Later it could be refined and alloyed by re-melting in crucibles. Objects were made by casting in moulds and by beating the material into shape. Hammering of the metal also improved its hardness, enabling it to be used for items such as nails.

The Sumerians and Egyptians used war captives as slave labour. This allowed some of their own workers to have the time to develop their skills as craftsmen in the manufacture of bronze artefacts. Ploughs, hoes and axes for farming; saw blades, chisels and bradawls for carpentry; helmets, spearheads and daggers for war and ornaments are just some of the items that were made out of bronze. Even after the use of iron became widespread bronze remained popular. Experience gained in alloying bronze with elements such as zinc and antimony ensured its continuing popularity.

As civilisation spread throughout the Middle East craftsmen such as metalworkers, potters, builders, carpenters, masons, tanners, weavers and dyers were all to be found. The more skilled artisans, for reasons of economy and supply, lived in the larger towns and cities. They usually lived in their own identifiable areas and later, in the Middle Ages, they would organise into craft guilds that became powerful political groups.

2.4 The Iron Age

Iron, which is harder and tougher than bronze, was first used in its naturally occurring form. Before smelting was feasible the original source was fallen meteorites, the ancient Egyptians calling it 'the metal from heaven'. The iron would be broken off, heated, then hammered into shape. The Hittites, who occupied the same area as the Sumerians had many years before, were the main workers in iron from about 1200 BC. It is likely that they smelted their iron from bog ore. This ore occurs freely in marshy ground and can be obtained by simply sieving it out of the water. The ore is then smelted and the resulting iron 'bloom' hammered to remove the impurities, termed 'slag', before quenching in water. After repeating the heating, hammering and quenching process several times an iron of useful strength is obtained. However, widespread use of iron did not occur until the Hittite Empire was destroyed and the ironsmiths dispersed along with their skills. The design and use of tools in general was continued in parallel with metallurgical developments. For example, primitive manually powered wooden lathes were in use around 700 BC for making items such as food bowls.

The use of iron was longer in developing than that of copper due to the difference in melting temperatures. Copper melts at 1083°C and iron at 1539°C, thus requiring much hotter and more efficient furnaces for smelting and casting. Eventually a primitive form of steel was produced by hammering charcoal into heated iron. The first iron ploughshare possibly came from Palestine around 1100 BC and by the time of Christ the use of iron was widespread.

In Europe, iron mining and smelting grew up in areas where the ore could be mined and there was a plentiful supply of wood from local forests. One of the main reasons for the Roman invasion of Britain in 55 AD was the attractiveness of the iron and tin mines. Iron was used for farm implements, armour, horseshoes, hand tools and a wide variety of weapons. Iron wrought into beams had also been used in structural work by both the Greeks and the Romans. With an increasing population, the need for manufactured items was also increasing. This meant that there was now a need for the 'mass production' of some products. The Romans, for example, had 'factories' for mass producing glassware but these were very labour intensive and had low production rates compared with today's expectations.

We now take a time jump through the Dark and Middle Ages to around the beginning of the Renaissance period. The quality of iron had remained relatively poor and it was not until about the fifteenth century that furnaces using water powered bellows were sufficiently hot to smelt iron satisfactorily. This allowed better control of the material properties of the finished castings. In cannon and mortar manufacture, for example, the iron had previously been so brittle that the cannon often blew up in the users' faces. The higher quality iron that could now be produced prevented this. In the sixteenth century water power was also used for other metalworking processes and rolling mills were developed for producing the metal strips from which coins could be stamped.

As well as technology for manufacture, organisation of manpower is important and progress also occurred here. Leonardo da Vinci, who lived between 1452 and 1519, apparently found time to turn his attention to the manner in which work is carried out. He implemented a study of how a man shovelled earth and worked out the time it should take to move a specified amount of material using the observed method and size of shovel. This was an early example of the scientific analysis of work, which is still important to the efficiency of modern manufacturing systems. Although Leonardo da Vinci and many others of this period were part of the great re-birth of interest in the arts and scientific discovery, it is not until the beginning of the eighteenth century that we find the manufacturing and engineering discoveries occurring that so changed the quality of life for the ordinary person.

The beginnings of the factory system were appearing in Europe during the seventeenth century in the woollen trade and cloth making. The 'factory system' is typified by the 'specialisation', or 'division', of labour. This entails individual workers specialising in one aspect of the total work required to complete a product. The 'commission system' had also arisen during this period. A 'commission merchant' was someone who bought a product from a manufacturer and then sold that product to an end user. For instance, he would buy wool from a sheep farmer and pay for it to be delivered to the wool cleaner. He would pay the cleaner, then pay for transport to the comber and pay him. This was repeated through the subsequent stages of carding, spinning and weaving. The commission merchant then paid for transport of the finished cloth to the market, either at home or abroad. The merchant financed his operation from borrowing money from a 'capitalist'. Capitalists' wealth grew considerably from this type of investment. Initially, these woolworkers were unlikely to be all together in the one 'factory' and were probably operating as individual 'outworkers' in their own homes. The division of labour could also be seen in woodworking, rather than general woodworkers we find carpenters, wheelwrights and cabinet makers beginning to appear.

It should be remembered that agriculture was still the main source of employment at this time. For example, it has been estimated that in England in 1688 there were 4.6 million people employed in agriculture and only 246 000 in manufacturing. Between 1750 and 1850 the population of Britain trebled, yet it remained self-sufficient in food, even though the number of workers in farming had dropped to around 1.5 million by 1841. This was due to the 'agricultural revolution', which was caused by new techniques such as planned crop selection and rotation, scientific animal breeding and the development of new farming tools. Probably the most important invention of this time was the mechanical seed drill, devised by Jethro Tull around 1701, which increased farming productivity. This was a good example of primary industry benefiting from a manufactured, or secondary industry, product. The advent of mechanisation on the farm would eventually lead to massive reductions in the numbers of labourers required; these were to provide a pool of workers for the blossoming Industrial Revolution.

2.5   The Industrial Revolution

A number of factors brought about what is sometimes called today the 'first' Industrial Revolution. It is generally accepted as having been started in Britain during the course of the eighteenth century. One of these factors, the division of labour within a factory system of working, already existed. Three other factors that were to contribute were cheap and efficient power in the form of steam engines, large supplies of good quality iron with which to make these engines and other products and the development of transport systems. The transport systems initially appeared due to the economic stimuli of the other factors but they also in themselves assisted further industrial development. A growing canal network and rail system facilitated trade and transport domestically, and increasing sea transport provided the essential ability to sell manufactured goods overseas.

Iron had always been produced in relatively small quantities, but in 1711 in Coalbrookdale, England, Abraham Darby began to use coke to smelt large quantities of ore. He produced good quality cast iron that was suitable for forging as well as for casting. Soon iron was to replace copper, bronze and brass for items such as pots and pans and other domestic goods.

Manufacturing industry at this time still relied on water power, even although Thomas Savery had installed a steam engine to raise water from a mine in 1698 and an 'atmospheric' engine had been developed by Thomas Newcomen and John Calley in 1705. Many of the improvements in manufacturing technology during this period were concerned with the textile industry. In 1733 water powered looms were made possible by the invention of the 'flying shuttle' by James Kay and in 1770 the 'Spinning Jenny' was patented by James Hargreaves. Replacing the spinning wheel, the Jenny was a frame containing a number of parallel spindles made to revolve by the use of a hand powered wheel and pulley belt. Where previously only one thread would have been spun by hand, eight could now be spun and this was later improved to 20 or 30.

The Industrial Revolution really got underway in 1776 when James Watt produced a much improved version of the steam engine. This was made a commercial possibility by the use of a boring machine made by John Wilkinson to produce the cylinders for the engine. Despite this, up until the beginning of the nineteenth century the relatively cheap Savery and Newcomen engines were often used in preference to the more expensive Watt engines. These were used on their own or as a supplement to water power.

Also at this time, the basis of modern economics was laid down by Adam Smith in his book *The Wealth of Nations*. In this he further developed the concept of the division of labour to improve productivity and stimulate economic progress. The general adoption of this principle contributed to the success of the Industrial Revolution.

In 1784, Edmund Cartwright built a loom capable of being driven by steam power, and in 1788 James Watt devised the centrifugal governor for automatically controlling the speed of a steam engine. Thus we have the beginnings of manufacturing automation. Looms had originally been located close to hills to make use of water power and the same type of equipment as corn mills, hence the name 'mills'. With the advent of steam power they were relocated to areas where coal was plentiful.

Since the source of power for the early machines was a large, relatively expensive steam engine, it was natural that all machinery driven by the engine should be located as close to it as possible. This also applied to associated activities. It was therefore now convenient for all workers to work within a short distance of each other and probably under the same roof. This was one factor that contributed to the rise of the factory. The other was social. People had been used to working to agricultural rhythms, but the new systems of work demanded tight organisation with workers expected to be available at specific times. Thus the time was ripe for the factory due to a combination of technological and social and organisational reasons. The factory was to peak in the early to mid-twentieth century with some employing around 70 000 people.

In 1794 in the UK Henry Maudsley created the first all metal screwcutting lathe (see Figure 2.2). Meanwhile, in the USA in 1792, Eli Whitney had produced the cotton gin that greatly increased the productivity of the southern cotton workers, who at that time were predominantly slaves. In 1796 in Britain Joseph Bramah developed the hydraulic press that provided a means of greatly increasing the force a manual worker was able to exert. In 1804 in France, we see Joseph Marie Jacquard introducing automatic control of a loom by the use of punched cards, see Figure 2.3. The use of punched cards to contain information for computers and automatic machine control was still being used more than 160 years later.

We have already seen the emergence of two of the main ingredients of modern manufacturing, that is, the 'factory' as a central manufacturing unit and specialisation of labour. At this point, at the beginning of the nineteenth century, we see the appearance of three further major concepts; that is, *mass production, standardisation* and *interchangeability*.

One of the first instances of the manufacture of interchangeable components was in the USA in 1798 when Eli Whitney used filing jigs to produce parts for a contract of 10 000 army muskets. Although the parts were almost identical, their manufacture did involve much manual labour in the filing to size of the components.

The manufacture of components using mechanised mass production techniques first occurred in 1803 in the Portsmouth Naval Dockyard in Britain. The components were pulley blocks for the Royal Navy. The machines for producing the blocks had

Figure 2.2 Maudsley's all metal screwcutting lathe 1794.

Figure 2.3 Jacquard punch card controlled loom 1804.

been designed by the famous engineer Marc Isambard Brunel and manufactured by Henry Maudsley. There were 45 machines and they were powered by two steam engines. Eventually, productivity was so improved that 10 unskilled workers could produce as many blocks as had previously been produced by 110 craftsmen. By 1808 the production rate had risen to 130 000 per year.

In the USA, two of the innovators in using machinery to make the close-toleranced components necessary for interchangeability were Samuel Colt and his chief engineer Elisha Root. In the Colt armoury in Connecticut they designed a turret lathe to facilitate the production of identical turned parts for the Colt revolver produced in 1835.

To achieve interchangeability of components not produced in the same factory, methods of *standardisation* had to be adopted for commonly used parts. One of the pioneers in this area was the British engineer Joseph Whitworth. Around 1830 he began developing precision measuring techniques and produced a measuring machine that utilised a large micrometre screw. These accurate measuring techniques and equipment were necessary for the inspection of components intended to be interchangeable. In 1841 he proposed a standard form and series of standard sizes for screw threads that were eventually adopted in Britain and although other countries began to adopt their own national standards these were not necessarily compatible on a world-wide basis. Even today, much effort still goes into achieving international standards for manufactured items.

In the textile industry, which was the first industry to experience the effects of the Industrial Revolution on a large scale, workers began to be displaced from their jobs by encroaching mechanisation. This was resented so much that the Luddite movement was formed in England in 1811. The Luddites' purpose being to riot against the machinery, they caused considerable damage to equipment but refrained from violence against people. Repressive measures, including the shooting of a band of Luddites in 1812, did not put an end to the movement. They appeared in waves in various textile manufacturing areas. However, as the prosperity of individuals continued to increase, so resistance to the machinery died out.

Greater possibilities for the improvement of working conditions in Britain were made possible by the repealing of the Combination Acts in 1824. This repeal allowed Trade Unions (combinations of workers) to be formed, thus providing the opportunity for organised resistance to exploitation. In the years 1833–1840, Industrial Commissioners were appointed to investigate poor social conditions. These measures were necessary as there were many major problems to overcome.

Large numbers of people were now coming to work in factories in close proximity to each other. This meant that when working a then normal 12-hour day, they had to live near to the factory. Thus factory towns and cities increased in size and population density. Congested housing, often in the type of houses known as tenements, forced men, women and children to live under very poor conditions. Rooms had poor lighting and ventilation, and were cramped and lacked privacy. There was also a 'ticket' system in operation for many workers. Under this system workers received tickets for their labour in lieu of money. To translate these tickets into the essentials of food and clothing, they had to be redeemed at the company store. Thus the company had a monopoly and was able to charge the highest prices for the poorest goods.

In the factories themselves, where many women and children were employed, working conditions were often poor. Low wages and long hours caused accidents, inefficient working and poor health. The 'piecework' system also caused hardship. Under this system, workers in their own homes, and eventually in what became known as 'sweatshops', worked on jobs sent out from the factories. They were payed per 'piece' produced and as they were forced to work at very low rates working conditions were often even worse than in the factories.

Considering the engineering aspects again, simple steam locomotives had been doing useful work in coalmines between 1813 and 1820, but it was not until 1825 that the first public railway was opened. This was between Stockton and Darlington in England. Improvements in manufacturing equipment and techniques were leading to stronger boilers and better fitting cylinders and pistons. Thus as the efficiency and power of the steam engines improved, so also did their ability to transport materials, products and people over greater distances at higher speeds. By 1848 there were 5000 miles of railway track laid in Britain. The first successful steam ship had been built on the banks of the river Clyde in Scotland in 1812. This was the 'Comet' and it was the forerunner of ships such as the 'Sirius' and the 'Great Western' that began regular transatlantic service in 1838.

In his book *Days at the Factories* published in 1843, George Dodd remarked:

> The bulk of the inhabitants of a great city, such as London, have very indistinct notions of the means whereby the necessaries, the comforts, or the luxuries of life are furnished. The simple fact that he who has money can command every variety of exchangeable produce seems to act as a veil which hides the producer from the consumer

And this is rather similar to today. Dodd differentiates between, 'mere handicrafts carried on by individuals of whom each makes the complete article', and the work that involves, 'the investment of capital and the division of labour incident to a factory'. He then goes on to describe the operation of various factories in the London of the early to mid-nineteenth century: these were apparently well organised, with relatively sophisticated cost control systems, machines utilising steam power and a variety of skilled, semi-skilled and unskilled workers.

The confluence of improvements in iron and steel production, the development of precision machinery, mechanised processes, division of labour, mass production techniques, interchangeability of components, standardisation, the application of steam to transportation giving access to many new markets and the beginnings of improvements in factory conditions, led into what has been termed the 'later factory period', beginning around 1850, see Figure 2.4.

The acceleration in industrial development during the second half of the nineteenth century can be seen by considering the production of iron and the expansion of the rail network in Britain. The annual production of pig iron had risen from just over 17 000 tons in 1740 to 1.25 million tons in 1840, and by 1906 it had reached more than 10 million tons. In 1850 there were 6500 miles of railway in England; by 1900 this had risen to over 25 000. Working conditions were improving and by 1872 in Britain a basic nine-hour

Figure 2.4 The pen-nib slitting room in the Hinks, Wells & Co. factory, Birmingham UK around 1850. Note the specialisation of labour.

Figure 2.5 Spencer's automatic lathe 1873. Source: Photo Courtesy of the Birmingham Museums Trust.

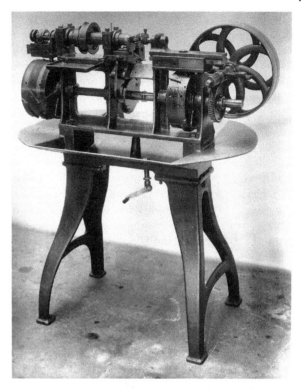

working day had been agreed, although the working week was of six days. Manufacturing equipment continued to improve in quality and capability. For example, in 1873 a fully automatic lathe was produced by Christopher Spencer, see Figure 2.5. The movements of the machine elements were controlled by drum cams, or 'brain wheels' as they were called, forerunners of today's computer control systems. By 1880 electric motors were being used to drive some machines. Coupled with the advent of the internal combustion engine at the end of the century, this meant that factories and other power reliant establishments would eventually be able to spread out and away from a central steam power source.

2.6   The Twentieth Century

Many of the advances made in science, medicine and engineering have been possible due to parallel advances in manufacturing technology. The understanding of how best to organise and manage people at work has also increased. Towards the end of the nineteenth and the beginning of the twentieth century, great progress was made in the realm of 'scientific management'. This was concerned with finding the best ways of doing work and became known as 'work study'; it was composed of 'work measurement' and 'method study'. As well as Leonardo da Vinci mentioned earlier, contributions had been made in this field by others. For example, around 1750 in France Jean R. Perronet carried out a detailed study of the method of manufacture of pins. Also Charles Babbage, the British inventor of the mechanical digital computer, published a book in 1832 titled

On the Economy of Machinery and Manufactures, in which he considered how best to organise people for improved productivity. However, widely recognised as the major contributors to work study are Frederick Winslow Taylor and Frank and Lillian Gilbreth at the end of the nineteenth and beginning of the twentieth centuries. Their work has provided the basis for modern work organisation, which is described more fully later.

Whereas in the eighteenth and nineteenth centuries the textile industry had been a spur to, and a beneficiary of, much of the advances in manufacturing technology, at the beginning of the twentieth century this role was taken over by the motor car industry.

One of the earliest examples of an assembly line was in the Olds Motor Works in Detroit, USA. In 'line production' the division of labour is fully utilised in that work is transported from worker to worker in a linear fashion, thus reducing work in progress and increasing productivity; in the Olds factory, which was rebuilt after a fire destroyed the original in 1901, the work was wheeled from worker to worker. But it is Henry Ford who is seen as the major pioneer of modern mass production techniques. It is at this point that the 'second' Industrial Revolution is sometimes said to have occurred.

Ford seized on the assembly line concept and together with interchangeable mass produced components, a production line utilising conveyors and specially designed assembly equipment, he made a tremendous contribution to manufacturing industry efficiency. Although the motor car engine was not made in a practical form until 1887, it has been estimated that Ford's Model T would have taken 27 days to build using the technology and techniques available at the time of the American Civil War (1861–1865). By 1914 Ford was producing a Model T in just over an hour and a half. By 1918 he had cut the price from 850 dollars to 400 dollars; this of course increased sales that led to a demand for higher productivity and by the 1920s the production time for the Model T had been reduced to 27 minutes.

Laws were passed in the early twentieth century to ensure tolerable working conditions in industry. For example, in the UK there were laws relating to hours worked, the minimum age of workers, the amount of floor and air space per worker and heating and lighting conditions. Also, laws regarding the safety of machines were created; many machines used dangerous belt and shaft transmission systems and these had to be guarded. Employers were also to pay compensation for any injuries suffered by workers while carrying out their work.

The First World War of 1914–1918 was a stimulus to technological progress. By the 1930s significant advances were being made in materials and manufacturing technology. For example, the quality of steel was improving, tungsten carbide was being used for cutting tools and new casting and extruding processes were developed. Plastics such as polyvinylchloride (PVC) and polyethylene were created. Automated machines for making screws and other items, and transfer machines capable of mechanically transporting parts from workstation to workstation were also in use.

During the Second World War of 1939–1945 industrial progress was rapid as the demand for manufactured goods for the armed forces mushroomed. For example, stick electrode welding was used for the first time on a large scale to fabricate ships. Production rates and quality and reliability of components had to be maximised. The integration of electronic control techniques with hydraulic mechanisms for gun and radar control was also to provide the foundation for the post-war automated machine tool industry. By 1947 the General Electric Company (GEC) in the USA demonstrated how it was possible to control a machine tool using a magnetic tape. This was termed the 'record playback control' system. The machine element movements necessary to produce an initial workpiece were recorded on the magnetic tape. When the tape was played back the machine movements were duplicated using a servo control system, so allowing further workpieces to be created automatically. Around this time a method of controlling a milling machine using punched cards was devised by J.T. Parsons, again in the USA. The coded pattern of holes on the cards was read automatically and the information fed to a controller to drive the milling cutter along the desired path. Through collaboration with the US Air Force and the Massachusetts Institute of Technology (MIT) the concept was further developed to produce a prototype three-axis milling machine in 1952. In the same year the GEC method was improved by MIT and launched on the market by the Giddings and Lewis Machine Tool Company this was the first commercial numerically controlled machine tool system. About the same time, Ferranti in Britain also developed a three-axis machine controlled by magnetic tape.

In the technical journals at this time there was much talk about the 'automatic factory' as the possibilities of applying mechanisation coupled with electronic control began to be understood. By 1954 the first large scale automated production line was built. This was named 'Project Tinkertoy', and was used to produce and assemble electronic products. The first industrial robot, made in the USA by Unimation, was installed in 1961, see Figure 2.6a. These examples of automation were not the computer controlled systems of today. They either used mechanical devices or 'hard wired' electronics to effect control and were therefore not easily reprogrammed to cope with changes of work. This only became possible with the advent of integrated electronic circuits.

The year 1967 saw the first large scale integrated circuit manufactured that contained hundreds of electronic components and around this the 'third' Industrial Revolution is suggested to have started. By 1970 the 'microprocessor' had been developed by Intel; this was a single electronic chip capable of interpreting and carrying out logical instructions. The cost of computing power was much reduced and 'minicomputers' could now be used for machine control, for example, the first computer-controlled industrial robot was produced by Cincinnati Milacron in 1973, see Figure 2.6b. Today, microprocessor-based

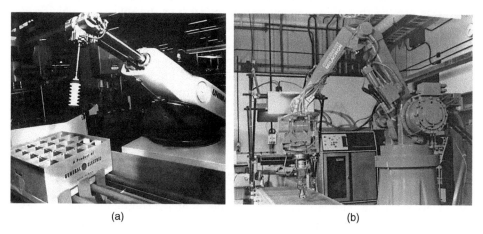

(a)   (b)

Figure 2.6 (a) 'Unimate'; the first industrial robot in 1961. (b) Cincinnati Milacron T3, the first microprocessor-controlled robot in 1973. Both were hydraulically powered.

Figure 2.7 A modern fully automated area within a factory where industrial robots build wrist units for more industrial robots. Source: Image Courtesy of FANUC Europe.

control, industrial robots and other programmable automation is ubiquitous throughout manufacturing industry.

The 'unmanned factory' was again a popular concept in the 1980s, and some fully automated factories were developed, but there are few of these today as they are hard to justify economically. However, many unmanned manufacturing cells and larger units have been created that utilise computer control, such as the robot assembly system shown in Figure 2.7; although to take humans out of the factory completely is still not economically feasible. In fact, in many countries where wage rates for unskilled and semi-skilled labour are relatively low, very large numbers of people can still be found employed in high volume assembly work. This is particularly the case where frequent design changes may occur, such as in mobile phone production.

2.7   The Twenty-First Century

Towards the end of the twentieth century the Internet began to appear in its modern form and now in the first quarter of the twenty-first century it has transformed much of the way we work and communicate. Various types of digital communication networks, whether they be industrial, government, commercial, domestic or social can now all interact, the advent of the World Wide Web, progress in advanced robotics, machine vision systems and artificial intelligence have all contributed to the 'Internet of Things' and what is now called the 'Fourth' Industrial Revolution. The impact of these factors on manufacturing industry is discussed later in the book.

The history of man's use of tools shows how manufacturing industry has shaped our society. These first two chapters have attempted to show the significance of manufacturing. The following chapters now describe the composition and operation of the manufacturing industry of today.

Review Questions

1 What feature made *Homo sapiens* different from his precursors, and how was this advantageous?

2 Why did Europe lag behind Asia in the development of metalworking?

3 Why did the full use of iron take longer than that of copper?

4 Why were improvements in the quality of items such as cannon barrels possible in the fifteenth century?

5 Explain the features that produced the 'factory system' of working in the seventeenth century.

6 Describe four factors whose combination produced the Industrial Revolution.

7 What was the principal reason for 'the factory' emerging as a central manufacturing unit?

8 Name and describe three major concepts in manufacturing that appeared during the initial years of the nineteenth century.

9 Briefly describe the production method, originally used by Olds, which Ford exploited to revolutionise car manufacture at the beginning of the twentieth century.

10 What device led to the emergence of 'reprogrammable automation' and why is this type of automation important in today's economy?

3

Typical Manufacturing Industries

3.1 Introduction

Manufacturing covers a broad spectrum of activity ranging from large multinational companies employing thousands of people to very small enterprises comprised of a few individuals. If every type of manufacturing industry was to be covered in this text, many volumes would be necessary. Therefore, the purpose of this chapter is to simply provide a brief overview of some of them.

In Chapter 1, it was shown that industry can be considered as primary, secondary and tertiary, and that manufacturing constituted the secondary category. In fact, in modern manufacturing industries the boundary between manufacturing as a secondary industry often merges with service industry activities. For instance, research, development, design, marketing and selling would appear to be tertiary industries but they are often an integral part of large manufacturing companies.

Manufacturing processes themselves can also be considered as composed of three categories; primary, secondary and tertiary. For example, in primary processes raw material such as iron ore is processed with other raw materials to produce cast iron or steel and so on. Similarly, crude oil can be further processed to produce petroleum or plastics. In secondary processes the materials produced at the primary stage are formed into more easily worked shapes or components. For example, components can be created by forming steel into rods then machining and plastics can be further processed by moulding or extruding. Finally, tertiary processes take the previously created components from the secondary processes and assemble them into a completed product. Primary processes are more fully explained in Part 2 and secondary and tertiary processes in Part 3.

The classification of manufacturing industries is often done to aid analysis of different aspects of each industry including performance, contribution to the national economy and employment figures. Different countries use various classifications for these subdivisions, for example, the North American Industry Classification System (NAICS) has over 350 different types whereas the UK Office of National Statistics (ONS) lists 14 categories but then subdivides each of these much further. However, for our purposes we will adopt a simpler approach. Thus the following list describes some typical manufacturing industries that create wealth as previously discussed in the introductory chapter.

- Aerospace
- Automotive

Essential Manufacturing, First Edition. Gordon Mair.
© 2019 John Wiley & Sons Ltd. Published 2019 by John Wiley & Sons Ltd.

- Shipbuilding and Marine Engineering
- Electronics and Electronic Products
- Household Appliances
- Pharmaceutical
- Food Processing
- Beverage Production
- Clothing
- Producer Goods
- Materials and Chemicals Production

It should be realised that the discipline of engineering is implied and inherent in each one of these. Specific manufacturing processes are examined later in the book.

An interesting aspect of manufacturing industries dealing with complex products and technology is systems engineering. The need for this is particularly acute in the aerospace, automotive and shipbuilding industries. Systems engineering is an interdisciplinary approach that addresses engineering and management aspects of an enterprise in an integrated manner. For example, in the case of an aircraft cockpit there are controls and displays that have to be designed. These controls may be interfaced with avionics systems for regulation of fuel supply to the engines and movement of the flight control surfaces, such as the ailerons and so on, also the displays will be connected to sensors to provide information on air speed and engine temperature and so on. As well as ensuring all of these subsystems work in harmony they will have to be suitably designed to enable the pilot to maintain situational awareness and understand the displays and operate the controls easily. Before all of this has occurred materials and components will have to be ordered, then assembled, installed and tested. It can therefore be seen that the disciplines of logistics, quality control, planning, product and engineering design and ergonomics are only some of the factors to be considered. Further, there is the need to consider the whole life cycle of the system including maintenance, repair and recycling. Therefore, systems engineering considers the entire system including design, hardware, software, human factors and management to ensure an optimum result is obtained for the final product. This is a very much simplified view of the discipline but it indicates not only the need for interdisciplinary cooperation but also the requirement for system engineers that are capable of grasping the holistic nature of modern complex product design and manufacture.

Important to all manufacturing industries is the supply of materials, components and services from sub-contractors. For the supply chain see Chapter 1 and for quality control purposes see Chapter 26; it is common to categorise these suppliers into 'tiers'. For example, a Tier 1 supplier will provide components direct to the producer of the finished product such as the seats for a commercial aircraft. A Tier 2 supplier will supply the seat manufacturer with the seat fabric and a Tier 3 supplier will provide the seat fabric manufacturer with the raw material to create the fabric. Industry today requires very tight control of quality and every supplier will require having their quality assurance procedures verified by the tier above it. This should ensure that the final product manufacturer can be confident of the quality of supplied parts.

3.2   Aerospace Industry

Aerospace is a very broad category and is an excellent example of a modern manufacturing engineering industry. Products are often highly complex, of very high added value and expensive to purchase. Typical products are commercial and military aircraft including unmanned aerial vehicles (UAVs), satellites, satellite launch vehicles and other space delivery systems, guided missiles and other defence equipment. The companies themselves are often big, multinational and large employers contributing significantly to the economies of the countries in which they are based.

The supply chains for these companies are also complex not only with many suppliers of components but also due to the distributed nature of the manufacture. For example, the Airbus series of aircraft have their components made in factories across the globe before they arrive at one of a number of final assembly plants. As noted in Chapter 1, the Airbus 380 has around four million manufactured components manufactured by around 1500 companies in 30 different countries.

The image in Figure 3.1 highlights the complexity of aerospace industry products. They are the result of a great deal of research, design, development and testing before the final manufacturing stages. This means that employees are usually well trained, very skilled and highly educated. These companies offer good apprenticeships for a broad spectrum of engineering careers as well as recruiting engineering graduates from universities and colleges. Product sales are usually to large airlines and governments rather than individuals. The aerospace industry additionally involves maintenance, repair and

Figure 3.1   Engineers working on a Rolls Royce jet engine. Source: Image courtesy of Rolls Royce.

the supply of replacement and refurbished parts for existing products. Divisions of the companies will also be involved in sales and marketing of their products, thus exhibiting the blending of the secondary and tertiary industries noted earlier.

3.3 Automotive Industry

The automotive industry is another large employer producing complex products but this time for direct sale to the consumer. Cars, trucks and buses are examples of products, and car manufacture in particular epitomises modern mass production techniques. For example, the supply chain has been carefully analysed, established and managed, automation using industrial robots is widespread, lead times between the design and supply of products has been minimised and response to customer requirements is quick and efficient (Figure 3.2).

A modern car manufacturing factory is mostly an assembly plant. Using the tier system mentioned earlier, a series of sub-contractors supply the plant with parts such fascia panels, electric window systems and seats. Subsidiary plants may produce complete engines. The car assembly plant itself will probably have presses for producing the car body elements from sheet steel and industrial robots for welding these together on a conveyor system. The welded body will then be surface finished with protective and cosmetic treatments including paint. The body will then join a conveyor system where the individual components from sub-contractors and subsidiary plants will be added in a carefully ordered and balanced sequence.

The lead time between concept and design of a new car model has been continually reduced over the years to ensure emerging customer needs are met. Also, the use of reprogrammable automation means that individual customer requirements such as style and upholstery can be accommodated quickly into the production system.

Figure 3.2 Car assembly by industrial robots. Source: Image courtesy of Jaguar.

3.4   Shipbuilding and Marine Engineering

This is an industry that demands a large amount of investment in space, materials and labour. Products created here include cruise, cargo and military ships, offshore oil platforms and submarines. Due to the large cost of manufacturing, these products' production is focused more and more in fewer and fewer countries. Currently more than 80% of the world's commercial shipbuilding occurs in China, South Korea and Japan. For security and employment purposes, military craft such as destroyers, aircraft carriers and submarines tend to be built in the countries that require them, providing the necessary expertise and manufacturing facilities have been retained.

In order to reduce costs and improve efficiency, large ships are often manufactured in a modular manner. For instance, an aircraft carrier like the one shown in Figure 3.3 was constructed in this way with prefabricated modules assembled in one location then transported to another where final assembly took place. The modules are often kitted out with all necessary services such as pipes and wiring in order to be easily connected to other modules at final assembly. Similarly, large cruise ships are also built in this way. Standardisation of the modules is adopted to reduce costs when building sister ships

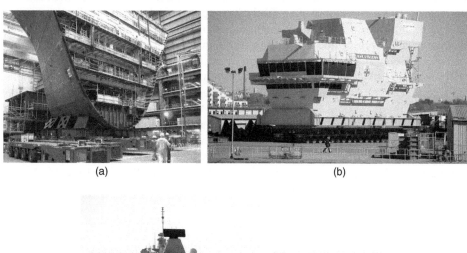

(a)   (b)

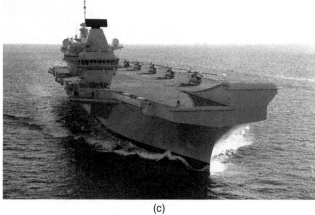

(c)

Figure 3.3 Queen Elizabeth class aircraft carrier. (a) Hull module, (b) bridge module and (c) completed carrier. *Source*: Photos courtesy of BAE Systems.

of similar design. The joining of the modules will be done by welding. Care has to be taken with the sequencing of the welds to prevent distortion, and the integrity of the welds has to be thoroughly checked to ensure metallurgical and mechanical properties are maintained. The lead time for a large aircraft carrier like the one shown from concept to placing contracts to design then manufacture and finally full service may be nearly 20 years. In contrast, the lead time for large cruise ships may be 5 years with attempts being made to reduce this to 3 years. For an aircraft carrier future strategic and political situations have to be considered as well as likely technological developments and this is extremely difficult since the ship may be expected to be in use for many decades, Whereas cruise ships need to be designed to cope with changing customer needs and demographics, these can change quite rapidly hence the requirement to reduce the time between design and service to a minimum.

3.5   Electronics and Electronic Products

The digital electronics industry has greatly contributed to the world's economic growth throughout the second half of the twentieth century and is continuing through the first decades of the twenty-first century. Today's computers, industrial robots, autonomous cars, communication and entertainment hardware – like mobile phones or flat screen technology televisions – and the supporting digital infrastructure – such as the Internet and wireless communications – all rely on digital electronics.

The building blocks of electronic products are electronic components including resistors, capacitors, diodes, transistors and integrated circuits such as microprocessors. Since the inception of integrated circuits the number of transistors on a chip has increased according the 'Moore's Law' (actually an observation rather than a real law), which states that they will double roughly every year, although recently limits are being approached. This means that research, design and development in the industry is essential and competition is fierce. Chapter 15 describes the actual manufacturing processes involved in microprocessor production.

Electronic products are ubiquitous, the industry is global and competition is just as fierce as in the components industry. For example, each manufacturer of mobile phones will not only produce a variety of models at different prices but will also bring out new and improved models on at least a yearly basis. The product life cycle, as described in Chapter 4, is extremely short. In the assembly of electronic products industrial robots may be used, however, much assembly work is done by humans. This is because although the design of the product may be carried out in a high wage country, it is often useful to employ the labour of a low wage country for the actual manufacture and assembly.

3.6   Household Appliances

Included in this category are what is often termed 'white goods' such as domestic kitchen washing machines, refrigerators, microwave ovens and cookers. This industry mostly involves assembly of individual components from other manufacturers. Taking a clothes washing machine as an example, the basic design will be a frame structure to which the other parts are fixed or hung. An electric motor, a pump, a rotating barrel for holding the washing, heating elements and an electronic module for wash timing and sequencing. Also, sheet metal parts to form the frame and the cosmetic panels that encase the whole appliance.

The structure of the washing machine means that it is not designed for automated assembly. This is because parts are being fitted within the frame from different angles and using flexible items like hoses and transmission belts. Therefore, assembly of large electrical appliances like this usually uses human labour although industrial robots can be used at the packing stages once assembly is complete. This type of manufacture, therefore, can employ relatively large numbers of people and the skill level required for the assembly work is not high.

3.7 Pharmaceutical Industry

In contrast to the household appliances industry, the pharmaceuticals industry is highly automated. However, in a similar manner to the electronics industry research and development is extremely important and occupies a large proportion of company expenditure. Globally, the cost of research into pharmaceuticals is in the order of many tens of billions of dollars and the expenditure on prescription pharmaceuticals is around a thousand billion dollars. It is a highly regulated industry and even after a new drug is developed it can take years before it can be approved and released for public use.

Pharmaceuticals are produced in various forms such as powders, capsules and tablets. The powder compounds use to create these delivery methods may be blended and they will be manufactured by milling to achieve a specific particle size, or by binding the small particles together into granules to aid flow and ensure the proper compound mix is maintained.

After the compounds have been pressed into tablets or filled into capsules they can then be packaged. This is often done in highly automated systems in clean environments. Various types of automation are used including vibratory bowl feeders, mechanical handling devices including industrial robots and conveyor systems; these are explained more fully in Part 4 of this book.

3.8 Food Processing

Food is available in most countries now as either fresh produce or as food that has been processed in some manner, it is this latter type that concerns us here in a manufacturing context. Products such as sausages, bacon, many cold meats such as corned beef, canned fruit, confectionary and convenience foods, like ready to eat meals for cooking in microwave ovens, are all processed. Baked products such as cakes and biscuits are also part of the food manufacturing industry. In this industry, the size of the companies and manufacturing facilities covers a broad spectrum. There are very small independent units comprising a few individuals engaged, say, in personalised wedding cakes, through to large multinational corporations employing thousands of employees.

The factories themselves vary in size and production methods. Some are very labour intensive with hundreds of workers carrying out operations analogous to those found on assembly lines, and some are highly automated using special purpose machines and industrial robots. The industrial robots are normally the very high speed types such as those described in Chapter 20. The use of machine vision is particularly important where automation is used to verify the quality of the product and this is covered in Chapter 21.

3.9 Beverage Industry

Bottled water, soft drinks, energy drinks, milk products and alcoholic drinks are all examples of beverage industry products. Water and milk are not as such manufactured but manufacturing industry is required to process them into a marketable form by bottling them or putting them into cartons. Tea, coffee and drinking chocolate are beverages but they are usually supplied to the customer in a dry form and the customer adds water, milk and heat to their personal taste.

A large proportion of soft drinks are sodas. These are comprised of syrups, also called concentrates, for flavouring and sweetening and they are usually supplied to the final bottling plant from a primary manufacturer. At the bottling factory they are mixed with filtered water in appropriate proportions and carbon dioxide gas is injected before pouring the mix into bottles or cans. Alcoholic drinks contain methanol obtained from fermentation and they come in a wide range of types such as wines, beers, ciders and spirits. The spirits have a higher content of alcohol obtained by distillation and maturation and they include whisky, gin, brandy and vodka. Once again the major manufacturing aspect occurs at the bottling and canning stages with highly automated high speed production lines.

3.10 Clothing Industry

The clothing industry, and in particular the fashion industry, is responsible for a large proportion of world economic output. Apart from bespoke haute couture almost all clothing is mass produced. Manufacturing facilities use not only the 'factory system' (see Chapter 2) employing large numbers of people, but they also use the latest computer controlled high speed automation technology such as computer controlled knives or lasers for cutting through thick stacks of material. Less commonly, reprogrammable automation can also be used to produce unique items to individual specifications.

Clothing manufacture is a large global industry that employs millions particularly in countries where wage levels are low. The movement of information and material through the supply chain in this industry is very fast. For example, fashion forecasters in one country may input ideas to garment range designers in another country who will be using computer-aided design technologies. The fabric required may then be produced in a third country, this material is then sent to a fourth for manufacture into the garments themselves, then shipped to the warehouse of a fifth for distribution to retail outlets around the world.

Also by automatically monitoring sales and stock levels in a large department store a particular fashion range and style can be recognised to be popular and stock running low identified. This will be noted by the system and an order triggered immediately to the manufacturing country to make and dispatch more stock.

The materials used may be natural such as wool, cotton and silk; or they may be synthetic such as polyester, nylon or acrylic. Much research goes into creating materials that can provide specialised properties for different applications. For example, moisture wicking for sport and active leisure wear, effective insulation for cold weather and material that protects the skin from the sun's rays.

Footwear can also be considered under this heading as many aspects of it are similar to that of clothing. For example, fashion changes and innovations occur frequently in the sports shoe industry. Synthetic materials are popular and rubber and leather are also used although natural leather is becoming increasingly expensive. The supply chain involved will again be similar to that of clothing from computer-aided design through to mass production in factories employing large labour forces with high speed automation also used where appropriate. Analogous to the bespoke fashion mentioned earlier it is also possible to produce unique footwear using additive manufacturing techniques as described in Chapter 13, but this is uncommon and expensive.

3.11 Producer Goods

Generally, any piece of equipment used to create components and products in the manufacturing industry is termed a producer good. For example, in the aerospace industry metal cutting tools are required to achieve high tolerances on machined parts; in the automotive industry industrial robots are widely used for spot welding; in the electronics industry, high speed machines for populating printed circuit boards are used; in the household appliances industry press tools will be needed to produce sheet metal parts; in the pharmaceuticals industry pills are blister packed using high speed machinery; in food processing ovens are required for baking; in the beverage industry complex automated conveyor systems are needed in bottling plants and in the clothing industry special purpose machinery is used for cutting cloth. The producer goods that manipulate material in some way such as by cutting or deforming are termed *machine tools* and the machine tool industry is often used as a manufacturing industry classification. However, there are other producer goods that are important such as the ovens and conveyors already mentioned, as well as machine vision systems for inspection and large scale items such as the equipment required for materials and chemical production.

This industry does not have such a high profile as others but is an essential one contributing significantly to the economic output and employment of countries where a producer goods industry exists. The design and manufacture of machine tools and other producer goods requires well-educated engineers, technicians and skilled craft workers. The need to continually improve the design of this type of equipment in a very competitive and sophisticated market is a good stimulant for research and development, much of which is carried out in technological universities and research establishments in collaboration with industry.

3.12 Materials and Chemicals Production

Finally, we briefly consider the industries that provide the raw materials for all of the other industries. Materials production is covered in Chapter 7 and this short section is

Figure 3.4 A massive oil refinery complex in Texas. *Source*: Reproduced with permission of Pixabay.

simply indicating that there is an industry that produces the basic materials for creating all subsequent artefacts from producer goods through to finished products.

Production of materials and chemicals is usually carried out in very large, expensive and complex plants covering many square kilometres. The equipment is highly automated with the movement and processing of the raw materials within the plant monitored and controlled through sensors and actuators such as valves and manipulators. The plants are often located close to rivers or railway lines where bulk materials can be easily transported to the processing areas (Figure 3.4).

This industry employs large numbers of skilled and well-educated technicians and engineers responsible for maintenance, repair, health and safety, quality control and process control within the plants. Environmental concerns are also addressed carefully as the processes are energy hungry and often are capable of producing high levels of pollution if not carefully monitored and controlled. They are also important producers of wealth for both the global and local economies.

Review Questions

1 Describe what you understand by the terms primary, secondary and tertiary as applied to *manufacturing processes*.

2 Explain what is meant by the term *systems engineering*.

3 Name 10 broad categories of manufacturing industry and, for five of these, describe the typical products manufactured and activities undertaken.

4

Designing for Manufacture

4.1   Introduction

Manufacturing companies may produce products that they have designed, or they may produce them on behalf of another company that has produced their own designs. For example, mobile phones may be designed in the USA but manufactured in China. If the product designer and the manufacturer are to survive, it is essential that well-designed products are created. Should a company's products not be designed to satisfy the needs of the consumer, then competitors' products will be purchased and the company will fail. This applies equally to domestic and export markets. Consider the contribution made by manufacturing to a nation's economy that was emphasised in Chapter 1, and that an inability of a company to sell its products leads inevitably to its demise. If this applies throughout a country's manufacturing industry it can be seen that good or bad product design can have a significant effect on the success or failure of the national economy.

Within the company, elements such as a good sales and marketing effort, tight cost controls and high productivity are useless if the design of the product is not what the customer wants. It must also be remembered that most of the cost of manufacturing a product is determined at the initial design stage: estimates of 70% in the car industry and 80% in the aerospace industry have been made. Materials and processes to be used are dictated by the product design. Therefore subsequent improvements in, for example, manufacturing efficiency, serve only to reduce the costs that have already been determined by the original design. For all of these reasons, the design process requires careful attention if a world class product is to be produced.

4.2   Computer Aided Design, Virtual Reality and Augmented Reality

Computer Aided Design (CAD) is ubiquitous; it is taught in schools, colleges and universities and is widely used by designers in every field, for example, architects, kitchen providers, artists, urban planners, landscape gardeners and of course product designers. In the case of product design, CAD is often integrated with other digital aspects of manufacturing such as computer numerical controlled (CNC) machine tools and additive manufacturing processes. Additionally, it is possible for designers of to acquire a 360° immersive experience of their CAD designs, for example, in car or ship design.

Essential Manufacturing, First Edition. Gordon Mair.

This is achieved through virtual reality (VR) and can be obtained by wearing a head mounted display (HMD) or by using a very wide angle screen or multiple screens to provide an immersive environment. Also, as opposed being immersed in a design, it is possible to observe the product design within a real environment. This is done by using augmented reality (AR) and this is achieved by wearing glasses on which an image of the product design is projected. By combining this with the ability to track the wearer's head the product can be examined as to how it will look from different angles within the real world. As an alternative to glasses, the same effect can be obtained by using a mobile phone or tablet with the appropriate software. However, the focus of this short book is not on the technology of design, such as CAD, but on manufacturing – therefore, the rest of this chapter is more concerned with the basic aspects of the design process itself and how it can be applied to the design of manufactured products.

4.3 Design for X

There is a wide range of factors that need to be considered in a product design. Features such as design for safety, function, aesthetics, maintenance, repair, ease of use, environmental considerations, sustainability and manufacture and assembly, are just some of them. Thus 'design for x' implies a consideration of all aspects relevant to the product in question.

Different products have different design priorities. For example, fashion clothing has as its main priority aesthetic appeal; functional characteristics are less important, it is the 'style' that sells the product. At the other end of the spectrum we find products such as the microprocessor. Aesthetic appeal is irrelevant as the chip is not intended to be seen in use; however, functional aspects such as processing speed will be critical in determining whether or not it will be a success. In the first type of product the designer with artistic flair will have the most important input, although he or she will also require a knowledge of natural and synthetic clothing materials and how they are produced. In the second type, it is the engineering design that is important and specialist circuit designers (using CAD systems) will be required. Most products today demand consideration of a full range of attributes. For example, a motor car must be aesthetically pleasing as well as providing performance and safety.

Ecological or 'green' issues continue to increase in priority. Everyday examples are the use of recyclable packaging, recycled paper and minimising the use of non-biodegradable or re-useable plastics in products. Product designers must also remember that recycling and 'environment friendly' products are only a partial solution to pollution and waste. Acid rain destroying forests and buildings and industrial effluent poisoning rivers and the atmosphere can be reduced by considering manufacturing methods and their energy and environmental implications. Good product design will ensure that the manufacturing processes used will cause minimal environmental damage without recourse to the very expensive methods that treat the industrial waste after it has been produced. These solutions involve chemically or physically separating the toxic elements of the industrial waste and then if possible converting them to something useful or harmless. Long term solutions to environmental damage must rely on products being designed to produce the minimum of environmental damage as they are manufactured. Designing for re-manufacture is also important. This is where products can be designed so that at the end of their life their individual components can be refurbished or replaced and then reassembled to the original product specification.

4.4   The Product Life Cycle

Design can be novel or incremental. Some novel designs have had important implications for society, the internal combustion engine and television being two examples. However, most design is incremental, that is each design is an improvement on what has gone before. Internal combustion engines today are much more fuel efficient and their power to weight ratios higher than that of their ancestors. Similarly the design of a modern television with a large, thin, high definition colour screen is quite a different product from the small low definition monochrome screen of the valve operated cathode ray tube (CRT) sets of the 1950s.

It is therefore apparent that the design process is not a once and for all event. If we consider each improvement as a new product then we can see that the initial and subsequent products will have a distinct lifetime. This is recognised as the Product Life Cycle and it is shown in Figure 4.1. In region A, the new product design is introduced into the market. In this area the marketing department has the major task of promoting the new product and ensuring that sales growth begins. In region B the new product is accepted by the market; it enjoys increasing demand and it experiences exponential growth. However, during this period competitors will have observed the success of the new product; this will stimulate them to produce their own competing design. In region C the product reaches maturity; it has already made its initial impact on the market, and it will probably be now competing with alternative designs by other manufacturers. In region D the sales of the product decline due to the availability of newer products possibly incorporating better technology.

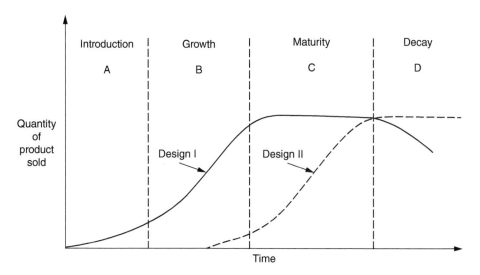

Figure 4.1 The Product Life Cycle.

It is therefore important for the company that designed and built the initial product to maintain technological progress through research and development (R&D). By doing this it can stay one step ahead of its competitors' designs and ensure that it maintains its market share. Products used to have life cycles of a few years, now they are much shorter; in the electronics industry some products have a life cycle of only a few months. The beneficial effect of incremental design improvements is shown by the dotted lines in Figure 4.1 where the company introduces its new product design. Naturally, the technology alone will not ensure success; other factors such as quality, price and after sales service will also be important in creating the right package for the customer.

4.5   The Design Process

Product design and product manufacture must be considered as being concurrent: the development of the design should evolve in parallel with the knowledge of how the product will be made. Only by doing this can the company ensure that the most competitive and profitable product will be manufactured.

Traditionally, a chronological development of a design through a number of stages has been recognised. The first stage is the identification of a market 'need'. This provides the initial idea for a product to satisfy that need. Second, a clear and unambiguous specification is created that fully describes all the attributes the product must have to satisfy the market need. Third, at the concept design stage a product concept is created, usually after having considered a number of alternatives, to satisfy the specification. Fourth, detailed design of the product is carried out. Finally, the production design stage occurs in which the design is modified to ensure the product can be manufactured economically. The problem with this traditional approach is that although it appears to follow a logical progression and each stage is an essential activity, the process becomes 'compartmentalised'. Thus, even if an apparently optimum design is obtained at each stage, an optimum for the overall design is lost.

The practical way to ensure the best overall design is to integrate all design activities and provide an environment conducive to good communications: this requires a team effort. For example, although a mechanical engineer involved in evaluating stress – strain relationships to maximise the strength of a product may produce a functioning design, they may not be fully aware of the manufacturing processes implied. A design must not only satisfy the product design specifications (PDSs), it must also be able to be manufactured economically. The mechanical designer may also not be aware of the implications of their design for the aesthetic appeal of the product. It is therefore necessary to ensure that he or she develops the design concurrently and in collaboration with the manufacturing engineer and, if appropriate, with an Industrial Designer. An approach using a team is adopted, this team comprising representatives from the whole spectrum of manufacturing including engineering, purchasing, manufacturing and even outside suppliers of components and raw materials. The approach attempts to integrate the product design with manufacturing process design to achieve a minimum total life cycle cost for the product.

Figure 4.2 traces the design process, highlights the design goals and illustrates the information flow necessary to obtain a well-designed product. The central spine shows the events that occur, the 'ribs' on either side show the information paths and the design goals that all exist concurrently.

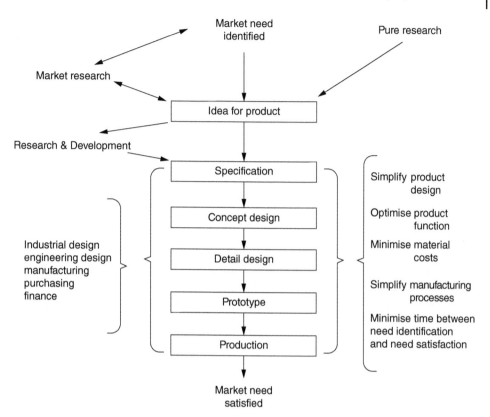

Figure 4.2 The design process.

4.6 Identifying the Market Need

In the beginning there is usually a perceived market need. This is observed and noted by the market research group in a large company, or by an individual such as an entrepreneur or inventor. From this need or 'market gap' springs the idea for the product.

In a large company the idea may also emerge from a research and development group. If a company has the financial resources, and wishes to continue being a leader in its field, then it will carry out pure or 'blue sky' research. Should it not be large enough to fund this on its own then it may work in collaboration with other companies and organisations, such as universities and government or private research laboratories. This type of research may not have an immediately obvious commercial worth, but it is conducted in the hope of future benefits and to ensure that an innovative advantage is maintained over the competition. 'Applied' research has more obvious and immediate commercial applicability; it has the short-term goal of producing a process or product suitable for immediate commercial exploitation.

4.7 The Product Design Specification

Once the basic idea of a product has been created, the next step is the formation of a product design specification, or 'PDS'. This document must be as comprehensive and as detailed as possible. It forms the basis of all the work that is to follow, and the success of the product depends on how well it is drawn up. The specification does not state what the design is to be, but it does state the 'boundary conditions' within which the design must be created. Some examples of the factors that would be included in a PDS are now considered.

- *Conformance to standards and specifications.* Most products sold in quantity today must conform to certain standards and specifications laid down by national and international authorities. For instance, there is the International Standards Office (ISO) that has its secretariat in Geneva, in the USA there is the American National Standards Institute (ANSI) and in the UK there is the British Standards Institute (BSI).
- *Product appearance.* Strongly influenced by the Industrial Designer, the shape, colour, texture and general 'style' of the product is an important part of the specification. This will have an influence on materials and manufacturing processes to be used.
- *Product performance.* Depending on the product, this may take many forms. Speed of operation, number of work cycles expected in product life, intermittent or continuous working, loads to be withstood and so on are typical considerations.
- *Life expectancy.* There will be a certain minimum product lifetime expected by the customer and a guarantee will normally have to be given with the product. This part of the specification will state how long the product should remain in working order provided reasonable care and maintenance is provided by the customer.
- *Maintenance.* How frequently and easily the product is to be maintained must be specified. If the product is to be maintenance free then costs may be increased due to the need for more expensive self-lubricating bearings and so on.
- *Working environment.* The conditions under which the product is likely to be used, stored and transported must be specified. The product may experience extremes of temperature, pressure and humidity. It may be subject to vibration, radiation or chemicals. It may have to operate in explosive or corrosive atmospheres. All of these factors will strongly influence the product design.
- *Quantity.* The total quantity of the product expected and, more importantly, the production rates and batch sizes required, should be specified. This will have implications for the types of manufacturing equipment and work organisation necessary.
- *Size and weight.* These factors obviously have a strong effect on the design. If the product has to be particularly small the cost may be decreased if material volume is reduced; conversely the cost may he increased if more precise manufacturing processes are demanded. Weight restriction will also influence materials to be used; this in turn will influence the manufacturing processes chosen.
- *Ergonomics.* This is concerned with how easy the product is to use by the targeted market. For example, the median anthropomorphic dimensions for the adult male and female population will be used when designing the driver's seating, instruments and controls in a car. This aspect is examined later in the book.
- *Safety.* The product must conform to all relevant safety standards within the countries where it will be sold. The specification should also state the possible abuse and misuse the product might be subjected to. Warning labels and instructions on safe operation of the product should be given, and these should be included in the PDS. It is important to remember that the designer can be held responsible for any accidents that may occur due to poor product design.

Some other factors to be included in the PDS are: likely methods of transportation; type of packaging necessary; quality and reliability expected; the time the finished product may lie around or be stored before use, that is the 'shelf life'; cost limitations and the testing procedures that will be necessary. Consideration of the company's existing manufacturing capabilities should be made, since these will determine if the product can be made 'in house' or if it will be made by a sub-contractor often in another country. This will influence cost and lead time. Also at this stage a search of existing patents, relevant literature and product data should be carried out. An analysis of competitors' products used to satisfy the same or a similar market need should be thoroughly investigated. Comparison of competitors' designs with the PDS will highlight any omissions, weaknesses or strengths of the forthcoming design. The PDS is an essential document, but it must not be immutable. Subsequent to its creation, at any part of the following design process, it should be changed should any opportunity for improvement appear.

4.8   Concept Design

The next stage is the 'concept design' stage, in which a design solution to the demands set by the specification is achieved. It is therefore essential that the specification exists, since it is only by using this as a datum that the adequacy of the conceptual design can be evaluated. Initially, there should be a number of alternative solutions obtained before a final selection of the 'best' one is made. The resulting solution should be a synthesis of all the characteristics and attributes expected of the product.

A concept design may come from an inspired individual; this does not happen very frequently and, for a large company, it is not a particularly reliable means of obtaining a steady supply of new products. Once again the importance of a synergetic team approach to design cannot be overemphasised. 'Synergy', in this context, is a term that implies that the effectiveness of a group of people working together is greater than the total effect of these people working individually. Teams, comprised of the individuals mentioned earlier in the chapter, can use various techniques such as 'brainstorming' to produce a number of possible design solutions. These will be presented for further consideration and discussion in the form of annotated drawings, textual explanations, physical 3D models and computer simulations including modelling and possibly virtual reality immersive environments.

All of the possible designs should now be critically evaluated. It may be that an aspect of one design does not exactly conform to the PDS. This design should not be discarded until the PDS has been checked to see if it should be changed to suit any advantages found in the product design. Each potential design should be fully analysed for performance and so on by carrying out all necessary calculations. This is usually done by computer to allow a speedy, efficient comparison of alternatives.

4.9   Detail Design

Following this there will be the detail design in which the individual components and sub-assemblies will be designed. Depending on the product, detailed calculations will be carried out on, for example, mechanical, electrical or thermodynamic aspects of the design. For some products such as aircraft, motor cars, ships and some of their sub-assemblies and components, for example engines, scale models will be made and tested and computer simulations carried out, to ensure that designs are optimised. These stages correspond to the development part of the 'research and development' function.

Design engineers with specialist knowledge will be involved here. Electronic circuit designers, thermodynamic engineers, mechanical stress analysts and dynamic vibration specialists are typical of the people who may be carrying out the detailed calculations and making decisions regarding the component design. Decisions affecting the details of the manufacturing process to be used will also be made here, for example, to what dimensions will the product be made, what tolerances can be allowed on the size and geometry of the components and what surface finishes are desired? In industry, around three-quarters of the total design activity is occupied by detail design and unless the PDS and concept design stages have been properly implemented, this time may easily be wasted.

4.10 Prototyping

Once the design has been completed the first prototype of the product can be made by model makers who may use various techniques including additive manufacturing. For large products only one prototype may be made but for smaller ones a number may be produced. These prototypes will undergo testing allowing design modifications to be made before production.

4.11 Production

Finally, a number of trial production runs will be made. This is necessary as the equipment and labour used for full production is quite different from that used for prototyping. Most production usually involves unskilled or semi-skilled labour operating special purpose and/or automated equipment. Trial production allows any last-minute changes to tooling and methods to be made. The tooling and production methods will have been designed in concert with the product design during the earlier stages. Although changes to the product design will still be possible at this late stage, they will be expensive.

The amount of design changes and modifications necessary as the design progresses are minimised by the integrated approach previously mentioned. Figure 4.2 shows that at all stages there must be a close collaboration between all functions. Depending on the complexity of the product, various individuals will be involved in this collaboration.

4.12 Contributors to the Design

The Industrial Designer often has the responsibility of finalising the general design and appearance especially in consumer products such as kettles, irons, radios, lamps and telephones and larger products such as cars. He or she should have a good engineering background coupled with artistic training and aesthetic flair. They will have an up to date knowledge of modern materials and their characteristics, and manufacturing processes and their capabilities. He or she will be aware of design trends, have a strong sense of 'style' regarding the finished appearance of the product, and will be concerned about 'ergonomics', that is, how easy the product is to use. Specialist Engineers from appropriate disciplines will be responsible for the functional aspects of the design, for example, electronic circuit design or strength of the product. The manufacturing or production engineer will be fully conversant with the details of modern manufacturing processes and materials. He or she will know such things as the precision of each process, the surface finish that can be produced, relative process costs and their suitability for different production rates and volumes. Provided all these individuals have an input to the design process from PDS to production, the time from start to finish of the project will be reduced and the total cost will be minimised.

Concurrent with these aspects, there will also need to be an input from purchasing and planning personnel. Purchasing can provide present and anticipated costs for materials and components. Planning Engineers will be aware of the availability of company equipment and the capabilities of sub-contractors. They will also be able to plan in advance any reorganisation necessary, plant layout changes and new tooling and machinery needed to manufacture the new design. Further comments on the need for collaboration between different roles can be seen in Chapter 5.

4.13   Some Principles of Product Design

The basic tenet here is that a good design is simple, that is, all functions of the product should be satisfied with the minimum of complexity. Following on from this guidelines have been developed that help to ensure that a product will be competitive in the market place; some of these are now considered.

- First, in general terms, the product function should be optimised. This involves getting the correct balance between all the operating factors. For example, in a car an optimum solution has to be found for each market segment regarding speed, acceleration, fuel consumption, load pulling ability and interior space. Different customers have different priorities; for example, one customer may want a sports car, another a family car capable of towing a caravan and yet another a small town car for commuting and shopping. The product must also be designed to be easy to use. Then material costs should be minimised; by selecting the appropriate material the designer can ensure that performance and manufacturability are optimal for the target market.
- Second, but no less important, the product must be 'designed for manufacture'. The design must allow the product to be made simply. This will minimise production costs by ensuring that low cost machinery and labour are used wherever possible. For example, if only one metal component is required an additive manufacturing process may be suitable, if 20 components are to be made then machining may be acceptable, however, if the number is to be 200 000 then the product should be designed to allow a process such as die casting to be used. An additive manufacturing process such as 3D printing is only suitable for single or extremely small quantities due to its slow speed. Machining is very flexible and can be used for a range of production runs. Die casting has a higher initial cost than machining due to the special tooling necessary,

but where large numbers are required the cost per part produced is much less. Also the labour required to operate the die casting machine is less skilled and hence less expensive than that for machining. These processes are explained in more detail later in the book. Some other goals are as follows:

1) Keep the number of individual parts in the design to a minimum; for example, if more than one part must be used then use integral clips for joining rather than separate nuts and bolts or screws.

2) Make components as similar as possible; this reduces the variety of manufacturing processes involved. Similarities may be in material (e.g. all plastic), shape, (e.g. all cylindrical or prismatic) or a specific process (e.g. shapes that can be extruded) and so on.

3) Avoid redundancy, that is, do not have two parts performing the same function (unless necessary for safety, e.g. a backup braking system in a bus or a backup computer system in a manned space vehicle).

4) Design for ease of assembly by allowing one component to be assembled to the next by vertical stacking movements. This also aids automation of the assembly process.

5) Avoid using floppy components in a design as they are difficult to handle, especially if automated manufacturing is to be used.

6) If possible, design individual components so that they can be made to a wide tolerance yet still function when assembled together; this greatly reduces manufacturing time and cost.

7) Use standardisation and modularisation wherever possible. This last goal is of particular importance to today's world manufacturing environment; it is therefore dealt with more fully in the following section.

4.14 Standardisation and Modularisation

It was at the beginning of the twentieth century that Henry Ford pioneered the mass production line methods that have brought the economies of scale essential to support today's consumer society. However the original techniques restricted the design of the product, for example, 15 million of the Model Ts were produced over a period of about 18 years, but they were all virtually identical. Today, customers demand greater variety in their product and the designer has the task of trying to achieve the economies of scale typified by the Model T era, yet satisfying a wide range of requirements for discerning customers. One means of doing this is by utilising the complementary concepts of standardisation and modularisation.

- *Standardisation.* This is the concept of using the minimum number of parts for the maximum number of purposes while considering the or overall cost and performance of the finished product. It has the overall effect of lowering the product design cost; removing the need for special tooling, for example, drills, reamers, milling cutters and so on; eliminating research, development and planning costs and removing drawing costs. As a very rough guide, a product that has been created by using 'off the shelf' components from a catalogue might provide only about 90% of the performance of a product that uses specially designed components; however, it will probably also have only about 50% of the cost. In products where performance is not critical this is likely to be a decisive competitive advantage. As an example, suppose an engineer calculates, for strength purposes, that a particular product requires six 4 mm, ten 5 mm and six 6 mm diameter bolts. The concept of standardisation would say that the design should be changed so that all bolts will be 6 mm diameter. This will mean that they are all of adequate strength and that, where large numbers of products are concerned, economies of bulk buying will be achieved. Also one size of drill and tap will be used for drilling and threading the bolt holes; this will reduce storage space and stock control costs, the purchasing department will have a simplified task and material handling will be more efficient. The principle of standardisation can be applied to all aspects of product design, for example, materials, components and processes.

- *Modularisation.* This is the broader concept of applying standardisation to produce modularised sub-assemblies. A car, computer or even a ship can be completely designed and assembled as a series of these modules. As well as bringing all the benefits of standardisation the concept also has other implications. For example, new modules can be designed on an individual basis to directly replace existing ones and if a number of these are introduced at the one time a new product is effectively created. Using the car as an example again – dashboards with associated instrumentation, sunroof 'kits' with glass and electric motors, door 'cassettes' containing the door shell, glass, winders, locks and stereo speakers, and seats including their frames and upholstery, are each produced as complete ready to install modules by factories specialising in their manufacture. They are delivered to the main car factory where they are assembled, often by industrial robots, to other modules on the production line. To avoid the final assembly company tying up its money in purchased modules, which also occupy valuable floor space, precise timing of delivery is important so that they arrive just as they are needed. This is called 'just in time' manufacture or JIT, and is discussed more fully in Chapter 23. Another implication of this is that if the module manufacturers wish to obtain maximum benefit from their investment in their manufacturing equipment then they must try to achieve economies of scale. This means that they may sell their modules to more than one end user, thus making it possible to find exactly the same modules in the products of different car manufacturers. Subsequently, end users will spend much of their design effort in making their products attractive through styling and the overall package presented to the customer.

4.15   A Design for Manufacture Example

Consider the section through the assembly design of a simplified car engine alternator as shown in Figure 4.3. This is used to illustrate a few of the design for manufacture (DFM) principles. The reader will appreciate the significance of the following exercise more fully after reading the rest of this book; however, by putting the example in here the importance of the designer having a knowledge of manufacturing processes and associated costs is highlighted. Since the alternator is a purely functional item hidden under the car bonnet no aesthetic considerations will be necessary.

Figure 4.3a shows the original prototype design and Figure 4.3b shows the design improved for manufacture. The following points explain the changes.

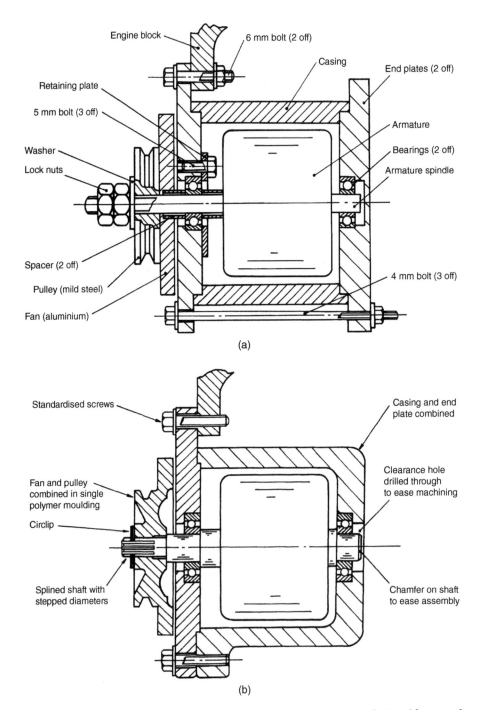

Figure 4.3 (a) Prototype design for an alternator assembly. (b) Alternator re-designed for ease of manufacture and assembly.

1) In Figure 4.3a the pulley and fan are separate items and are made of different metals. The pulley was machined from mild steel bar material and the fan was pressed from aluminium strip. By designing the pulley and fan as one item, as in Figure 4.3b, to be made from one polymer moulding a number of cost savings are made.
   - The costs of the mild steel bar and machining for the pulley are saved.
   - The aluminium strip and presswork tooling costs for stamping out the fan are saved.
   - The costs of holding separate stocks of finished pulleys and fans are reduced, as are the costs of transporting and assembling the parts since only one component is now involved.

2) In Figure 4.3b the use of lock nuts, a washer, and a threaded armature spindle to hold the pulley and fan assembly in place have been replaced by a simple circlip to retain the composite pulley/fan on a splined spindle. The pulley/fan will have a mating internal spline as part of the mould design. This system will be sufficient to prevent rotational slippage of the pulley/fan and also retain its axial position on the spindle. The new arrangement reduces the number of parts and makes assembly much quicker.

3) The need for the retaining plate and associated bolts has been removed by adding stepped diameters to the shaft. As well as removing the need for four parts, assembly of the whole product is much improved since a 'stacking' sequence can now be followed. Previously the left hand end plate assembly would have to be completed as a 'sub-assembly' before completing the final assembly of the product. Removal of the retaining plate allows the right hand section of the alternator to be used as the 'base' for assembly into which the other components can be stacked sequentially. This means that only one fixture needs be used to hold the work, and also that automatic assembly of the product becomes economically attractive.

4) The use of stepped diameters removes the need for the two spacers, again reducing the number of parts, simplifying assembly, reducing assembly time, and lowering handling and storage costs.

5) A chamfer has been added to the right hand side of the armature spindle to ease assembly.

6) The right hand end plate can be combined with the casing into one casting. As well as reducing the number of parts, this type of design change also reduces the effect of tolerance build up, that is, the mating faces of the end plate and casing no longer exist therefore machining of them to within specified sizes is no longer required.

7) The 4 mm nut, bolt, and washer arrangement for holding the assembly together is no longer necessary once step 6 is accepted. Thus cheaper hexagonal headed screws can be used for assembly, again reducing material and labour costs. This principle is also applied to the 6 mm bolts holding the alternator to the engine block. In practice, a check would need to be made to ensure that the clamping forces remained adequate and that vibration would not loosen the screws.

8) By standardising the size of all the screws to 6 mm diameter and making the lengths the same, savings are again possible by introducing the opportunity for reduced costs due to high quantity buying, and by simplifying storage, material handling and assembly. An additional advantage to the customer is that maintenance is easier since only one size of tool is now necessary for removal and disassembly.

4.16   Conclusion

This chapter has indicated the importance of, and means of achieving, a good product design. The significance of DFM has been highlighted. It is now worth mentioning in closing a few specific historical techniques that have been developed to achieve an optimal design.

- Value Analysis was originally developed by H. Erlicher and L.D. Miles in the USA after the Second World War. It is an organised and critical approach that questions the function of each part of a product with respect to its cost. 'Value analysis' is the term used when examining existing products, its aim being to achieve the same performance as the original design at a lower cost without affecting the quality or reliability of the finished product. 'Value Engineering' is the term used when applying the technique to new products.
- DFA, or Design for Assembly, is based on work by Boothroyd and Dewhurst and aims to minimise the cost of assembly by reducing the number of parts, and then ensuring that those remaining are easy to assemble. Another method, pioneered by Genichi Taguchi, uses statistical design of experiment theory to analyse the product design.
- Finally, Professor Stuart Pugh developed an approach that aims to take into consideration the complete commercial and technical environment within which the design process is taking place. This approach is termed Total Design and Pugh defined it as 'The systematic activity necessary, from the identification of the market/user need, to the selling of the successful product to satisfy that need – an activity that encompasses product, process, people and organisation'.

In conclusion, no matter what specific technique is used, the end result must be a product the cost of which over its whole life cycle will be such that it at least holds its own in the market place and returns an adequate profit to the company that sells it.

Review Questions

1   Explain why good product design is essential for the survival of a manufacturing company and why it is important economically at the national level.

2   What is meant by the term 'Product Life Cycle' and what are its implications for the product design activity?

3   Compare the traditional approach to product design with that of team based engineering; discuss both differences and similarities.

4   How might the market need for a product be identified?

5   Why is it so important to get the PDS correct?

6   Using the guidelines in Chapter 4 write a PDS for one of the following: a domestic electric food mixer; a car jack; a mobile phone.

7 Discuss briefly what takes place at the 'concept' and 'detail' design stages of the design exercise.

8 Describe the job functions that should be involved in creating a successful product for manufacture.

9 In *general* terms, name three criteria that a product designer should attempt to satisfy.

10 List six different criteria that should be satisfied when *designing for manufacture*.

11 Explain fully the advantages of standardisation and modularisation.

5

Manufacturing Concepts

In Chapter 1, it was shown that the purpose of manufacturing industry was the creation of wealth. This can be used to improve the quality of life both of individuals and the population in general. What, then, is the purpose of an individual manufacturing organisation within this industry? In line with the industry goal the individual company must also have wealth creation as its target. Just as wealth at a national level can be measured by the gross domestic product (GDP) per head, so the success of a manufacturing company is measured by how well it performs at making money. In practice, this is measured by the profit it makes, by the ratio of this profit to the value of the resources it has to employ, that is, its 'return on investment' (ROI), and by other factors such as the percentage share it has of the total possible market for its products. The fruit of a company that successfully makes money should be secure and well paid employment for the workforce, good working conditions, stimulation of the local economy around the manufacturing plant, incentives for further investment by shareholders and others to make the company 'grow' and of course a contribution towards the national economy. It has to do all of this within the context of caring for the environment including factory emissions, use of renewable energy and materials where possible, and producing environmentally friendly products designed for recyclability or ease of remanufacture.

The manufacturing company is therefore an organisation whose measure of health is its ability to make money. To do this the company, which comprises people, equipment, materials and finance, must be managed well. This chapter considers the individual elements of the organisation and how they perform in concert to achieve the corporate goal. It is evident that manufacturing is concerned with the efficient use of resources to provide useful goods for customers.

5.1 The Manufacturing System

For the manufacturing organisation to fulfil its function, it must make and sell products. These products will sell at a high enough price to make a profit only if they are available at the right time and they are of the right cost, quality and type to suit the market place.

The essential stimulus is therefore the demand, or potential demand, for a product. Thus the organisation must be fully aware of the detailed requirements of the customer, otherwise competing companies will capture the market with their own products. If this happens there will be no inflow of cash from sales and the organisation will die. Figure 5.1 shows the manufacturing system in operation. The input to the system is the market

Essential Manufacturing, First Edition. Gordon Mair.
© 2019 John Wiley & Sons Ltd. Published 2019 by John Wiley & Sons Ltd.

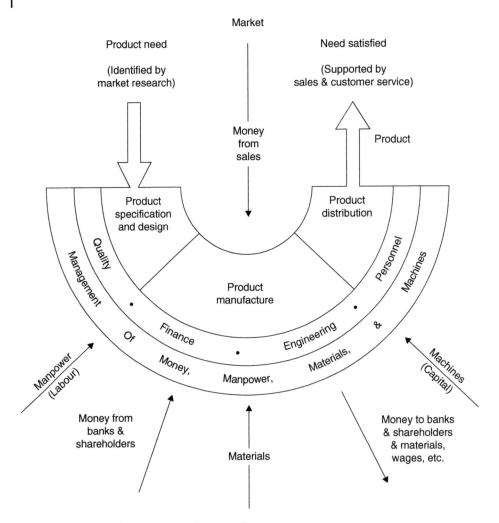

Figure 5.1 The manufacturing system inputs and outputs.

need and the output is the satisfier of that need, which in the case of manufacturing is a product.

Subsequent to the identification of the need comes the specification of the product; this requires careful attention as all subsequent work will be useless if the specification is wrong. Using this 'spec' the designer is now able to form the concept of the product as noted in Chapter 4. This conceptual design is verified by marketing before a detailed design is produced. At the detailed design stage the designer will select appropriate materials and consider what manufacturing processes will be necessary.

When the design has been thoroughly checked the next stage is the construction of a prototype of the product. How this is done will depend on the nature of the product, but it will usually involve craftsmen using general purpose machines and hand tools. This prototype will be thoroughly tested to determine any problems that may necessitate design changes. Possible production difficulties should also be identified at this stage.

The final specification of material and processes to be used should now be possible. Special tools needed for production will be designed and manufactured and the materials ordered. The actual production methods can also be planned, sequences of operations, the layout of the plant and scheduling of the work can be initiated. How the product is to be inspected should also be determined, that is, what dimensions and functions are critical, and at what stages inspection should be carried out.

Assuming manpower, machines and materials are all available the actual process of production can begin. If all foregoing steps have been properly implemented, the product should now be able to be manufactured to the customer's requirements and produced ready for on-time delivery. The individual elements of this production stage are covered later. Throughout the whole process quality of the product must be maintained; this is done by ensuring the quality awareness of all personnel, the provision of proper manufacturing methods and strict adherence to inspection procedures. After the products are completed they will be given a final inspection and, if necessary, test before packing and dispatch to the customer.

Money will have to be made available for various items such as rent and rates, wages and the purchase of equipment and materials: also, costs incurred within the facility must be monitored and controlled; these are the responsibilities of the finance department.

As can be seen from Figure 5.1, the sales and marketing functions form the interface between the company and the market: It is their responsibility to keep up to date with market demands and to ensure the product is sold.

5.1.1 Manufacturing Sub-Systems

From the previous description of the manufacturing system it is apparent that there must be various sub-systems working within it. The situation is analogous to a motor car. The car is a system composed of sub-systems, for example, a transmission system, a braking system, a suspension system and an electrical system. These systems can themselves be broken down into further systems, for example, the electrical system can be broken down into the low tension and high tension systems. The same concepts apply in the manufacturing organisation. For example, there is the management system, the production control system, the financial control system and the production system. The production system could be further broken down to describe individual production areas, groups of machines within these areas, individual machines, machine control systems and eventually the individual hardware components used. This idea of 'systems within systems' is useful when considering control, and in particular computer control, which will be considered in Part 4.

5.2 Lean Manufacturing and Added Value

The concept of 'adding value' in manufacturing was introduced in Chapter 1. Lean Manufacturing, based on the Toyota Production System, includes the concept of ensuring that the added value during the manufacture of a product is maximised, and any wasted effort, time or material is minimised as this adds no value to the customer but adds cost to the company. Lean Manufacturing covers many aspects of production but with regard to waste it focuses on a number of areas, usually seven, where waste can occur, these are noted in the following. The first three problem areas are addressed by the concept of Just In Time Manufacture (JIT) mentioned briefly in Chapter 1 and discussed more fully in Chapter 23.

Unnecessary production quantities. Occurring when more product is made than is required by the customer, this is particularly bad as it will result in some of the other waste producing factors listed next.

Unnecessary inventory or stock. This can be materials and components held in stores, material in progress through the factory but not being worked on, and finished product held in the factory but not delivered to the customer.

Unproductive or waiting time. Occurring when materials are not being worked on, or components are not being assembled into a sub-assembly or finished product. This waiting time is not adding value for the customer but represents a cost to the company in the investment tied up in the materials or components. It is important to note that this does not necessarily apply to manufacturing equipment. For example, if an expensive machine is idle because of lack of work then it should not be used to produce components that may be required in the future, this will simply add to unnecessary inventory.

Unnecessary internal transport of materials. While materials or components are being moved around in the factory no value is being added, that is, they are not being transformed from one state into another of higher value. Thus careful consideration to the layout of machines, stores and work flow is required at the factory planning stage with frequent reviews to cope with changing conditions.

Unnecessary movement of people and equipment. This happens when workers movement, either locally at the workplace or generally when walking between tasks is not optimal and also when equipment is inefficiently used. Method Study, as noted in Chapter 24, can be used to improve human efficiency and machines that combine operations such as multitasking computer numerical controlled (CNC) machines, as shown in Chapter 20, can reduce the need for automated guided vehicles and fork lift trucks to transport parts different machining operations.

Unnecessary work on the product. There should be precise instructions as to how the product should be made with regards to such aspects as dimensional tolerance, surface finish and so on. These instructions should enable the product specification to be met (see Chapter 4). Making the product to a higher specification simply adds unnecessary cost and must be avoided.

Non-conformance of manufactured parts or product. This entails reworking of the components until they conform to what is specified, or actually scrapping the parts and making new ones. In both cases waste has been produced and the additional work required to correct the situation adds cost.

5.3   Integrating the Effort

Throughout this chapter, and indeed throughout the whole book, the manufacturing organisation and the functions, jobs and processes involved, are described as objectively and unemotionally as possible. However, it is worth noting that because people are involved in every area mentioned, subjectivity and emotion are factors that greatly influence much of what happens in a factory. In moderation, these factors add colour and excitement to the environment, but too much is a recipe for chaos. The personal ambitions of individuals, the different ways each person sees a situation and attitudes, beliefs and personalities all lead to some inevitable disharmony in any work situation. This can be advantageous if not allowed to get out of hand, but excessive arguing, internal politics and stubbornness must be avoided as the company would soon lose its way and become uncompetitive. Companies therefore look for ways in which they can structure their organisation to minimise human failings and optimise the functioning of the organisation as an entity in itself. Two examples are now considered.

5.3.1 Simultaneous or Concurrent Engineering

The systems mentioned earlier will not naturally operate in sympathy with each other. A great deal of effort has to be applied by the management and other individuals within the organisation to achieve full integration. Good communication between departments is essential to ensure the correct products are being produced to the correct specification at the correct time. This is not as simple as it sounds, and a conscious and determined effort must be made to create the proper environment for this communication. The concept of simultaneous engineering, sometimes used synonymously with concurrent engineering, is an example of such an effort. The principle is shown in Figure 5.2 in comparison with the traditional sequence of events.

Simultaneous engineering tries to ensure that all relevant problems associated with the design and manufacture of a new product are tackled in parallel, that is, concurrently. This means that the design can be modified dynamically as manufacturing constraints and requirements are considered. Also, the time between concept and actual manufacture is much reduced as problems are identified quicker and machinery, materials and labour can be organised well in advance of the production start date. A good team structure is particularly helpful here.

5.3.2 A Common Goal

The efforts and aims of each individual and group must be such that they act in concert. This can be achieved by the use of techniques such as 'Management by Objectives' (MBO); a term popularised more than half a century ago by Peter Drucker in his 1954 book *The Practice of Management*. Despite some notable detractors of the concept it is still being very effectively used today in many successful companies although it is now usually referred to by different names. The concept provides a structure within an organisation that ensures that the work done by each individual has as its aim a target that contributes directly towards the organisation's ultimate goal. Therefore, since we have said the ultimate goal of the manufacturing organisation is to make money, the sub-goals could be stated as an acceptable profit, ROI and market share. These sub-goals will be quantified to enable specific objectives to be set annually. A breakdown of these can then provide monthly objectives for the company and individual entities within the company. In the case of manufacturing, these will be factories. In each factory these objectives would be broken down further into objectives for the individual managers, supervisors, factory floor workers and other staff. Obviously, the objectives set at the level of individuals will not be specified as profit, ROI or market share. Rather, they will be objectives that, if achieved, will contribute towards the ultimate goal. For individuals, they

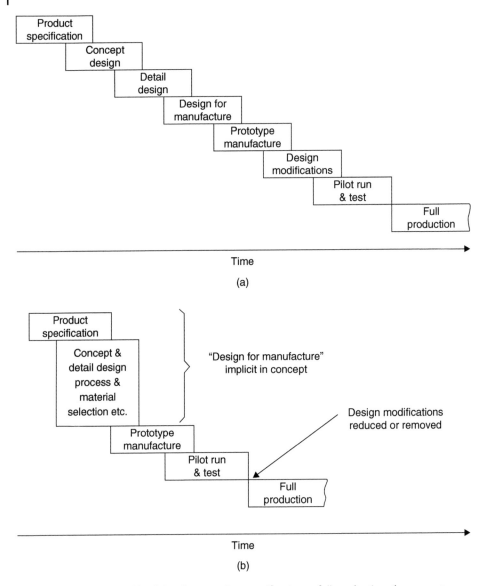

Figure 5.2 Comparison of lead time from product specification to full production, the concept can also be applied for the full product lifetime through to its eventual disposal or remanufacture of the product. (a) The traditional event sequence. (b) Simultaneous or concurrent engineering sequence.

will probably take the form of budgetary objectives, productivity performance statistics, quality levels and sales targets. These targets can be set using the widely used mnemonic SMART, which stands for:

- Specific – they must be precise, clear and unambiguous
- Measurable – they must be quantifiable or otherwise able to be confirmed
- Agreed – they must be agreed by all parties to be achievable
- Relevant – they must be appropriate and conform with the overall objectives
- Time – they must have a target time for completion

The workers themselves must participate in the setting of their own objectives in order for this type of management to work effectively. An example of the process would involve a production manager, based on customer requirements, agreeing with the factory manager on, say, the number of products to be produced within a given time, the costs involved in producing them, the amount of labour to be used, and the maximum scrap, reject and rework costs to be incurred. The factory manager himself would have overall cost and productivity targets to achieve for the whole factory, whereas the individual production operator on the shop floor would have targets relating to his own efficiency and quality of work. The principle is shown in Figure 5.3. The system also provides a means of monitoring performance since each set of objectives can be compared at a later date with what has been achieved.

Figure 5.3 Hierarchy of objectives from an individual employee to company level.

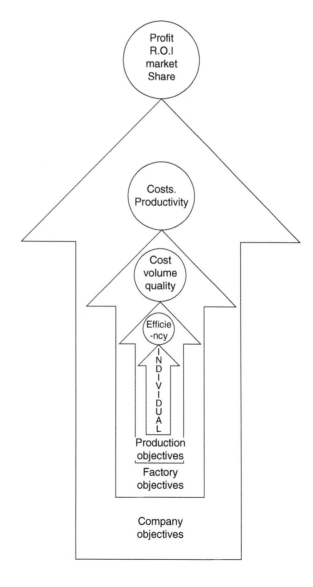

5.4 The Formal Organisation

5.4.1 The Structure

Manufacturing organisations are usually 'bureaucratic' in structure. The word 'bureaucratic' is not used disparagingly but rather in its true sense, which implies a hierarchy of authority with a division of labour into bureaus or departments. A pyramid type structure exists with each post at one level subordinate to a post at a higher level. Each person in the structure is responsible to only one person in the higher level. The division of labour, as explained more fully in Chapter 2, allows the advantages of *specialisation* to be obtained, that is, each person has a specific, and often unique, task(s) to perform. Another characteristic of a bureaucratic organisation is the existence of an established system of rules to cope with various situations and to ensure impartial and just treatment of individuals.

Sometimes there is a distinction made in an organisation between 'line' and 'staff' personnel. For example, in a large manufacturing company authority is clearly defined as running in lines down from the Managing Director, through the factory manager, production managers, foremen to the shop floor operators. All of these people are directly responsible for production. However, they all require the services of other people such as industrial engineers to select the appropriate methods of manufacture and quality assurance personnel to ensure that the product is of an acceptable quality. These may be referred to as 'staff' functions, since they are an aid or support to the production function. In practice, the principles outlined earlier concerning good communication and integration must be followed by all concerned otherwise artificial barriers can arise and 'protectionist' attitudes occur between departments.

5.4.2 The Organisation Chart

For a number of reasons, it is advantageous to have an organisation chart constructed for the company. Assuming adequate descriptions or footnotes are appended, it allows everyone within the company to know who has responsibility for what function and who should be approached to get something done. It can be shown to newly appointed personnel, or even interviewees, allowing them to appreciate their relative position within the structure. It is also useful to potential customers or quality approval bodies to prove that a recognised structure of authority and responsibility exists.

Some disadvantages of organisation charts are that they can quickly become out of date if not regularly revised and if constructed retrospectively individuals may feel aggrieved at being placed lower on the chart than they had expected. Also, formal charts cannot communicate informal relationships and roles that develop within the company due to personalities and individual knowledge and experience.

Organisation charts, or 'trees', are therefore useful but their limitations should always be remembered. A simplified chart for a manufacturing company is shown in Figure 5.4. It puts the job functions described in Section 5.4.3 into perspective. It should be noted that the actual titles used to describe the jobs, and even the jobs themselves, vary from company to company; however, the actual tasks implied by the titles have to be done by someone no matter what terminology is used. In a large factory this pyramid type of structure becomes unwieldy. One way of overcoming this is to split the factory organisation into smaller units with as much autonomy for their actions as is practically possible.

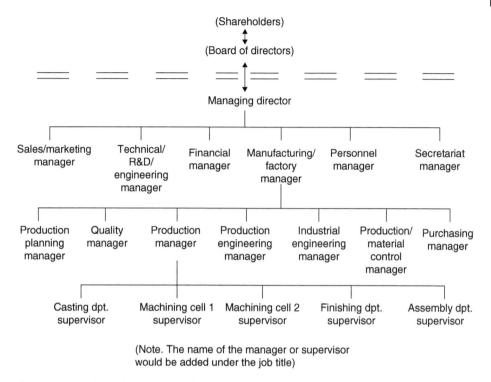

(Note. The name of the manager or supervisor would be added under the job title)

Figure 5.4 A traditional organisation chart for a large manufacturing company although the shareholders and Board of Directors would not normally be included.

Another very effective means of improving efficiency, speeding up decision making and reducing response time, is to remove many of the management layers. This has the effect of 'flattening' the pyramid. The use of computers and electronic communication networks has been an important factor in making this possible.

5.4.3 The Major Functions

In a manufacturing organisation, a number of major functions are performed by a variety of personnel. These personnel have specific jobs and titles, as is evident from Figure 5.4. It is the norm in an organisation to specify the job, then find an appropriate person to fill the post. The major functions and jobs within a manufacturing organisation are now listed in the following subsections.

5.4.3.1 The Board of Directors

This body is at the apex of the organisational structure of the company; it is appointed by the shareholders and represents the company ownership. In the company's Articles of Association and Memorandum of Association, all authority is delegated by the shareholders to the Board of Directors. The Board's fundamental duty is to protect the interests of the shareholders; this implies protection of the company. It is the policy-making body of the organisation, deciding broad objectives and the future direction(s) of the company. It is responsible for the appointment of company officers, for example, the Chairman, Managing Director and Company Secretary.

The Board of Directors delegates its authority to the Managing Director for the everyday running of the organisation. He or she is responsible directly to the Board for the successful achievement of the company's broad objectives. The 'MD' must be of high intelligence and integrity, and in a large company will probably have a wide business experience rather than being a specialist in any one area.

5.4.3.2 Manufacturing

This function has the typical 'line' structure mentioned earlier. The largest number of people will probably be employed under this heading in either supervisory or direct and indirect operative roles. Activities such as production engineering, production planning, production control, work study, maintenance, quality, purchasing and stores and so on will be part of this function. The main objectives of Production are to produce the products required by the customer, at the correct time, at or below the budgeted cost and at the agreed quality levels.

The *factory manager* (or manufacturing or works manager), is responsible to the Managing Director for the efficient operation of the manufacturing facility. He or she will have a large span of control, that is, a large number of people will report directly to him. Under him will be the managers of production, production engineering, industrial engineering, production control, purchasing, maintenance and so on.

The *production manager* is responsible for ensuring that the product is completed as scheduled, within the budgeted costs and at the specified quality levels. They have direct authority over the production labour force, which remains the largest grouping of people in many factories. Reporting to them will be various supervisors. Depending on the type of operation there could be further layers of foremen, section leaders and so on before arriving at the pyramid base composed of the shop floor production workers or operatives.

The *production engineering manager* has the responsibility of ensuring that all production machinery and other equipment is appropriate to the task in hand and is working efficiently. He or she must also keep up to date with the latest technology and processes to ensure that the product is being made at a competitive cost.

The *industrial engineering manager* is responsible for the Work Study department. The task of this department is to create efficient work methods and determine accurately the times required to carry out the various jobs involved in making a product. This enables jobs to be costed properly, workforce requirements estimated and productivity measured quantitatively. The manager will also be responsible for making sure that the layout of the factory produces the minimum of movement of materials and labour.

The *material control manager* coordinates the flow of material, components and products within the production facility. It is his responsibility to provide all the necessary material for manufacture at the appropriate time and place. He or she must keep stores, work in progress and stock to a minimum while making sure that no shortages occur on the production lines. This function is important since costs associated with parts and materials probably account for around 50% of the total manufacturing cost of a product, whereas direct labour constitutes only about 10%.

Each of the following functions will have its own manager or director depending on the company size and structure.

5.4.3.3   Finance

This department has the responsibility of ensuring that funding is available for the smooth operation of the organisation. It also must compile budgets for the other departments; these will be set in cooperation with the managers concerned in a similar manner to other objectives. The department operates the cost and budgetary control systems by gathering and analysing costs and other financial data, before redistributing this information in the form of performance reports. Other activities are the keeping of all accounts including income and expenditure, the payment of wages and participation in costing and pricing decisions.

5.4.3.4   Marketing

In large organisations the marketing function will be composed of a number of sub-functions such as Advertising, Sales, Service, Distribution and Market Research. Marketing has the responsibility of providing a steady flow of orders to the manufacturing facility to keep it producing at optimum production rates and to ensure that profits are maximised by striving to increase the company's share of the potential market.

5.4.3.5   Human Resources or Personnel

With a broad range of responsibilities this department attempts to relieve the other functions of the difficult and time consuming human problems, concerns and decisions that have to be made on both a short- and a long-term basis. Grouped under this heading might be found the responsibilities for welfare, industrial relations, training, safety, job evaluation, wage negotiations and 'hiring, firing and retiring' of personnel.

5.4.3.6   Technical

Another broad departmental title that covers a variety of functions all related to the technical and engineering aspects of the organisation. For example, basic research of the scientific and technical factors that lead to new products, subsequent design and development of products resulting from this research, construction of prototypes, quality control and the setting of product specifications and standards are all part of the technical function. There will also be close liaison between the technical and marketing functions where customer service and complaints are concerned.

5.4.3.7   Secretariat

Under the Company Secretary this function is concerned with the legal aspects of the company, the official recording of the work of the Board of Directors and generally the handling of corporate level correspondence and books. The Company Secretary's office may also be responsible for the administration of pension schemes, insurance, patents, trademarks and the handling of share issues and transfers of stock.

5.5   Types of Manufacture

Manufacturing industry produces a broad range of products that range in size from microchips to oil tankers, and in production volumes from one to one million or more.

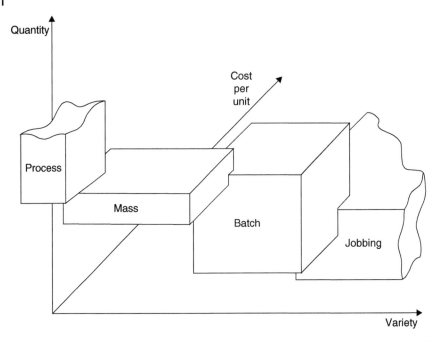

Figure 5.5 The four classes of production showing the relationship of quantity or volume of product produced, cost per unit and variety of product produced.

While the size of the product has an obvious effect on manufacturing methods, it is the variance in volume that has the greatest influence on the type of organisation, equipment and labour employed. There are a number of ways in which manufacturing can be classified; the terminology we will use here is Process, Mass, Batch and Jobbing production. Figure 5.5 shows the relationship of these four types with respect to volume, variety and cost per unit. Process and mass production are used for the highest volumes and least variety, they also produce the lowest unit cost; batch production has a medium unit cost for medium volumes and variety; jobbing production has the highest unit cost, greatest variety and is used for single units. Before describing them further it should be noted that there is often a mixture of types used to create a finished product. For example, although motor cars are mass produced, many of their components such as shafts, engine blocks and conrods and so on will be produced in economic batch quantities (EBQs). A large ship may well be the only one of its kind ever made, which implies jobbing production, but the steel plates that form its hull will have been produced by continuous casting and rolling that are really 'process production' techniques. There is also a 'blurring at the edges' where one type of production gives way to another. The classifications used here should therefore be simply regarded as useful labels to attach to various production systems when considering their organisation.

5.5.1 Process Production

This may also be referred to as 'continuous production'. It relates to that type of production in which the plant operates 24 hours per day for weeks or months without a halt. This does not always happen in practice due to equipment breakdowns and planned maintenance. The product being made will be required in 'bulk' and the plant output is likely to be measured in volume or weight rather than in numbers of discrete products produced. Oil refining, chemical manufacture, food processing and steelmaking are all examples of processes that should ideally run continuously. The investment in the capital equipment will be high as it will probably be automated and specially designed and constructed for the product. Although plant and equipment costs will be high, the labour costs are likely to be low. The type of people employed here would be mostly technical and supervisory.

5.5.2 Mass Production

Again high volume production is involved, but this time it is discrete items that are produced rather than bulk quantities. Typical products made by this type of manufacture are mobile phones, laptop and tablet computers, motor cars, refrigerators, television sets, clocks and radios. Individual components and sub-assemblies of these products may also be mass produced, for example batteries, refrigeration compressors, transformers and transistors. As in continuous production, mass production utilises expensive dedicated special purpose equipment. The variety of the product is kept to a minimum with standardisation and modularisation applied wherever possible. The workforce will have a large proportion of semi-skilled and unskilled workers for machine operation and assembly. In fact, the work will be deskilled as much as possible to allow easy interchangeability of labour between jobs. Work is broken down into the simplest of operations to make operator training quick and simple. Because much of the work is tedious the use of industrial robots has become increasingly popular, especially in car manufacturing. Equipment, labour and material supply are all highly organised to ensure a smooth flow of work through the plant and minimise the cost of each product unit.

5.5.3 Batch Production

Here, specific quantities of a product are required either in a single production run or in batches to be repeated at given times. Batch sizes may range from two or three to a thousand or more. Throughout its useful life each piece of manufacturing equipment will therefore be used for making many different products. This means that the equipment must be much more flexible than that used in mass production. Also as the labour force has to cope with a larger variety of work it will tend to be more highly skilled and higher paid. In a large factory many batches of different products of varying quantities, scheduled for different delivery dates and customers, will be being processed at any one time. As many of these products, though following different routes, will require to be worked on by the same machines and personnel, a complex control problem often arises. Since batch production manufacture contributes significantly to the national economy much effort has been given to ways of improving control and, hence, efficiency. The use of computer controlled machines and computerised production control and management systems has greatly increased in an attempt to bring unit costs down towards those experienced in mass production. Examples of products produced in this type of manufacture are components and spares for aircraft, cars, buses and construction equipment, machine tools, valves, pumps and compressors for the process industries, furniture and so on.

5.5.4 Jobbing Production

Single unit quantities, usually made to a single customer's specification, are manufactured using this method. General purpose equipment, additive manufacturing techniques and hand tools are used. The labour force is commonly composed of highly skilled craftsmen and technicians as well as the normal supervisory and administrative personnel. This type of manufacture produces the highest unit cost for a product and typical examples would be product prototypes, communication satellites, ships, oil rigs and special purpose manufacturing equipment.

5.6 Types of Manufacturing Equipment

It is possible to create a general classification for the types of equipment used in manufacturing; specific types of machine will be covered in later chapters in more detail. The process production industries use specially designed plant dedicated to the particular product being made. The massive investment required is apparent when oil refineries and chemical manufacturing plants are considered. The process industries require specialised treatment and, apart from the production of metals and plastics, this book does not examine them in detail.

Special purpose, dedicated equipment may found in very high volume mass production facilities. This type is built specifically to suit the product and is completely dedicated to the manufacture of that product. The equipment has therefore to be specially designed and built, making it very expensive; thus, very long production runs are needed to justify the expense. High production rates will be achieved and lower cost labour can be used. The main disadvantage of this equipment is that when the product becomes obsolete or has major design changes, the equipment hardware becomes redundant and has to be scrapped or broken down.

Reprogrammable equipment overcomes some of the limitations of the special purpose type. In today's environment, product changes occur frequently as customer requirements change and new technology creates new product possibilities. This means that special purpose equipment cannot be used since production runs will not be sufficiently long to recoup expenditure. Examples of reprogrammable equipment in mass production can be seen in the industrial robots used for spot welding in car assembly lines. When a car model or style changes the robots can simply be reprogrammed and then resume working on the new design. Reprogrammable machines are also widely now used in batch production, as they are ideally suited to short production runs repeated at regular or irregular intervals. The programs for each batch can be stored away and reused when required.

Automation is the term used to describe equipment that can operate on its own without continuous human control and monitoring. Most of the special purpose equipment used in process and mass production is highly automated. The reprogrammable equipment in the previous paragraph is automated. Sometimes the terms hard and soft automation are used. Hard automation refers to equipment, machines and systems that are made to operate automatically but are not easily reconfigured to cope with changes in product or product design. Special purpose machines, machines controlled by cams and 'hard wired' electronically controlled machines all usually fall into this category. Soft automation refers to equipment controlled by computers or other easily reprogrammed devices. Industrial robots, numerically controlled machine tools and vision systems all come under this heading. The actual operation of these machines does not need highly skilled personnel. Automation is dealt with more fully in Part 4.

General purpose equipment is used for making prototypes and in jobbing production. Although electrically or hydraulically powered the machines will often be manually operated and some may be numerically controlled. This is the most flexible type of equipment and can easily be used to produce any type of product. The handheld 'stick' welding system used in shipyards is an example of general purpose equipment; so is the traditional centre lathe used for producing cylindrical type components. The skill levels of the equipment users are highest here as considerable training and experience needs to be gained before a good quality product can be produced.

Additive Manufacturing Equipment is used to create a wide variety of products, is extremely versatile, but due to the time needed to create each item it is normally only economically viable for single and small numbers of items. It is covered more fully in Chapter 13.

There are also different ways of laying out a manufacturing plant, for example 'line' or 'cell' layout, or layout by product or process. These aspects of 'plant layout' will be covered in Chapter 22.

This chapter concludes the introductory section of the book. An appreciation of the economic importance of manufacturing, its history, the importance of good design and some of the basic concepts of the organisation for manufacturing has been given. The rest of the book explores more fully the manufacturing processes used to create products and how people, material, machines and money need to be organised to optimise the creation activity.

Review Questions

1 Briefly describe how the success of a manufacturing company is measured.

2 State the major elements that must be properly managed in a manufacturing company.

3 List the main events that take place between the conception and the sale of a manufactured product.

4 Explain what is meant by the term 'manufacturing system'.

5 Explain what is meant by the term 'simultaneous engineering'.

6 Explain what is meant by the term 'management by objectives' (MBO).

7 Do you agree that a bureaucratic method of management is suitable for a manufacturing organisation? If so, why?

8 List the major job functions that need to exist within a manufacturing company.

9 Describe what is meant by the terms 'process', 'mass' batch and 'jobbing' production.

10 Discuss what is implied by the terms 'hard' and 'soft' automation.

Part II

Manufacturing Materials

6

Materials for Manufacture

6.1 Introduction

A design and manufacturing engineer must have a good knowledge of the types and properties of materials from which a product can be manufactured. The material must not be so expensive as to make the product uncompetitive in the market place, yet it must possess all the characteristics necessary for the functioning of the product. Factors such as cost, strength, hardness and how easily it can be worked must be considered carefully by the designer. Although there is already a wide range of materials available, research and development ensure that new ones are constantly being created. In plastics alone it is estimated that hundreds of new variants are developed each year. Advances in various areas of technology can often stimulate the need for new materials. For example, the possibility of building spacecraft created a requirement for materials, including metals, which could withstand extremes of temperature, pressure and vibration.

We will mainly concern ourselves with metals and polymers that are the most common engineering materials. Other materials, such as composites and ceramics, are also noted. These materials are gaining in importance as more is learned about how to manipulate their structures to obtain desired properties.

One way of grouping materials is shown in Figure 6.1, that is, into metals and non-metals. Metals are essentially chemical elements, such as iron, copper, gold and aluminium, or alloys of elements, such as steel and bronze. They are often lustrous when smoothed or polished, are usually good conductors of electricity and heat and are likely to have a good strength to weight ratio. They can be further classified into the iron-based metals, that is, the ferrous and non-ferrous metals. The ferrous metals include the steels widely used in the construction of a broad range of products from pins and paper clips, through to motor cars, ships and bridges. The non-ferrous materials are also used in a variety of familiar applications, for example, electrical wiring made from copper and strong light-weight aircraft components from aluminium.

The non-metallic materials can be split into organic and inorganic materials. Originally the organic materials were only those that had their origin in living organisms, for example, wood and leather. Now organic materials are regarded as those based on carbon compounds generally, though it is interesting that the basic ingredients for even the synthetic polymers or 'plastics' are obtained from oil, natural gas or coal, which in turn come from long-dead living matter. The inorganic materials come

Essential Manufacturing, First Edition. Gordon Mair.
© 2019 John Wiley & Sons Ltd. Published 2019 by John Wiley & Sons Ltd.

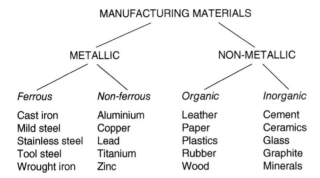

Figure 6.1 A simple materials classification.

originally from the earth in the form of minerals. In the past they were used more in the construction industry than in manufacturing, but developments in ceramics and cements are producing many new manufactured products.

Where strength at relatively low cost is required metals are still by far the most common material used. Other materials, especially the plastics, are steadily gaining in ground in properties and popularity. For instance, it is claimed that the volume of plastics sold is now greater than that of metals. Very approximately, as the percentages can vary, a typical family car will be comprised of over 75% metals and around 15% plastics with the remainder made up of rubber, glass, leather/fabric, ceramics and fuel, oil, coolant and so on. The metals will mostly be steel used in the body of the car, with cast iron or aluminium alloy used in the engine block, lead in the battery and copper for the wiring. More expensive cars will have aluminium used in the body and the most expensive may have carbon fibre bodies for very high strength to weight ratios. However, the predominant material in manufactured products where strength at relatively low cost is required remains metal, and in particular steel, in the form of various alloys. We will now therefore consider how metals are structured as this is necessary for an understanding of how they can be manipulated to manufacture products.

6.2 The Structure of Metals

6.2.1 Atomic Bonding

At this fundamental level we find that there are two ways in which atoms can be held together, that is, primary and secondary bonding. Secondary bonding is a weak bonding, which is not necessary to consider here. The primary bonding is important as it is strong. Primary bonding has three types: (i) the *ionic* bond in which electron transfer occurs, (ii) the *covalent* bond in which electrons are shared and (iii) the *metallic* bond in which there is a structure of positive ions surrounded by universally shared wandering electrons. These electrons provide metals with their relatively high thermal and electrical conductivity. The structure can be deformed without the bonds breaking; this allows metallic bond materials to be changed in shape yet retain their original strength. The metallic bond is therefore of most relevance here; the covalent and ionic bonds will be mentioned later when ceramics are discussed.

6.2.2 Atomic Arrangement

This tells us how the atoms arrange themselves in a material. There are three types of structure: (i) *molecular* as in water, (ii) *amorphous* as in glass and (iii) *crystal* as in metals and most minerals. In crystal structures the atoms are arranged in a regular geometric array known as a space lattice. When metals solidify by cooling they adopt a crystalline structure and the atoms group themselves into one of these lattices. The three main types of lattice are body centred cubic (BCC), face centred cubic (FCC) and hexagonal close packed (HCP), see Figure 6.2. This structure helps determine how easily the material can be worked or deformed, for example, FCC metals can usually be easily deformed without fracturing, whereas HCP metals are difficult to work.

6.2.3 Grain Formation

As described before, when a metal solidifies atoms arrange themselves geometrically to form a crystalline structure. The initial lattices that appear become the nuclei or seeds from which the crystals of metal will grow. Many of these nuclei form in the initial stages of solidification but the direction in which each lies is random. As the crystals grow, the lattice pattern of the source seed is maintained as successive lattices align themselves with their predecessors. Eventually, when one growing crystal comes into contact with another of different orientation, growth of both will stop. The surfaces where they meet will be irregular in nature and will form part of a 'grain boundary'. This process is illustrated in Figure 6.3. It is interesting to note that some high quality jet turbine blades are made from metal consisting of a single large crystal. This unusual and expensive material gives good performance at high temperatures.

6.2.4 Recrystallisation

This is an important feature in manufacturing with metals. When a metal composed of many crystals, that is, a 'polycrystalline metal', is deformed the crystals are twisted and strained. If the metal is now heated to a high enough temperature new equiaxed and unstrained crystals will be formed from the original distorted grains. This process is known as *recrystallisation* and is illustrated in Figure 6.4. The temperature at which it occurs is different for each metal and varies with the amount of cold deformation that has previously taken place; that is, the more deformation the lower the temperature at which recrystallisation will occur. The recrystallisation process tends to produce uniform grains of comparatively small size. As properties of metal tend to diminish as grain size increases, good control is important to keep the grain size small or at the optimum level for the application. Generally, if metals are allowed to cool slowly after being taken above their recrystallisation temperature, large crystals will form; if they are cooled rapidly, small crystals will result.

When metals are deformed below their recrystallisation temperature, cold working is said to take place. The structure consists of distorted grains and the metal is strain hardened; this can make it difficult to work the metal further. When deformation takes place above the recrystallisation temperature, hot working occurs. A recrystallised structure continually forms and no strain hardening is present. As metals may fracture if deformed too much it is common practice to recrystallise metal at intervals during cold working processes. This restores ductility and prepares the metal for further deformation; this process is known as recrystallisation annealing.

Bcc or body centre cubic

E.g.   Chromium
        Tungsten
        Vanadium
        Iron (at room temperature)

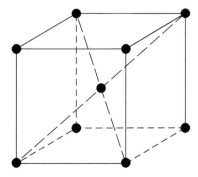

Fcc or face centre cubic
Usually ductile

E.g.   Aluminium
        Copper
        Gold
        Silver
        Platinum
        Nickel
        Lead
        Iron (at elevated temperatures)

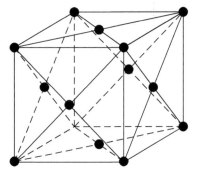

Hcp or hexagonal close packed
Usually exhibits poor formability

E.g.   Beryllium
        Cadmium
        Magnesium
        Titanium

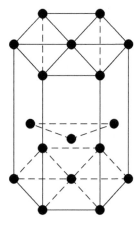

Figure 6.2 Three main types of crystal structure.

6.2.5   The Importance of Grain Structure

The grain size of a metal depends on the rate at which it was cooled and the extent and nature of the hot or cold working process. A metal with small, fine grains will have better strength and toughness compared to the same metal with large, coarse grains. This is due to the atoms being closer together in the smaller grained metal and causing more interference in the lattice structure when a force is applied. Larger grained metals are

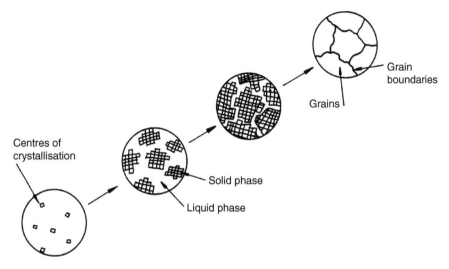

Figure 6.3 Crystal formation and grain growth.

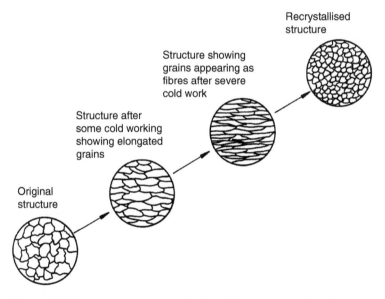

Figure 6.4 Effect on grain structure of cold working and recrystallisation.

characterised by easier machining, more uniform hardenability during heat treatment, but with a greater tendency to crack when cooled by quenching. Additives can be added to a molten metal to promote a specific grain size, for example, aluminium may be added to steel to promote fine grains.

Both hardness and grain size are affected by the temperature history of the metal. Quenching a hot metal quickly from a high temperature will usually harden it, whereas cooling it slowly will give it maximum softness. Annealing is the slow cooling of a metal from high temperature to increase the softness, toughness and ductility while also removing internal stresses.

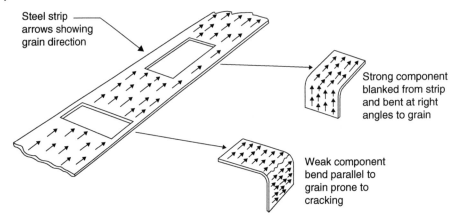

Figure 6.5 The effect of grain direction on component strength.

When a metal is deformed the grains become elongated in the direction of metal flow. This gives the metal the appearance of having a fibre structure, similar to the grain structure in wood. Due to the creation of strain hardening and the fact that the intergranular boundaries will no longer be randomly oriented, the strength and other mechanical properties of the metal will not be the same in all directions, see Figure 6.5. This fact is exploited by engineers, with processes such as rolling and forging being used to impart such properties.

6.2.6 Alloys

Alloys are formed when metals combine with other metals and, occasionally, with non-metals. The metals involved usually dissolve in each other in the liquid state to form a completely homogeneous liquid solution. In engineering, metals are normally used in the form of alloys. The most useful alloys contain a large quantity of one metal combined with a much smaller quantity of one or more added elements. These alloys are said to be based upon the metal that predominates in amount, for example, iron or copper based.

Each alloying element has its own unique effect on a metal, the whole effect of two or more elements generally being greater than their individual sum. Carbon is an alloying element with iron that forms steel. The properties of steel are significantly affected by varying the quantity of carbon. Plain carbon steels are therefore referred to as being low (0.05–0.3% C), medium (0.3–0.6% C), or high (0.6–1.4% C) 'carbon'. Below 0.05% C are the wrought irons and above 1.4% C are the cast irons. In the steels, tensile strength, yield strength and hardness increase with carbon content, whereas impact strength and ductility are decreased.

6.2.7 Some Typical Engineering Metals

- *Wrought iron.* Though first made as long ago as the late eighteenth century, wrought iron is still a useful material where toughness and resistance to shock is required at relatively low cost. It usually has between 0.02% and 0.03% carbon, and may be used for such things as anchor chains and crane hooks.

- *Cast irons.* These are relatively cheap materials, made with various alloying elements, to give a range of types with differing properties. The carbon content varies from 2.0 to 3.6%, depending on the particular cast iron being made. Alloying materials, such as silicon to improve fluidity and hence ease of casting, and manganese to increase hardness and strength, are used. In comparison to steel, cast irons are usually of lower strength and more brittle, but easier to cast. They absorb vibrations well and are often found as parts of the structure of machine tools.

- *Mild steel.* This is the most popular ferrous material and is widely used in the manufacture of motor cars, ships and household goods such as washing machines. It is relatively cheap, is easy to machine, and is available in a wide variety of shapes and sizes. It is normally in the 'low carbon steel' category described previously. 'Medium carbon steel' provides higher strength and may be used for parts such as gears, axles and other mechanical parts.

- *High carbon steels.* These steels can be hardened and tempered to give precise hardness, strength and wear resistance characteristics. Cutting tools, screw drivers, press tools, cutlery, chisels, drills and saws are some uses of this material.

- *Alloy steels.* Although all steels are alloy steels, that is, they all contain carbon and other elements, such as molybdenum that increases strength and hardness, the 'alloy steels' have had a number of additional alloying elements added. Control of the manufacturing process is more precise than in the previous types to ensure a high quality metal that closely conforms to specification. Only a few of the possible alloying elements are noted here. The addition of nickel improves toughness and impact resistance. Chromium increases resistance to wear, abrasion and corrosion; it also improves hardenability. Molybdenum increases hardenability and toughness. Vanadium improves impact and fatigue resistance. At high temperatures added tungsten will form hard tungsten carbides. When tungsten and vanadium are combined in steel the resulting material is known as High Speed Steel (HSS), which is used for cutting tool materials.

- *Tool steel.* This is an alloy steel able to be heat treated to give very good hardness for use in metal forming and cutting dies. Such steel may contain over 18% tungsten and will retain its hardness at high temperatures.

- *Stainless steel.* This is an alloy steel with more than 12% chromium and usually also containing nickel. It has very good resistance to corrosion and is used for cutlery, food processing equipment, sink units, valve and pump components, dies and so on.
  These are only some of the wide range of ferrous materials commonly available. The non-ferrous materials also provide a wide variety of characteristics. Like the ferrous materials they are most useful in engineering in their alloyed forms.

- *Aluminium alloys.* These are used where a light weight is required. In its pure form aluminium has good electrical conductivity and corrosion resistance. The alloyed form is much stronger but less corrosion resistant. Alloying elements such as copper, silicon, manganese and zinc are commonly used. For special applications, such as satellite construction where lightness and stiffness are required, lithium has been used. However, lithium reacts explosively with water thus necessitating more expensive manufacturing procedures.

- *Copper.* Copper has extremely low electrical resistance and is corrosion resistant. However, it is both soft and expensive in its pure form. It is therefore usually alloyed to provide bronze, brass and other materials.

- *Bronze.* Bronze is an alloy of copper with tin, aluminium, manganese or silicon. It is corrosion resistant and relatively strong.
- *Brass.* Brass is an alloy of copper with zinc. It has good corrosion resistance, is easily machined and cast, but can be less strong than bronze.
- *Titanium.* Titanium and its alloys, though expensive to produce, have generally high strengths up to 500°C and are highly resistant to corrosion. This makes them a cost-effective solution to many problems in the aerospace and chemical industries.

6.2.8 Heat Treatment of Metals

The final mechanical properties exhibited by a metal product are usually the result of three factors, that is, the alloying materials, the way in which the metal was worked and the heat treatment processes used. The first factor has already been explained, the second will become apparent as the various processes are studied later; heat treatment is briefly considered here. Heat treatment is usually employed to relieve internal stresses built up during cold working of the material, or to harden or soften the material to a specific value to suit a particular application.

- *Annealing.* The main purposes of annealing are to restore ductility and softness to a metal and relieve internal stresses after it has been hardened by cold working or rapid quenching from a high temperature. It is carried out by slowly heating the metal to an appropriate temperature, keeping it at this temperature for a specified time then allowing it to cool slowly. The annealing process follows the stages of stress relief, recrystallisation and grain growth, as was shown previously in Figure 6.4.
- *Hardening.* Steels with a carbon content greater than 0.3% can be hardened by raising them to a high temperature, then rapidly cooling them in a liquid such as cold water. The temperature to which the steel needs to be raised depends on its carbon content and ranges between 720 and 1100°C.
  If the steel has a carbon content lower than, say, 0.3% then its surface can be hardened using a process known as carburising. Here the steel is heated to above 900°C in contact with a substance rich in carbon. The carbon in the substance will diffuse into the surface of the steel, forming a skin containing around 0.8% carbon. The carburising substance may be gas, liquid or solid. When the steel is quenched it will be said to have been 'surface hardened'. It should now have a tough core and hard outer shell that will be wear resistant. In this process, surface areas of the steel that do not require hardening can be protected against carbon penetration by a surface coating such as copper plating.
- *Tempering.* When steel is fully hardened throughout, it is brittle and is likely to contain internal stresses. It is therefore necessary to reduce this hardness to that required for the application; this 'tempering' process will also restore toughness to the structure. Steel is tempered after hardening by reheating it to a specific temperature, usually below 550°C and then cooling it. The exact temperature to which it is reheated determines the final hardness of the metal. For example, a fully hardened chisel made entirely from high carbon steel would be too brittle to use, its edge would crack easily if dropped and, when hit by a hammer, the head would be liable to chip and cause an eye injury from flying particles. The chisel is therefore tempered, the edge is raised to a temperature high enough to remove the possibility of cracking

yet ensuring that sharpness is maintained in use, and the head is raised to a higher temperature producing a softer but tougher composition that will not fracture.

Heat treatment of non-ferrous materials is often restricted to annealing to remove the effects of cold working. Some aluminium and copper base alloys can be hardened using a process called 'precipitation hardening'.

Where high strength is required at reasonable cost, metals are still the most likely engineering choice. Where special properties of light weight, heat resistance and high strength to weight ratios are required materials such as polymers, ceramics and composites are becoming more popular. However, metals continue to be improved and new forms are always emerging from the world's laboratories. High stiffness steels capable of being useful in very thin light-weight sheets for car bodies and ultra-high carbon superplastic steels capable of 1000% deformation have been developed.

6.3 Plastics

6.3.1 Plastics and Polymers

A 'plastic' is an engineering material that can easily be moulded into a desired shape, usually at an elevated temperature. After cooling the plastic retains its new shape. In the case of thermoplastics, reheating will allow the plastic to be remoulded. Thermosetting plastics retain their moulded shape even when reheated. Plastics can be transparent, translucent or opaque. They can be produced in any colour or finish desired. They can be used for wrapping chocolate bars, housing computers, replacing human organs or armour plating a tank. On average one new plastic is being created almost every day. In fact, plastics are now so widely used and the variety is so great that major producers provide computer programs, free of charge to users, to assist with the appropriate plastic selection.

Plastics are generally synthetic polymers. Natural polymers have been used for many years. Latex is the sap of 'rubber trees' that, when mixed with certain chemicals and allowed to coagulate, can be processed with sulfur and placed in a mould. When the mould is heated a chemical reaction takes place called 'vulcanisation'; the product is a rubber product that has considerable mechanical strength and has the shape of the mould. Horn is another example. Horn when heated becomes soft and can be moulded. It was used for buttons and in thin translucent sheets it was used as we now use glass. In fact the word 'lantern' comes from the term 'lanthorn', an early application of the natural polymer.

6.3.2 The Chemical Structure of Plastics

Plastics are synthetic polymers made from organic molecules using the process of synthesis. An organic molecule has carbon atoms as its base; an example of this is the ethylene molecule shown in Figure 6.6a. In the production of plastics this ethylene molecule is known as a 'monomer', which means 'one part'. In the process of synthesis used to manufacture the plastic, conditions of high temperature and pressure may be created, causing one of the links in the central double bond to break. This allows the individual monomers to link up and form a chain, or 'polymer', as shown in

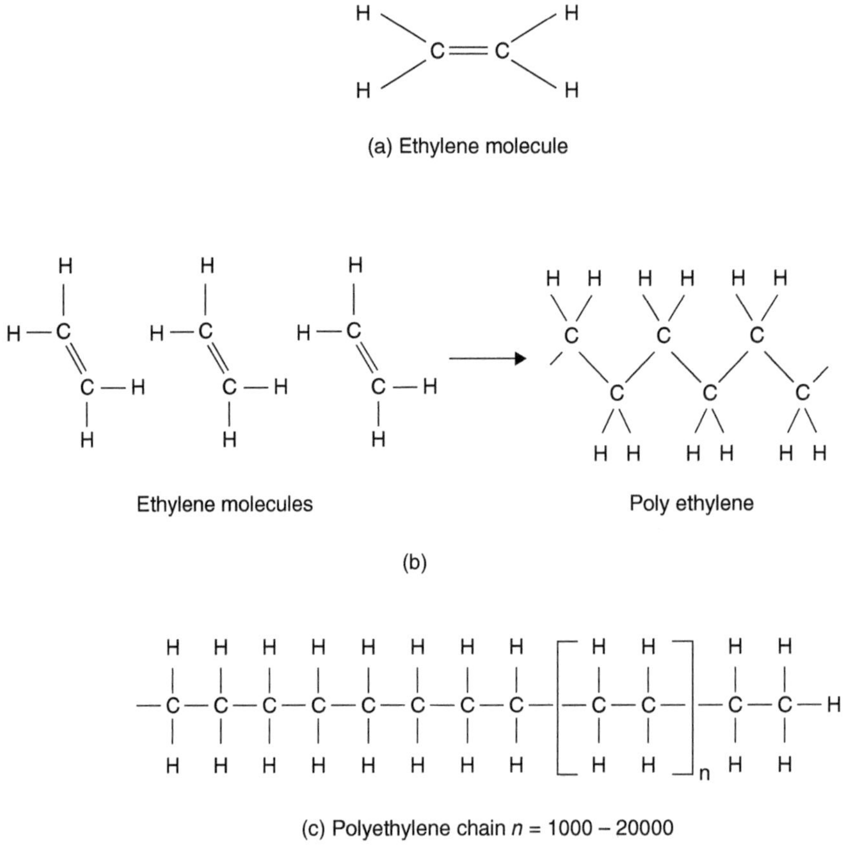

(a) Ethylene molecule

Ethylene molecules   Poly ethylene

(b)

(c) Polyethylene chain $n = 1000 - 20000$

Figure 6.6 Organic molecules.

Figures 6.6b and c. Around 1000–20 000 of these monomers combine to form one polymer of 'polyethylene', which is a simple plastic composed only of carbon and hydrogen atoms. By introducing other atoms or groups of atoms plastics with different properties can be created. For example, chlorine is used to produce polyvinylchloride or PVC and fluorine can be used to create polytetrafluoroethylene or PTFE.

The manner in which the polymer chains arrange themselves with respect to each other influences the properties of the final plastic. In the thermosetting plastics the molecular chains are designed so that further chemical linking can occur between the chains themselves. This produces a three-dimensional network structure that forms as the plastic is being moulded under heat, see Figure 6.7a. This produces a very strong plastic with good hardness and stiffness but is usually brittle. Such plastics do not soften on reheating.

In the thermoplastics there is considerable scope for organisation of the chains to provide different properties. Plastics such as PVC, polyethylene and the acrylics and nylons have linear structures, see Figure 6.7b. Since the chains are linear they can slide over each other, thus providing a certain flexibility to the plastic. Polyethylene is so flexible it can be used as wrapping film, whereas PVC is relatively rigid unless a 'plasticiser' is used. Plasticisers are liquid or semi-liquid additives that tend to separate

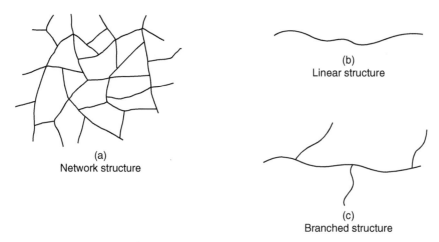

Figure 6.7 Polymer chains: network, linear and branched polymers.

out the polymer chains, thus allowing them to slide over each other more easily. Thermoplastics can also have branched structures, see Figure 6.7c, which are less dense and have an apparently higher strength than the linear ones. Both linear and branched structures often exist together in the one plastic. It is also possible to introduce a crystalline structure into plastics during processing. This improves hardness and stiffness, increases density and decreases ductility.

6.3.3 Some Common Plastics and Their Properties

Generally, plastics provide light-weight, corrosion resistance, low electrical and thermal conductivity and a low cost to weight ratio. It is also possible to produce plastic products directly in any colour or degree of transparency and with any type of surface finish desired, with no need for secondary operations such as machining or painting. As mentioned earlier, new plastics are being introduced at the rate of a few hundred per year; we therefore concern ourselves with only a few basic types that highlight typical properties available.

Thermosetting plastics now include a wide range of materials. For example, there are the epoxies. These have very good mechanical and electrical properties, good elasticity, toughness, resistance to heat and chemicals and strong adhesive qualities. They are often fibre reinforced and used in structural components such as tanks for holding chemicals. Another group are the phenolics. These are relatively strong and hard but brittle. Widely used for products such as telephone cases, handles and electrical insulators, they are available in many forms; for example, sheet, rod or tube. There are many other thermosetting plastics, for example, alkyds, aminos, polyamides and silicones, all exhibiting their own individual combinations of properties.

Thermoplastics are available in an extremely broad range and some of them are considered next.

- *Acrylics.* These plastics have good strength properties, especially impact strength, and have very good optical transparency. They can be used for vehicle windshields, goggles, lenses, windows and so on.

- *Polyamides.* These include the nylons and aramids. Nylon is tough, has good abrasion resistance and is self-lubricating. It is therefore used for gears, bearings, fasteners and so on. In monofilament form it is used for fishing lines and climbing ropes. Aramids have high tensile strength and stiffness and are used in bullet proof vests and pneumatic tyres.
- *Polycarbonates.* These have high strength and toughness and good impact resistance. Safety helmets, bottles and machinery guards are some applications.
- *PVC.* A relatively inexpensive plastic, PVC has a wide range of properties and can be made either rigid or flexible. It does not have high strength or heat resistance. Rigid PVC can be used for items such as signs and pipes. Flexible PVC can be used for flexible tubes, cable insulation, floor tiles and so on.
- *Polyethylene.* This has good electrical and chemical resistance with mechanical properties dependent on the particular type used. Two of the types are low density, LDPE, used for litter bins, toys, bottles and so on, and high density, HDPE, used for products as diverse as canoes and machine components that require good wear resistance.

Just as with the thermosetting plastics, there are many other thermoplastics. Examples of some are polypropylenes, polystyrenes, polysulfones and polyesters. The variety of types and properties of all plastics is now so vast that selection of an appropriate material for a specific application is best done using the material supplier's own updated computer program or product information catalogues.

6.4 Ceramics

Ceramics, first developed over 7000 years ago for making clay pots, are now used today to protect the surface of spacecraft re-entering the atmosphere. With the demand for materials that can function at increasingly high temperatures and speeds and yet retain their properties of strength, hardness and electrical and chemical resistance, there has been a resurgence of interest in ceramics. Even at normal temperatures ceramics can provide a combination of hardness, lightness, stiffness and resistance to corrosion that most other materials could not better. The main disadvantage, however, is that they are brittle.

Ceramics have been widely used for some time now as electrical insulators in electrical power systems and in items such as sparking plugs where high temperature strength is required. Other applications have been in cutting tools using tungsten carbide and grinding wheels using silicon carbide as an abrasive. Now many new applications are seen due to improved understanding of ceramic constituents, and strict quality controls during manufacture produce less brittle ceramics.

The structure of a ceramic material is a compound of metallic and non-metallic elements. The covalent and ionic bonds that hold the atoms together in a ceramic material are much stronger than metallic bonds, thus giving the ceramic greater hardness and thermal and electrical resistance than, for example, steel. The structure of a ceramic may be single crystal, or polycrystalline where the smaller the grain size the better the strength and toughness.

Examples of more recent ceramic products are ball bearings and turbine blades. Motor car manufacturers are particularly interested in using them. Ceramic exhaust liners, coatings for pistons and catalytic converters are already in wide use, but in the future it is hoped that much more of a car's engine will be made out of ceramics. Conventional piston engines that can run at high temperatures without the need of a radiator or ceramic gas turbine power units are some of the possibilities.

6.5 Composites

The composites are probably the materials with the highest strength to weight ratio of all the types previously considered. They are relatively expensive compared to metals but in many applications the extra cost is acceptable. One of the first and least expensive composites is glass fibre, which is widely used for boat hulls and less frequently now some car components. The simplest form has short fibres of glass randomly oriented in a matrix of plastic. By using longer fibres and arranging them all to run in the same direction within the matrix greater strength is obtained. By using sheets of these, and laminating them so that each layer has fibres running in different directions, structural aircraft components can be made. Further strength and stability can be obtained by using, within the plastic matrix, fibres that have been woven into a three-dimensional pattern; this is used in products such as skis. The plastic matrix is usually an epoxy or a polyester. This supports the fibres, protects the fibres from damage, acts as a crack arrestor and transfers stresses to the fibres. The fibres themselves are usually high strength stiff materials, but are sometimes brittle. A composite therefore exhibits the best properties of the plastic matrix and the integral fibres to give a tough, strong and light-weight structure.

Other examples of other typical fibres used are carbon and the organic aramid 'Kevlar'. Carbon fibre reinforced plastics are widely used due to their high strength to weight ratio. Applications can be found in aircraft, cars, boats, windmills, satellites, golf club shafts, sports racquets and fishing rods. Kevlar, either on its own or combined with other materials, is used to reinforce many products such as bullet proof vests, cut resistant gloves, bicycle and car tyres, walking boots and military helmets.

Some composites use a variety of materials such as metals, polymers and ceramics. They may be used as the fibres or as the matrix. For example, silicon carbide fibres in a matrix of titanium, a metal matrix composite, is suitable for high speed aircraft structures. Advanced ceramic, metal and polymer matrix composites are expensive, but they are attractive materials for high performance machines, motor cars and aircraft.

6.6 Properties and Testing of Materials

Engineers are normally interested in the physical, chemical and mechanical properties of materials. Typical physical properties are (i) density, this is important for weight, for example, a good strength to weight ratio is imperative in aircraft structures; (ii) thermal and electrical conductivity; (iii) melting point, which is important in manufacturing as it determines the ease with which the material can be cast and also the amount of energy required for the process; (iv) magnetic properties; (v) colour and (vi) coefficient of thermal expansion. Chemical properties such as the ability to resist corrosion are also important. However, it is the mechanical properties that often have the greatest influence on the manufacturing methods used to work the materials. Tensile, compressive and shear strength, hardness, ductility and impact and fatigue resistance are all relevant. They are described more fully next.

6.6.1   Stress, Strain and the Strength of Materials

It is essential to understand the meaning of these the terms 'stress' and 'strain' before proceeding further. When a material is subjected to an axial load, as shown in Figure 6.8, two things happen to it: (i) it becomes deformed; this deformation is termed 'strain' and is defined quantitively as the change in length divided by the original length; (ii) internal forces are set up within the material to resist the applied forces; this is called 'stress' and it is defined quantitively as the force exerted by the load divided by the cross sectional area of the material.

A simplified stress strain diagram for a low carbon steel is shown in Figure 6.9. Up to the proportional limit the material obeys Hooke's Law, which states that stress is directly proportional to strain; this ratio is known as Young's Modulus or the modulus of elasticity. Either on or just above the proportional limit, the elastic limit occurs; beyond this point increases in strain do not require proportionate increases in stress. Elongation is now unrecoverable and is known as plastic deformation; usually when this happens the metal is said to have 'failed'. It is in this plastic region, before rupture

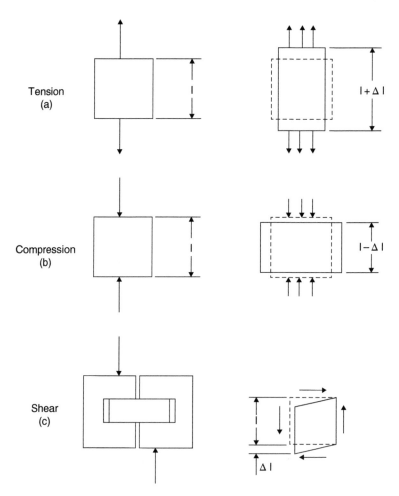

Figure 6.8 Tensile, compressive and shear loading and the resulting strain effects.

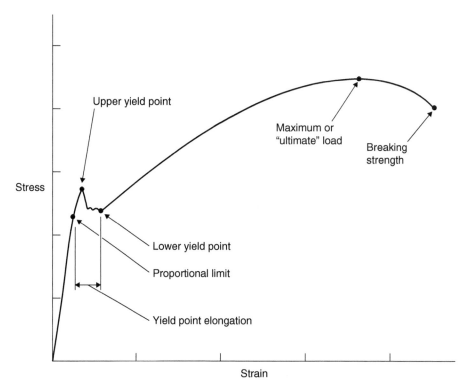

Figure 6.9 Stress/strain diagram for a low carbon steel.

occurs, that plastic deformation is used to shape a metal product in many manufacturing processes.

6.6.2 Tensile Strength

This is the strength exhibited by a material when it is being pulled apart from two opposing directions. Tensile strength is determined by pulling on the two ends of a specimen machined, as shown in Figure 6.10. When the specimen is pulled the smaller diameter section necks down from an area A to an area A', and the gauge length increases from L to L'. For most engineering purposes the area A is used in all calculations since A' is difficult to measure. From the data collected while pulling the specimen, a curve can be plotted from the two values of stress and strain where:

Stress = Force/Area A

Strain = (L' − L)/L

The value of stress where necking begins is called the Ultimate or Tensile Strength of the material.

6.6.3 Compressive and Shear Strengths

Figure 6.8b,c shows the effects of compressive and shear forces on a material. The compressive strength of a material shows its ability to resist squeezing forces without

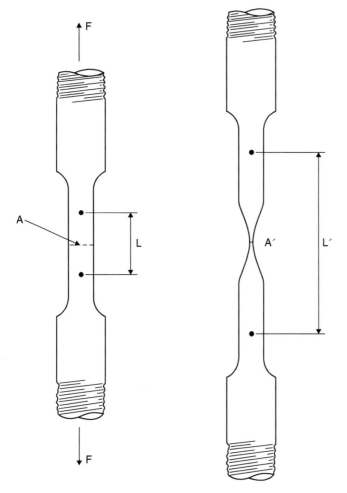

Figure 6.10 A tensile test specimen.

crumbling or cracking. Shear strength is exhibited by a material when resisting slightly displaced opposing forces, tending to cause adjacent parallel planes of the material to slip over one another; shear strength is often about 50% of the tensile strength. There is also torsional strength; this is exhibited when a material resists a twisting force and is often about 75% of the tensile strength. To ascertain these strengths the material can be subjected to compressive, shear and torsional stresses and strains, and in each case the appropriate load values noted at point of failure.

6.6.4 Hardness

There are several techniques used for determining the hardness of a material, but most industrial methods measure the resistance of the material to penetration of a small sphere, cone or pyramid. Initially, the penetrator and material are forced into contact with a predetermined initial load. An increased load is then applied to the penetrator and the hardness reading is obtained by noting the difference in penetration caused by

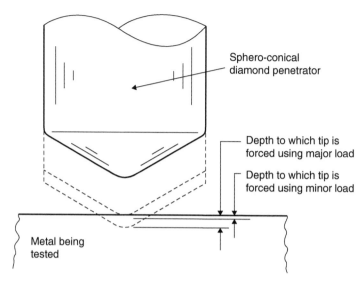

Sphero-conical
diamond penetrator

Depth to which tip is
forced using major load

Depth to which tip is
forced using minor load

Metal being
tested

Figure 6.11 Rockwell hardness test.

the final load as compared to the initial load. One of the more common scales used is the 'Rockwell' test in which the load applied and shape of the penetrator are specified, see Figure 6.11.

6.6.5 Ductility

The ductility of a material indicates how much it can be bent, drawn, stretched, formed or permanently distorted, without rupture. Normally a material that has high ductility will not be brittle or very hard. Conversely, hard materials are often brittle and lack ductility. The tensile test can be used as a measure of ductility by calculating the percentage elongation of the specimen upon fracture.

6.6.6 Impact Resistance

A material may be hard and have a high tensile strength, but it may still be unsuitable for an application that requires it to withstand impact or sudden load. A number of tests are used to determine this impact resistance. Two common tests are the Izod test and the Charpy test, in which a notched specimen is struck by an anvil mounted on a pendulum. The energy required to break the specimen is an indication of the impact resistance of the material. For some common engineering materials this energy can vary quite dramatically with temperature, even at temperatures close to ambient. The Izod test is illustrated in Figure 6.12.

6.6.7 Fatigue Resistance

The yield strength is useful for designing components that are subjected to a static load, but for cyclic or repetitive loading the endurance or fatigue strength has to be known. This is found by loading the part and subjecting it to repetitive stress. Usually, a number

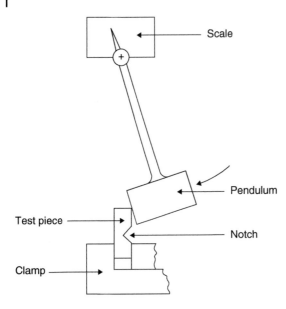

Figure 6.12 Izod impact test.

Scale

Pendulum

Test piece

Notch

Clamp

of specimens of a material are tested at various loadings and the number of cycles to failure are noted.

6.7 Conclusion

The variety of materials available to the engineer is vast. He or she must make optimum use of them if his product is to be competitive in the market place. A knowledge of materials is therefore essential not only for product design but also for consideration of how the product will be manufactured. The next chapter now describes how some of these materials are produced from natural resources.

Review Questions

1 Give an example of each of the following types of material together with a typical use: (a) ferrous, (b) non-ferrous, (c) organic, (d) inorganic.

2 What type of atomic bonding is exhibited by metals and what implication has this with respect to their properties?

3 What is meant by 'grain formation' in metals?

4 Discuss the significance of grain structure and the importance of recrystallisation.

5 What is an alloy?

6 In steels, how are properties generally affected by carbon content?

7 Describe two ferrous alloys and their uses.

8 Why are ferrous metals hardened and tempered?

9 What are polymers and why are they useful?

10 What is the main difference between a thermoplastic and a thermosetting plastic?

11 Describe two polymers and their uses.

12 What combination of properties makes ceramics attractive for today's product designers?

13 Why have 'composites' become desirable for products such as the structural elements of aircraft?

14 What do the terms 'stress' and 'strain' imply when materials are subjected to an axial load?

15 Briefly review the types of material properties of interest to product designers and manufacturers.

7

Materials Production

7.1 Introduction

Manufacturing processes are the means used to change a material from one state to another state of higher value. For example, iron ore is smelted to make pig iron that is converted to make steel. Next, the steel might be continuous cast before rolling to make sheet steel. This can be blanked to make car body frames; these frames are then assembled to other components to produce a finished car.

Although manufacturing is a secondary industry, it relies on primary industries such as mining and quarrying to supply the raw materials. Some materials used by manufacturing industry were examined in Chapter 6; in this chapter, we will look at the production of the metals iron, steel, copper and aluminium, and also basic polymer production. Most metals require to be mined since they are found in naturally occurring mineral deposits known as 'ores'. In the natural state the metals are usually combined with other, undesired, elements. This 'gangue', is removed to leave the concentrated ore ready for the metal extraction process.

7.2 Ferrous Metals Production

7.2.1 Pig Iron

Pig iron is the initial raw material for all ferrous metals. The composition of the pig iron will determine how it will be used. Along with iron the alloys usually contain between 3 and 4% carbon plus a total of about another 3 or 4% of the elements silicon, manganese, sulfur and phosphorous.

Figure 7.1 shows the blast furnace, which is used for smelting the ore along with coke and limestone. The ore may be of different types depending on its source – magnetite, which is 72% iron; haematite, 70%; limonite, 60–65%; siderite, 48%; and taconite, which although containing only 20–27% iron is normally pre-processed close to where it has been mined to produce pellets that are 63% iron and suitable for the blast furnace. Together with heat, the coke and the limestone produce the necessary chemical reactions in the ore.

The actual process is as follows. The blast furnace is composed of an outer shell of steel plates encasing a lining of refractory bricks, thus creating a hollow cylindrical chamber

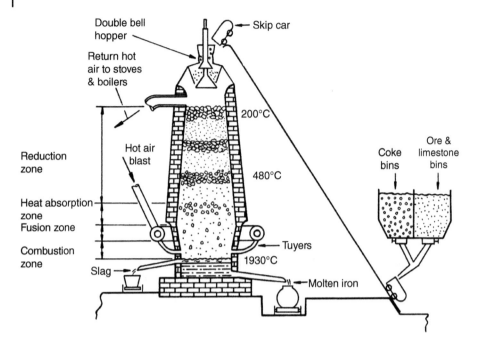

Figure 7.1 Blast furnace for making iron.

approximately 60 m high and 8 m in diameter. The process is a continuous one with the furnace operating 24 hours per day. The daily capacity of a typical furnace ranges from 1000 to 4000 tonnes. To produce 1000 tonnes of pig iron the total charge might consist of about 2000 tonnes of ore, 800 tonnes of coke, 500 tonnes of limestone and 4000 tonnes of hot air. The heated air is blasted into the furnace through water cooled nozzles called *tuyers*; these can be seen at the base of the furnace in Figure 7.1. Passing through the incandescent coke the air causes large volumes of carbon monoxide to be produced; this together with the carbon in the coke causes a chemical reaction in the ore called 'reduction', a term for the removal of oxygen from a substance. Thus, the iron oxides are reduced to iron. The limestone promotes the reduction process and additionally combines with the undesired oxides of calcium, magnesium, silicon and aluminium to form a 'slag'. This slag is lighter than the molten iron and therefore floats to the surface where it can easily be drained off. The ore, limestone and coke are fed in constantly at the top of the furnace while the molten pig iron at the bottom is tapped off about every five hours.

7.2.2 Cast Iron

Using a combination of pig iron and scrap, cast iron is produced in a furnace called a 'cupola' using coke as fuel. In a manner similar to that used for pig iron, the molten cast iron is tapped off at the bottom. Cast iron composition, discussed earlier in Chapter 4, is also similar to that of pig iron. The major difference between the two types is in the form of supply. Pig iron is supplied as cast bars called 'pigs', whereas cast iron is supplied in the form of castings of a design specified by the customer. Even this distinction is blurred in practice, as extremely large castings are sometimes made by using pig iron direct from the blast furnace.

7.2.3 Steel

Steel is an alloy of iron with a little carbon plus other alloying elements to provide specific desired properties. The major problem with early steel was the slag waste from the ore that, especially when trying to make large volumes, would remain in the finished structure of the steel and so weaken it. However, in the second half of the nineteenth century the Bessemer process was developed, which allowed large volumes of steel to be produced as cheaply as cast iron had been. The principle of the process was to force air through the melt and so oxidise the excess carbon. A few years after its introduction another process, the open hearth furnace, was introduced. This also allowed the production of good quality steel by melting the ingredients of the charge in such proportions that the excess carbon and oxygen were driven off in the form of carbon monoxide. These two processes have now been superseded by the basic oxygen process and the electric arc furnace.

7.2.3.1 The Basic Oxygen Process

Between 65 and 80% molten pig iron from a blast furnace is used plus scrap, lime and fluorospar. The production rate is around 300 tonnes of steel every 45 minutes. The scrap is first loaded into a vessel lined with refractory material, then the pig iron is poured in. The vessel is held vertically, as shown in Figure 7.2, and a water cooled oxygen carrying lance is lowered to a height of between 1 and 2 m above the molten charge. When the oxygen is blown through the lance and over the surface of the bath the metal immediately ignites and the temperature rises close to the boiling point of iron, which is around 1650°C. Carbon, silicon and manganese are oxidised and the lime and fluorospar added to collect various impurities such as phosphorous and sulfur in the form of slag. Unlike pig iron production steelmaking is not a continuous process. When a batch of steel is complete, the oxygen is shut off and the lance is retracted through the hood. The furnace is then tilted in one direction to pour off the slag, then after testing the melt it is tilted in the opposite direction to allow the steel to be poured into the ladle transfer car.

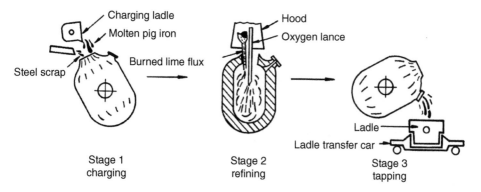

Figure 7.2 Stages in the basic oxygen steelmaking process.

7.2.3.2 The Electric Arc Furnace

A sketch of the furnace is shown in Figure 7.3. Instead of pig iron it is charged with carefully selected steel scrap and alloying materials. The production rate is around 150 tonnes every 3 hours. Typically, it will be used to produce melts for ingots and castings of stainless steel, tool steel, heat resistant steels and other general purpose alloy steels. The recycled scrap is loaded through the charging door or the top of the furnace. Three graphite electrodes are held in the roof and arranged to sit just above the scrap heap. A three phase current arcs back and forward between the electrodes and the charge creating the necessary heat for the process.

7.2.4 The Integrated Steel Plant

The most efficient way to produce steel is to create an integrated plant. These complexes require many square kilometres of land, kilometres of roads, expensive equipment, power and labour. Typically the raw materials of ore, coal, limestone and alloying materials arrive by rail or sea and are stored and blended. The coal is changed to coke in a coking plant; the materials are then loaded into the blast furnace for the production of pig iron. This can then be transferred in its molten state to the basic oxygen furnace for conversion to steel. The molten steel can then be poured into the mould of a continuous casting plant for direct production of steel slabs or into an ingot mould (see Chapter 8). The ingots are then transported to a rolling mill (see Chapter 10) for rolling into slabs. The slabs from either the rolling mill or the continuous casting process are then taken to other rolling mills for rolling into strip, plate, bars or other forms. The advantages of an integrated plant are savings in energy, transport and organisational costs. Global demand for steel can vary depending on the world and local economies. At the time of writing there is overcapacity in steel production with China producing as much steel as the rest of the world's output combined. Also, the utilisation of other materials has led to a reduction in the requirements for steel in many products. These factors

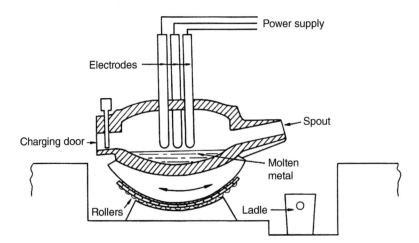

Figure 7.3 Electric arc furnace for steelmaking.

have led to the demise of steel plants in countries where steelmaking was previously a major industry.

7.3 Non-Ferrous Metals Production

Non-ferrous metals are seldom used in their pure state since they lack physical strength, in fact less than 20% of metals used in industrial products are non-ferrous. However, since they do have useful properties such as resistance to corrosion, high electrical conductivity and malleability, they are often used as alloys with other materials. The natural colours of metals such as aluminium, copper, tin and their alloys also provides a selection of materials that are aesthetically pleasing and can enhance the appearance of a product. Casting of these materials is usually simple but welding is often difficult especially with those of lower density. Machining of some, such as aluminium and copper alloys, is easier than that of steel whereas titanium and nickel are more difficult.

7.3.1 Copper Production

The major sources of copper are the sulfide ores such as copper pyrites, chalconite and bornite. When mined these minerals are found mixed with waste so that only about 4% of the mined material is copper. To obtain the copper from the raw material a number of production stages can be identified, as show in Figure 7.4.

In the first stage the ore is crushed to reduce it to a fine powder. It is then concentrated by a flotation process in which a tank is filled with a suspension of powdered ore in water. Small quantities of frothing agents are added and air is bubbled through the suspension. The desired particles are carried to the surface where they form a froth that can be removed by skimming, the undesired residue remaining in the tank. In the second stage the concentrate is heated with other materials called fluxes that allow a molten mixture of copper and iron sulfides to form under a slag that contains most of the remaining waste. The slag is run off continuously and the metal sulfides are periodically tapped and transferred in the molten state to the next stage. The furnaces used are termed 'reverberatory', because the heat from the melt is reflected back downwards by a low roof, and they are fuelled by gas, oil or pulverised coal. In the third stage the melt is poured into a large cylindrical vessel lined with refractory material; this is termed the

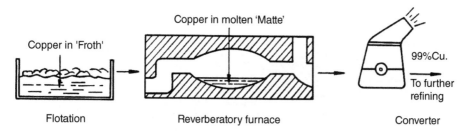

Figure 7.4 One method of copper production.

'converter'. Air is blown through tuyers and into the melt. This causes the iron to oxidise and when silica is added to the melt they combine to form a slag; this can then be tapped off, thus effectively removing the iron. Continued blowing removes unwanted sulfur as sulfur dioxide gas is created. After a period of approximately 10 hours crude molten copper, which is about 99% pure, is left in the converter. This can either be east into slabs called 'blister copper' or while still molten it can be transferred to a fourth stage for further refining. Using electrolytic refining, copper up to 99.99% pure can be obtained.

7.3.2 Aluminium Production

The main source of aluminium is bauxite. This is a naturally occurring mixture of gibbsite and diaspore containing 45–60% aluminium. The impurities present are typically iron and titanium oxides and silica. The main producers of the ore are countries like Jamaica and Australia. However, production of aluminium demands large amounts of electricity, around 13–18 kW h kg^{-1}. High volume production therefore started historically near sources of hydro-electricity such as in Scotland, Norway and in the Niagara Falls area in the USA. Main users of aluminium today are countries like Japan and the USA who purchase the ore and then do their own refining. Two stages can be identified in aluminium production.

In the first stage, alumina, that is aluminium oxide, is obtained by removing the water from the ore, crushing it and placing it in a hot solution of caustic soda, that is, sodium hydroxide, in a pressure vessel. Under high temperature and pressure the alumina is dissolved and the undissolved impurities precipitate out and settle as a red mud (which has to be carefully disposed of as it has been a troublesome pollutant). The separated liquid solution is cooled and aluminium is precipitated as hydroxide. Pure aluminium oxide is obtained by heating at about 1000–1500°C to drive off the combined water.

In the second stage (see Figure 7.5), the electrolytic extraction of the aluminium is practicable only if the alumina is first dissolved in some other substance to form a liquid solution that is capable of conducting electricity. Cryolite, that is natural sodium aluminium fluoride, is used for this purpose. The solution, which is red hot and around 950°C, is held within a cell composed of steel plates and lined with carbon. Suspended in this are thick carbon anodes. Direct current is passed via the anodes through the electrolyte to the lining of the cell. This causes the alumina to split into aluminium and oxygen. The oxygen burns the anodes to form carbon monoxide and carbon dioxide. Molten aluminium is produced and, as its density is greater than that of the alumina and cryolite solution, it sinks to the bottom of the cell and forms a layer. This layer is

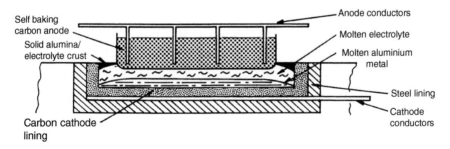

Figure 7.5 Electrolytic reduction cell for aluminium production.

then periodically tapped to provide aluminium that is 99.8% pure. It is sometimes refined further in a subsequent cell to provide aluminium 99.99% pure. It may also be noted that increasing amounts of 'scrap' aluminium are being used; this recycling is much more efficient as it uses only about 5% of the energy required for production from bauxite.

7.4 Forms of Material Supply

After the metal has been produced to sufficient purity and alloyed with other metals to provide desired properties, it then undergoes further processing to produce a form suitable for further working. These forms are created by the 'primary processes' that include casting, rolling, forging and extrusion. These processes, and others, are described later in the book. Figure 7.6 illustrates some of the standard forms in which material is supplied to manufacturing companies.

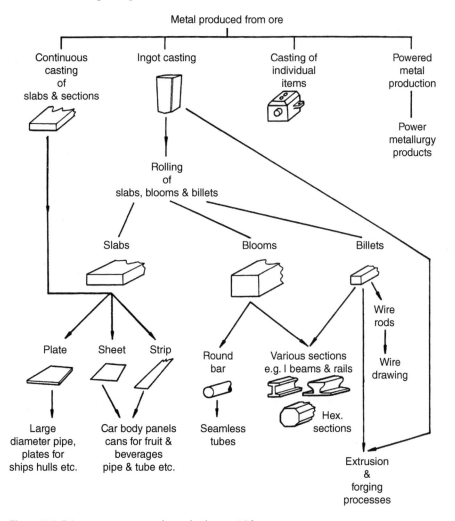

Figure 7.6 Primary processes and standard material forms.

7.5 The Primary Production of Plastics

Plastics are mostly synthetic polymers and they are produced today, along with synthetic rubbers, from petroleum products. However, although about 90% of the world's polymers come from petroleum a brief description of the manufacture of the natural polymer, rubber, is given first.

Plantations of rubber trees are grown in a suitable climate; Thailand, for example, is at time of writing the world's largest producer. A helical cut is made around the trunk of the rubber tree and the watery sap, that is, latex, which is a polymer of isoprene, is tapped off. This is then passed through filters to remove impurities. It is next coagulated into a relatively weak, soft and inelastic solid by the use of acid and a squeeze drying process or by centrifugal action. By further heating the rubber and mixing it with sulfur the rubber is cured, that is, vulcanised. In this process the sulfur atoms form cross links between the long polymer molecules, thus restricting their ability easily to slide over each other. This produces a harder, stronger rubber more suitable for engineering purposes. If the amount of sulfur is sufficiently large, then a hard rigid material called ebonite is produced.

Synthetic rubbers are mostly produced from petroleum industry products. For example, SBR or styrene butadiene rubber, which is used for tyres, transmission and conveyor belts, shoe soles, cable insulation and so on, is a copolymer made from the monomers styrene and butadiene. The ability to control the chemical process allows a variety of rubbers to be made to suit various applications, for example, for resistance to chemicals neoprene and nitrite rubbers are used, or silicon based types are produced that are resilient to extreme thermal cycling making them suitable for seals in aerospace products. Combinations of natural and synthetic rubbers may be used to obtain materials with any desired combination of properties.

Just as in the primary production of metals, the initial stages of plastics production are carried out in large expensive complexes called refineries for processing crude oil. The oil is composed of a mixture of hydrocarbons, that is, compounds of hydrogen and carbon. These are separated into their components, called fractions, in a fractionating column (see Figure 7.7). Each hydrocarbon has its own boiling point; this means that the mixture can be separated by fractional distillation. In this process the oil is boiled by passing superheated steam through the column; the fractions condense and separate at different levels in the column; each is drawn off at the appropriate height. The bubble caps allow the vapour to rise freely through the column but prevent condensed material from running back down. As can be seen from the sketch the naphthas, with boiling points around 120°C, can be drawn off somewhere between the kerosene and petrol levels; it should be noted that all boiling points shown in Figure 7.7 are approximate.

Continuing with the example of the naphthas, the fractions can then be 'cracked' to produce the gases ethylene and propylene. 'Cracking' is the name given to the process of breaking down larger molecules into smaller ones. This is usually done with the assistance of a catalyst to accelerate the process. The ethylene is then polymerised to form polyethylene, the structure of which was shown in Figure 6.6b, or PVC, polyester or synthetic rubber. The propylene is polymerised to produce polypropylene, which is the basis for polyurethane, acrylic fibres, nylon and some foam plastics. Chemicals other than the naphthas are also used to produce plastics, for example, toluene is used to produce benzene that can be further processed to give a range of useful plastics.

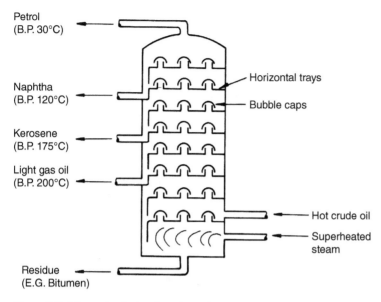

Figure 7.7 Schematic of a fractionating column.

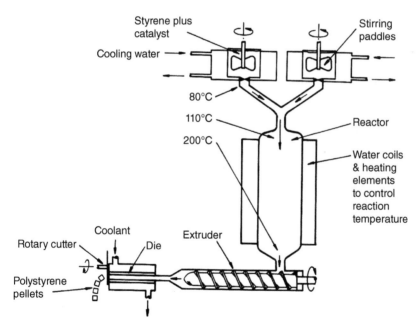

Figure 7.8 Manufacture of polystyrene.

Ethylene is also used to produce styrene, which polymerises to produce polystyrene. Figure 7.8 shows schematically the production of polystyrene; this serves as a typical example of the plastic production process. A partial polymerisation by mixing the styrene with a catalyst in tanks begins the process. The valves to the preliminary mixing tanks can be opened and closed as necessary to maintain a continuous feed to the reactor. It is in the reactor that full polymerisation takes place, the temperature of this heat generating process is closely controlled. By the time the material reaches the base of the reactor it is a hot liquid plastic at around 200°C. It then passes into a screw extruder that forces the material through a water cooled die. As it emerges from the die the cooled solid plastic is cut up into pellets. As polystyrene is a thermoplastic the pellets will be re-melted when subsequently used in one of the component moulding or forming processes.

Review Questions

1   What are the main ingredients necessary to produce pig iron?

2   Briefly describe the process of pig iron production (no sketches are necessary).

3   Describe one modern method of steel production.

4   What is an 'integrated' steel plant and what advantages are to be gained from the integration?

5   What is 'blister' copper and how is it obtained?

6   Discuss why large scale aluminium production first started in areas where cheap electricity was available.

7   Sketch five different types of standard material form.

8   Natural rubbers are vulcanised during their manufacturing process; briefly state what vulcanisation is and why it is necessary.

9   What is the major source of the world's plastics and synthetic rubbers?

10   What is the purpose of a 'fractionating column' and how does it work?

11   Briefly describe how polystyrene is produced.

Part III

Manufacturing Processes

8

Casting

8.1 Introduction

Casting is the process of pouring or injecting molten metal into a mould and then allowing it to solidify. Products weighing many tonnes or just a few grams can be produced in a variety of surface finishes and accuracies; internal cavities are also possible. It may be that casting is only the first process in a series of operations that will lead to the finished product, or the desired item may be cast with such precision that it can be fully utilised immediately in the 'as cast' condition.

The process is probably one of the earliest, dating from about 4000 BC. However, in recent years much refinement has taken place and a variety of techniques have been devised to satisfy different needs. There are still some problems experienced generally with castings due to their metallurgical structure, as relative to other processes they tend to have lower toughness and ductility and porosity can occur.

As the metal cools in the mould from the molten state and solidifies, it also begins to contract. This means that, if parts of the casting cool before others, depressions can appear on the surface and cavities can occur internally. The problem can be decreased by ensuring that there is always molten metal available to fill the spaces as they begin to grow. This is achieved by providing a 'head' of metal in reservoirs in the case of gravity fed moulds and injecting the metal under pressure in the die casting processes.

8.2 Ingot Casting

This is a preliminary process in which the metal is cast in the form of ingots. These are usually further worked by rolling, forging or extrusion to produce sheet, strip, rod, tube or other forms such as 'T' beams for the construction industry. Any type of metal may be cast into ingots but steel is primarily considered here.

Steel ingots are often cast into large iron moulds. These hold several tonnes of metal and are tapered slightly so that the mould can be lifted clear of the cooled ingot. A phenomenon that appears here is 'piping'. This happens because of the relatively rapid cooling of the metal in contact with the mould surface; as the metal cools from the outside in so the shape shown in Figure 8.1a is formed. These 'pipes' are undesirable as

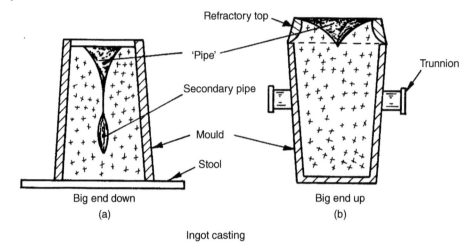

Big end down
(a)

Big end up
(b)

Ingot casting

Figure 8.1 Ingot casting and effect of mould orientation on piping.

impurities tend to gather in the vicinity of the pipe surface; this surface also tends to oxidise. When the ingot is further worked, say by rolling, then this surface may become an internal feature of the rolled component and thus be a source of weakness in the finished product. Slow pouring of the metal can lessen the problem, or for smaller ingots reversing the mould and mounting it on trunnions to allow it to be rotated for ingot removal and using a 'hot top' as shown in Figure 8.1b also provides good results. In both methods the top region must be cut off and rejected due to the high impurity content.

8.3   Continuous Casting

Continuous lengths of slabs, bar and other sections are produced in ferrous and non-ferrous alloys using this process. The product may be used as cast or further worked to give stronger directional properties. Continuous casting is a process that is popular for producing steel slabs for rolling work. It is more efficient for this than casting ingots, transporting them to energy hungry 'soaking pits' that reheat them to hot working temperature before being rolled by a massive rolling mill into slabs. It also removes the problem of rejecting the top part of the casting because of impurities, the problems of piping and mould spatter and the cost of the ingot mould.

Continuous casting installations often comprise part of an integrated steel plant as described in Chapter 7. The molten metal is poured into a water cooled mould open at the top and bottom, as shown in Figure 8.2. There are a number of variations in the process but the two methods shown illustrate the basic principle. The retractable base is drawn downwards at a rate that allows the metal to solidify and yet keeps pace with the metal being poured. As the metal passes through the water cooled moulds it is transported along a path using rollers. Careful control must be maintained over the cooling rate and speed of casting as the material uses its solidified skin to support itself. The mould is often made of copper and to ease the movement of the casting it is usually vibrated and lubricated with a graphite type material.

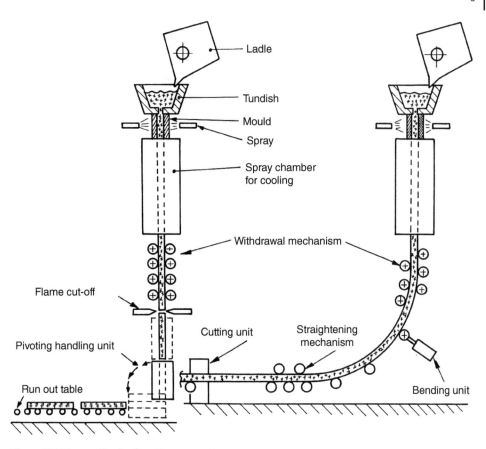

Figure 8.2 Two methods of continuous casting.

8.4 Sand Casting

There are essentially two types of sand casting; these are 'green sand' and 'dry sand'. In the former the moist sand contains between 2 and 8% water. This is the most common method used to produce castings weighing from under half a kilogram to around 4 tonnes, and where high precision and surface finish are unimportant. It can be used for ferrous and non-ferrous metals and the easy collapsibility of the mould reduces the stress and strain induced in the casting, so making it suitable for intricate work. The dry sand technique is used for large and very heavy castings. The sand is strengthened by giving the mould surfaces a refractory coating, which is dried before pouring the molten metal. This section is primarily concerned with green sand casting.

Figure 8.3 illustrates the principal elements necessary for the green sand casting of the pipe shown in Figure 8.3a. First a wooden pattern is made by a patternmaker. This pattern, shown in Figure 8.3b, includes an allowance for contraction of the metal in the mould and if necessary a slight taper or 'draft' on the surface to allow easy removal of the pattern from the sand. As a hollow section is required a 'core' must be made. This necessitates the construction of the core box shown in Figure 8.3c.

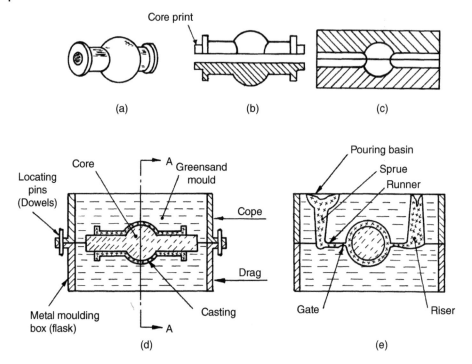

Figure 8.3 Sand casting elements. (a) Required component, (b) upper and lower patterns, (c) core box, (d) an assembled mould and (e) view of mould section A – A.

The sand for the main mould must have the following properties. It must be able to withstand high temperature, that is be refractory. It must be able to retain its given shape, that is, be cohesive. And it must be able to permit gases to escape, that is, be permeable. Most of the sand, up to 90%, is composed of silica that provides the refractoriness, between 4 and 8% is clay to provide cohesiveness and the remainder may be composed of iron oxide, coal dust and water, which contribute to the permeability once subjected to the heat of the molten metal. The core produced in the core box is composed of similar sand to the main mould, but it is further strengthened by additional bonding agents.

Figure 8.3d and e shows an assembled mould. The moulding box containing the green sand mould has an upper portion, the 'cope', and a lower, the 'drag'. The two parts are held in precise relation to each other by means of the locating dowels.

As well as the mould cavity it can be seen that there is also a pouring basin, a sprue, runner and gate system and a riser. The pouring basin provides a facility for efficient entry of the molten metal, the sprue provides a reservoir and 'head' of metal, the runner is designed to carry ample supplies of molten metal rapidly to all areas of the mould cavity, and the gate is designed to provide a controlled flow of molten metal into the cavity and a point at which the casting can easily be broken off from the rest of the sprue and runner material. The riser accepts the overflow of metal from the cavity and again acts as a reservoir and 'head' of metal.

To produce the mould as shown, the cope and drag halves are initially made separately. The bottom pattern half is laid flat side down on a moulding board with the drag half of the moulding box around it. Moulding sand is then riddled over the pattern and rammed down sufficiently for the particles to adhere together; the sand is then cut level with the edge of the box. The box is turned over and the pattern removed. This process is repeated with the top half of the pattern and wood forms to make the runners, risers and so on. After making the sand core in the core box the core is laid in the impressions formed by the core prints in the drag section. A fine layer of dry, clay free, parting sand is sifted onto the sand surface; this prevents the cope from adhering to the drag. The upper half is then placed on the lower and the pouring of the molten metal can begin. When the casting has cooled and has been removed the sand core is broken up and shaken out, thus leaving the completed hollow pipe. Although a simple product has been shown here the same technique is followed for most products made by this casting process.

The advantages of sand casting are that almost any metal can be used, there is virtually no limit on the size and weight of product, high complexity is possible, tooling costs are low and the route is direct from pattern to mould. The disadvantages are that some machining is usually necessary, surface finish is poor, it is difficult to achieve close tolerances and it is not practical to cast long thin sections.

8.5 Centrifugal Casting

Long, hollow, cylindrical castings are commonly produced without the use of a central core using this process. The permanent cylindrical metal mould is rotated at a high speed, usually between 300 and 3000 rpm, while molten metal is poured into it. The centrifugal force created by the spinning, which pushes the molten metal against the cylindrical surface of the mould; this produces a hollow cylinder of uniform wall thickness. A good dense structure is obtained with all of the lighter impurities concentrated on the inner face, thus allowing them to be easily removed by machining the bore.

The advantages of the process are that the centrifugal force facilitates complete filling of the mould, the gases and impurities concentrate on the inner surface for easy removal, a good solid outer surface is obtained and the mould is relatively simple. The main disadvantages are that the process is limited to symmetrical products and that if alloys of separable compounds are being cast these compounds may not be evenly distributed.

8.6 Shell Moulding

This is essentially a sand casting process in which the clay used for bonding in the green sand process is replaced by a synthetic material of the phenol- or urea-formaldehyde type. Figure 8.4 shows the five stages of the process. The patterns must be made of metal due to the relatively high temperature required for the setting of the bonding material. The process is carried out as follows. (1) The pattern and plate are heated to around 250°C and coated in silicone oil to aid stripping. The dump box containing the sand and the thermosetting resin mixture is attached. (2) The box, which is mounted on trunnions, is rotated and the sand mixture falls over the pattern. The resin melts and hardens, the resulting thickness of the shell depending on the pattern temperature and the length of time the sand is allowed to remain in contact with the pattern. (3) The box is again inverted and the uncured sand falls back into the box; the partially cured material is left

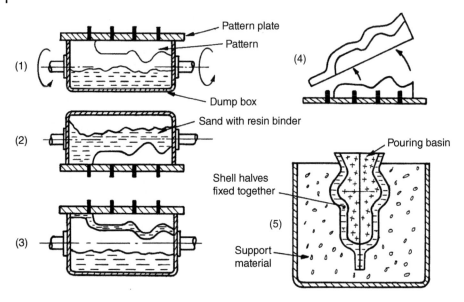

Figure 8.4 Stages in shell moulding.

adhering to the pattern. The shell thickness is usually about 3 mm. The shell and pattern are then removed from the box and placed in an oven for about two minutes at 315°C; this finally cures the shell. (4) The hard shell is stripped from the pattern. (5) Two shells are fixed together to form the completed mould. They are usually placed in a pouring jacket and supported by sand or shot. The molten metal is then poured in. This process provides a rapid production rate, a good grain structure and a good surface finish and accuracy, thus minimising finishing operations. However, it does require expensive patterns, equipment and resin binder, and there is a limit to the size of part that can be made.

8.7 Full Mould Process

A consumable polystyrene pattern is used in this process. Complex product shapes, together with the pouring basin, sprite and runner systems are formed in foamed polystyrene. The patterns are packed in sand and when the molten metal is poured they instantaneously vaporise. Though originally a one off type process suitable for prototype production, the process has been applied for large volume production using metal dies for mass producing the polystyrene patterns.

The process has the advantages that expensive wooden patterns do not have to be made and that as the pattern does not have to be withdrawn, the need for taper or draft on the pattern is eliminated. Some machining is usually required to achieve the desired precision and surface finish, though this has been reduced due to improvements in the process.

8.8   Investment Casting

Expendable patterns are again used here. However, in contrast to the full mould process high precision products with good surface finishes are obtained in large volume production. The 'lost wax' investment casting process is described visually in Figure 8.5.

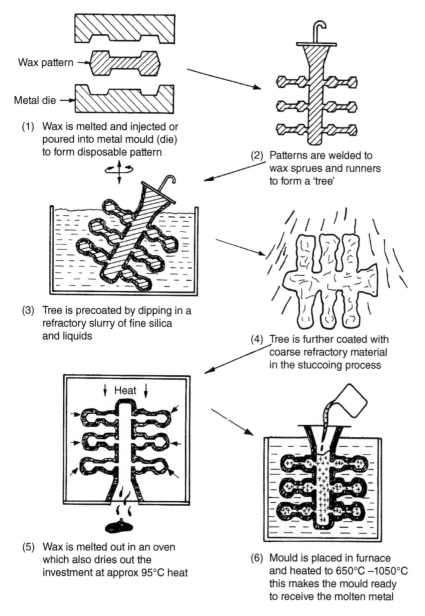

Figure 8.5 Investment or 'lost wax' casting process.

1) A metal die is made with a cavity conforming to the desired shape of the component. Wax is injected or poured into the die cavity. When the wax has cooled the die is opened and the wax pattern is removed. A large number of patterns are usually made.
2) The wax patterns are welded onto a wax sprue and runner system to form a 'tree'.
3) The wax tree is dipped into a fine slurry of refractory material and plaster. This will provide the smooth interior surface of the finished mould, which is capable of replicating intricate detail.
4) The tree is now further coated by dipping or spraying with a coarser refractory material to give the mould additional strength. This step is repeated a number of times to increase the wall thickness.
5) The tree is placed upside down in an oven at a temperature of approximately 95°C. This melts out the wax and dries out the investment or coating.
6) The mould is now preheated in a furnace to a temperature of between 650 and 1050°C. This allows the poured molten metal to flow freely into all corners of the mould. It also promotes sympathetic contraction of the mould and casting thus providing better dimensional control. When the casting has cooled the mould is broken away leaving the finished components attached to the sprue and runner system.

The process provides high dimensional accuracy and an excellent surface finish, components of extreme intricacy can be produced in almost any metal, a good size range is available from a few grams to around 40 kg. The main disadvantages of the process are the high cost of the metal dies and the time consuming manufacture of the moulds.

8.9 Die Casting

For high volume production of non-ferrous and some ferrous items, die casting is the most likely casting process to be used. In this process molten metal is injected under high pressure into precision made metal dies. These dies have sprue, runner and gate systems just as in the other casting processes. The metal is held in the die under pressure until solidification takes place; the finished item is then automatically ejected. The machines and dies are expensive and are therefore financially viable only where large numbers of castings are required. The machines come in two forms, that is, 'hot chamber' and 'cold chamber'; these are shown schematically in Figures 8.6 and 8.7. In both sketches the hydraulic rams and toggle arrangement for clamping the moving portion of the die to the fixed die during metal injection would be located to the right of the chamber.

In the hot chamber process a 'gooseneck' duct is partially submerged in the molten metal held within the crucible. There is an intake port opening from the crucible into the duct. A plunger is shown operating vertically. On the downward stroke of the plunger the molten metal is forced into the die. When the plunger returns to its original position the chamber is refilled with the molten metal flowing through the intake port. This produces a fast operation and short cycle times are obtained. The direct injection of the molten metal from the melting pot into the die chamber makes this a relatively efficient process. It is not used with the higher melting point metals but is commonly used for the zinc and tin based alloys. It is also suited to automated operation.

For the higher melting point alloys, such as those of aluminium, magnesium and copper, the cold chamber process is used. In this process the molten metal is ladled into the 'shot' chamber; it is then pushed forward by the plunger and injected into the die cavity.

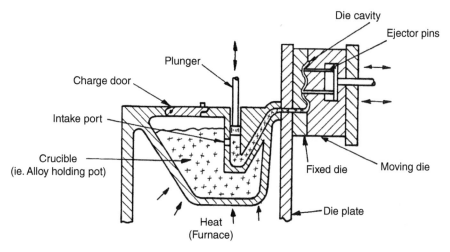

Figure 8.6 The hot chamber die casting process.

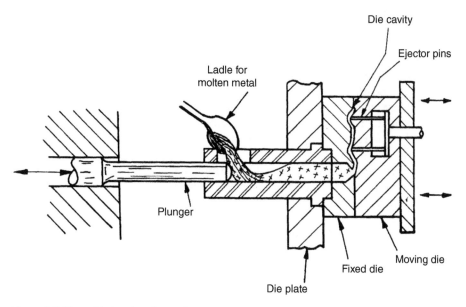

Figure 8.7 The cold chamber die casting process

Due to the high pressures involved, for example, in the hot chamber process, the metal is injected at about 15 MPa and in the cold chamber between 20 and 70 MPa; large forces must be applied by the machines to keep the dies closed during operation. Thus, die casting machines are rated according to the force they can apply, typically they range from about 25–3000 tonnes or approximately 200–27 000 kN.

The advantages of die casting are: high production rates, high precision, high quality surface finishes, high integrity castings with low porosity and good grain structure, and complex castings with thin walls can be made. These advantages have to be paid for in the expensive machines and tooling that are required. It is therefore not a viable process to select if quantities and production rates are not high enough to produce an adequate return on investment.

8.10   Defects in Castings

A number of defects can occur in casting; the more common are listed next with a comment on their likely causes.

1) *Shrinkage.* This is evidenced as internal porosity or cavities, or as depressions on the surface of the casting. Improvements and additions to the risers in the mould design should remove the problem.
2) *Scabs.* These are rough lumps of excess metal on the surface of sand castings. Probably caused by poorly rammed sand or sand with insufficient binding material.
3) *Fins.* This is the term given to the excess metal occurring along mould parting lines. It is caused by poor fitting of the mould halves and possibly other mould components such as cores and inserts.
4) *Blow holes.* Internal cavities in the casting caused by gas. The gas is originally present in the molten metal and as the metal solidifies it is rejected from solution so producing the cavities. Commercial degassing agents should be added to the melt.
5) *Blows.* Cavities on or near the casting surface. They are caused by poor mould venting and, in the case of sand casting, caused by gases emanating from the mould due to excess moisture and poor permeability.
6) *Inclusions.* These are slag, oxide or sand particles evident in a finished sand casting. They are caused by poor skimming or fluxing of the melt prior to pouring, and poorly made sand moulds lacking cohesiveness.
7) *Cold shuts.* 'Seam' like discontinuities in the casting. They are formed when two metal streams meet inside the mould but have insufficient fluidity to allow them to break the oxide films that separate them. Improved mould design is necessary to ensure that the runner and gating system carries the molten metal rapidly enough to all areas of the mould cavity.
8) *Misruns.* Incomplete castings caused by the molten metal not penetrating throughout the mould. Usually caused by the metal being poured or injected at too low a temperature or pressure.
9) *Hot tears.* Cracks in the casting. Caused by stresses set up during contraction of the casting through poor mould design and in the case of sand casting through the use of cores lacking in collapsibility.
10) *Porosity.* This is said to exist if fluids can be transferred through the metal even though the pores are invisible. May be caused by contamination of the metal.
11) *Warping.* This is distortion of the finished casting due to poor mould design.

8.11   Cleaning of Castings

When the casting is removed from the mould it is attached to the sprue, runner and riser system. It must be separated from these and in addition any 'fins' or other surface protrusions must be removed. The sprue and runner system is removed at the gating points by hammering, flame cutting, grinding or sawing. Pneumatic chisels may be used on larger castings. This initial process is termed 'fettling'. The sand left adhering to sand castings can be removed by vibration or shot/sand blasting. For polishing and cleaning the casting surface and removing fins and rough edges, tumbling the castings in a cylindrical steel drum is sometimes used.

8.12   When to Use Casting

There are a number of indicators that would suggest when it would be advantageous to select a casting process in preference to another manufacturing technique. These are outlined next.

1) When the product required has large heavy sections of complex shapes, a process such as sand casting may be more efficient than fabrication and machining.
2) When using materials that are difficult to machine, for example refractory materials, casting to fine tolerances using investment casting may be an attractive solution.
3) When large production volumes of small to medium sized complex components are required in zinc or aluminium alloys, then pressure die casting is often appropriate.
4) When the desired component has a complex structure, possibly with re-entrant angles and internal cavities, then casting could be the only technique possible.
5) When vibration effects have to be absorbed, for example, the damping of machine tools to decrease the effects of mechanical vibration and noise, then sand casting of components in grey cast iron is often carried out.
6) When it is necessary to produce items, such as machine pedestals and base plates, in which masses of metal have to be strategically placed, casting is usually preferable to fabrication.
7) When single items are required quickly, as in the case of prototypes, then the ability to produce a pattern rapidly, as in the full mould process, makes casting attractive.
8) When directional strength properties are not desired in the finished component.
9) When valuable or precious metals are to be used, casting provides a technique that minimises wastage.
10) When it is possible to design or redesign the product in such a manner that components that were originally separate can be integrated into one unit.

Review Questions

1   What is the phenomenon called 'piping' in ingot casting and why is it undesirable?

2   Continuous casting is now more popular for making steel slabs than the rolling process. Discuss why this is so.

3   What is the difference between 'green sand' and 'dry sand' casting?

4   Describe three requirements for a good moulding sand.

5   State one further requirement of the moulding sand necessary during and after the casting has cooled.

6   What is the function of a 'core' in sand casting and how is it made?

7   Explain the function of the following casting elements: sprue, runner, gate and riser.

8   State why the structural integrity of products made by centrifugal casting is generally good.

9   Describe, with the aid of sketches where necessary, the stages of the shell moulding process.

10   Give two reasons why the full mould process is particularly suited to prototype manufacture.

11   Describe, without the aid of sketches, the lost wax process.

12   Discuss the differences between the hot and cold chamber die casting process, and state under what conditions each may be used.

13   List six defects commonly found in castings and discuss how they could be avoided.

14   What is meant by the term 'fettling' and how is it carried out?

15   State the casting processes you would select to produce the following items and give reasons for your answer in each case:
1. Small complex steel components, 50 mm × 40 mm × 10 mm; a tolerance of ±0.05 mm and a good surface finish is required; quantities will be 2000 per week.
2. Motor car wing mirror mountings. A tolerance of ±0.1 mm and a special textured surface finish is required; the material will be a zinc alloy and 2000 per week are required.
3. Bowl shaped stainless steel pump casings for nuclear reactors. The outside diameter is 1 m and the wall thickness is 150 mm. The tolerance is ±2.5 mm as the component will be finished by machining; 4 per month are required.
4. A seamless, steel, pressure vessel 2.5 mm long, 0.5 m diameter, with a wall thickness of 10 mm; the tolerance required is ±1 mm; 500 are required and a good sound casting with no internal flaws is necessary.
5. A pedestal to hold a prototype cutting machine; accuracy and surface finish is unimportant, the material will be grey cast iron.

9

Cutting Processes

9.1 Introduction

There are a number of ways to cut materials and in this section we will consider some of the most popular. The first to be considered, and the most widely used in manufacturing industry, is *machining*. This generally involves using a cutting tool that removes material by chip formation. Another method is *thermal cutting*, this involves separating metal by localised heat using for example gas, electric arc or laser. Both of these methods are considered here in relation to metal cutting although machining by chip formation is also used for a wide variety of other materials. Finally, water jet cutting will be considered; this is popular for cutting polymers and composite materials when using industrial robots.

9.2 Sawing and Filing

Both sawing and filing involve cutting the material by chip formation. In sawing, a narrow strip of metal called the blade has a series of cutting teeth arranged along its edge. Although hand saws of various types have been used in woodworking for hundreds of years, modern industrial power saws used for cutting metal in the manufacturing industry are normally of three types. First, the hacksaw has a blade of limited length and is only slightly flexible; it is often used for cutting rods and bars to specified lengths. Second, the bandsaw has a flexible blade formed into a continuous band. The blade is usually made mainly of flexible high tensile strength alloy steel with tungsten carbide or high speed steel teeth bonded to one edge. The blade passes round two wheels, one of which is driven while the other maintains the tension. This blade passes through a slot in a worktable on which the material to be cut rests. The bandsaw can be used for cutting shapes from sheet and plate materials. Finally, the circular saw is a rigid disc with cutting teeth around the circumference. This is used for fast cut off operations and it usually leaves a clean smooth cut surface.

In filing, the cutting operation is similar to that of a saw except that the cutting teeth are much broader and arranged in series on a relatively broad rigid surface. Because only small amounts of material are removed, filing is not used to separate or cut off material but is used to obtain precise shapes and dimensions on a workpiece. Files can vary greatly in size and shape. Large coarse files are used for removing large amounts of material quickly; at the other end of the spectrum small fine toothed needle files are used

Essential Manufacturing, First Edition. Gordon Mair.
© 2019 John Wiley & Sons Ltd. Published 2019 by John Wiley & Sons Ltd.

for delicate work, and a wide variety of file cross-sections are available, for example, flat, round, square, triangular and so on. Filing machines use file segments joined to form a continuous band analogous to the handsaw, circular files are used in disc filing machines and straight solid files, similar to those that are hand held, are used in reciprocating die filing machines.

9.3 Basic Principles of Machining

In machining, a 'cutting tool' is held in a 'machine tool' such as a lathe, drill or milling machine. The machine causes relative movement of the tool with the work such that the material is cut. As the tool enters the work and moves through it a chip is produced, hence the term 'chip removal'; the operation of a single point tool is shown in Figure 9.1. Cutting tools may have more than one cutting edge; for example, a drill usually has two, a milling cutter may have 22 and the small hard particles that are the cutting edges on a grinding wheel can be numbered in their millions.

9.3.1 Advantages and Disadvantages of Machining

Advantages of machining are as follows. Generally speaking, machining is the best manufacturing process for achieving high precision components with specific surface finish characteristics. It is possible to achieve very high production rates with automated machines yet batch sizes from single units upwards can be economic. For manually operated machines the capital cost of equipment can be relatively low. Complex shapes can be made by machining particularly when multi degrees of freedom computer numerical controlled (CNC) machines are used. Disadvantages of machining include the production of scrap due to the chip forming nature of the process and the limited life of cutting tools. Skilled, and hence expensive, labour may be required. Machining may require more energy and some geometric shapes may take much longer to produce than other processes. Also unless cutting conditions are carefully controlled, machining can adversely affect the surface properties of the workpiece.

9.3.2 Single Point Cutting Tools

In lathe work a single point cutting tool is often used; it is also useful here for illustrating a tool's basic geometry, see Figure 9.2. It is apparent that the shape of a metal cutting tool

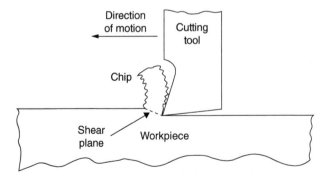

Figure 9.1 Two-dimensional (orthogonal) cutting using single point tool.

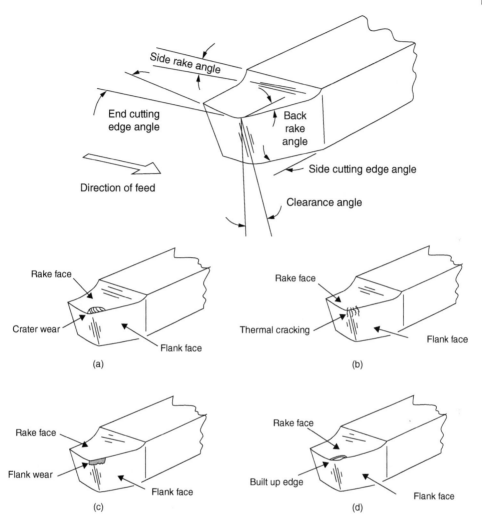

Figure 9.2 Single point cutting tool for lathe turning. (a) Crater wear. (b) Thermal cracking. (c) Flank wear. (d) Built-up edge (BUE).

is quite different from a knife or a woodworking chisel, that is, the metal cutting edge appears much thicker. This is for two main reasons. First the tool is subject to very high loads when cutting strong and often hard metals, therefore the tool itself must be strong. It also has to be harder than the metal it is cutting, this makes it relatively brittle, hence the need for support behind and below the cutting edge.

9.3.3 Chip Formation

Referring again to Figure 9.1 it can be seen that as the cutting tool moves through the workpiece material a strong force is directed by the tool into the workpiece in a direction at right angles to the rake angles. This force increases until the material shears in the direction of the force and the area over which this occurs is called the 'shear plane'. Depending on the machinability of the workpiece material, the cutting tool material, the speeds and feeds used, the depth of cut and the rigidity of the machine and tool, different types of chip can be formed: two are considered here.

Discontinuous chips. These are usually formed when the material being cut is brittle and relatively hard. One of the reasons chip discontinuity occurs is that the material cannot resist the high shear forces mentioned previously and therefore separate chips or even a fine shower of material is created. Other causes are large chip thicknesses resulting from coarse feed rates and low cutting speed, negative or small positive rake angles on the cutting tool, inefficient lubrication or cutting without a cutting fluid. An advantage of the discontinuous chip is that chip disposal is easier. Disadvantages are that the fluctuating cutting forces can cause tool chatter and vibration, thus spoiling surface finish, this can be minimised by ensuring the cutting tool, tool holder and material being cut are all rigid.

Continuous chips. These are usually formed when the material being cut is ductile and relatively soft. Surface finish is usually better since tool chatter is less likely to happen than with discontinuous chips. They often occur with high cutting speeds, large cutting tool rake angles, sharp cutting edges and where there is low friction between the chip and tool face – normally due to good lubrication and a smooth tool face. However, a disadvantage is that they can be very long and sharp coiling around the workpiece and other parts of the machine, thus being very dangerous to handle. A way of overcoming this is to use a 'chip breaker'. This is a small component clamped to the top of the tool that forces the chip to curl so tightly that it hits the workpiece and breaks off into smaller pieces.

9.3.4 Cutting Tool Wear

Due to the forces acting on the tool during cutting, the abrasive nature of chip production and high temperatures, the cutting tool will experience wear. This wear will eventually affect the cutting process such that the surface finish on the material being cut will degrade, the workpiece may go out of tolerance and the cutting forces will rapidly increase. The tool should therefore be changed or resharpened before any of these damaging effects occur.

Crater Wear is found on the rake face of the tool just behind the cutting edge, see Figure 9.2a, and it is created by the rubbing of the chip across the face, by high temperatures at the tool chip interface and a chemical affinity between the tool and workpiece materials.

Thermal Cracking is caused by high temperatures, see Figure 9.2b. These cracks occur at right angles to the tool cutting edge and can lead to chipping of the tool; this will spoil the surface finish of the workpiece as well as reduce the life of the tool.

Flank Wear is found on the clearance angle of the tool and is caused by the rubbing of the tool along the surface of the workpiece, see Figure 9.2c. A result of this is the creation of undesirably high temperatures, adhesion and abrasive wear, all of which reduce the properties of the tool.

A *Built Up Edge (BCE)* is formed by particles of the workpiece material that are gradually built up in welded layers onto the rake face of the tool while cutting, see Figure 9.2d. The BUE can increase until it dislodges from the tool and is partly carried away by the chip and partly left adhering to the surface of the workpiece. Therefore, large BUEs are very undesirable since they can spoil the surface finish of the workpiece and reduce cutting tool life, a thin BUE, however, is seen as desirable since it protects the rake face of the tool and reduces frictional wear.

9.3.5 Cutting Tool Materials

A cutting tool requires to be hard, tough and wear resistant. Until the early years of the twentieth century high carbon alloy steel was able to satisfy these requirements. However, as the need for higher cutting speeds increased so the need for cutting materials that held their hardness, toughness and wear resistance at high temperatures also increased. This led to the use of 'carbide' cutting tools in the 1930s. The tool composition was in fact particles of tungsten carbide bonded into a matrix of cobalt for cutting non-ferrous metals and cast iron, or titanium carbide for use at higher cutting speeds when cutting harder materials such as steels. These tools are still used today in the form of disposable tool inserts, see Figure 9.3. Many other cutting tool materials are now available. Because of their cost they are used in the form of indexable inserts. Ceramic, cubic boron nitride, which is almost as hard as diamond, and diamond itself are examples. Suitable materials can also be used to coat the surface of a base material, for example, titanium nitride can be used to coat high speed steel tools.

9.3.6 Cutting Parameters

Three critical parameters in machining are the speed of cutting, the depth of cut taken and the rate at which the cutting tool is fed into the work; this is illustrated for a turning operation in Figure 9.4. These are determined by considering the rate at which material is to be removed, the type of material being cut and the material of the cutting tool being used, the surface finish required, the power of the machine tool and the amount of support given to the work. For a fast material removal rate the cutting speed, feed rate and depth of cut all need to be maximised. Tough or hard materials being cut will reduce the material removal rate and hard and tough cutting tool materials will allow

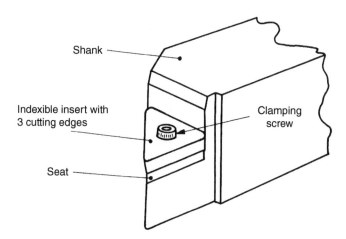

Figure 9.3 Indexable insert.

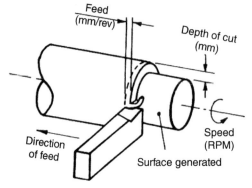

Figure 9.4 Generating a surface.

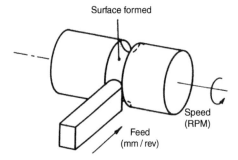

Figure 9.5 Forming a surface.

it to be increased. Generally speaking, high feed rates with single point tools produce coarse surface finishes, smooth surfaces being obtained by using high cutting speeds, low feed rates, shallow depths of cut and cutting tools in good condition.

Cutting tools can be used to create the shape of a workpiece by either 'generating' or 'forming' the surface. In generating, the shape of the surface produced is determined by the nature of the movement of the cutting tool relative to the workpiece. Thus in turning, a cylindrical shape is *generated* when the tool moves parallel to the rotational axis of the workpiece, see Figure 9.4.

Using turning again as an example, a surface can be *formed* as shown in Figure 9.5. In this case, the tool is fed straight into the work at right angles to the rotational axis and the final shape of the surface is determined by the shape of the cutting tool.

9.3.7 Machinability

The machinability of a material is an indicator as to how easily it can be cut to provide a good surface finish and precision. With metals the metallurgical structure and chemical composition; the hardness, ductility, yield and tensile strengths; susceptibility to 'work hardening' and the presence of inclusions, will all affect the machinability. A metal that has good machinability, called a 'free machining' metal, is good economically and environmentally since it will (i) minimise the cost of machining by being able to be cut quickly, (ii) need relatively low power therefore requiring less energy than a more difficult material and (iii) it will produce less wear on the cutting tool.

Softer materials such as aluminium are easier to machine than, say, stainless steel. However, if a material is too soft problems can arise. For example, low carbon steels are softer than high carbon but have a tendency to stick to the cutting tool; this results in the built-up edge noted earlier that shortens tool life. Low ductility makes cutting easier since chip production is facilitated but increased hardness will cause increased tool wear and demand higher power for cutting. A compromise between hardness and ductility is therefore often necessary. High thermal conductivity in the material being cut is good since the heat generated at the cutting tool tip is carried away quickly thus increasing tool life. Substances included in the material in trace amounts are called inclusions. These may increase or decrease the machinability. For example, aluminium oxide, which is hard and abrasive, will decrease machinability whereas manganese sulfides, which reduce the steel's strength, will increase machinability.

9.4 Machine Tools

The variety of machine tools available is large; there are those that are manually operated, those that are computer controlled and those that are specially designed for specific operations – these may be controlled mechanically, pneumatically or electronically. The ones we will examine in this section are the universal manually operated types. These might be found in a toolroom where possibly only a single component is to be made at a time. Today lathes, milling machines and machining centres are often 'numerically controlled', that is, a computer is incorporated in the machine control system. No matter what the method of control, the basic principles of operation are the same as discussed here.

9.4.1 The Lathe

The lathe is the most widely used general purpose machine tool. Machining operations on a lathe involve rotating the workpiece and generating a surface. When the surface is external and parallel to the rotational axis the cutting process is termed 'turning', when cutting an internal surface the term 'boring' is used and when creating a surface at right angles to the rotational axis a 'facing' operation is being carried out, when cutting a component off from the material held in the lathe workholder, the process is called 'parting off'. Screw threads are often cut using a lathe and this operation is termed 'screwcutting'. These operations are shown in Figure 9.6.

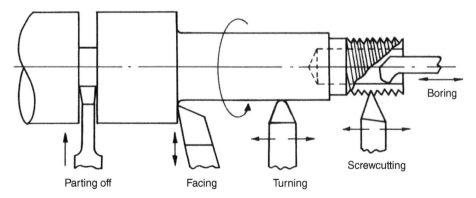

Figure 9.6 Some machining operations carried out on a lathe.

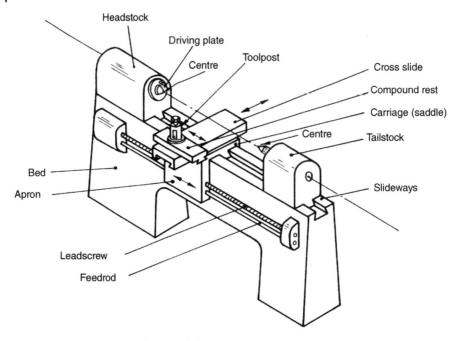

Figure 9.7 Basic elements of a centre lathe.

There are many different types of lathe, each suited to different production volumes and types of product. Three are outlined here to explain the essential principles of operation. The basic type is called a centre lathe since the workpiece may be held between a revolving powered centre in the lathe headstock and a free centre mounted in the tailstock. This type was also known as an engine lathe, the term arising from when they were first used in factories and driven via pulleys and shafts from a common steam engine. The basic construction of this lathe is shown in Figure 9.7.

The main body of the lathe is termed the *bed*. This supports and contains all the other lathe components; a large portion of it may be made from cast iron or other materials such as concrete or a granite-epoxy composite to absorb vibration. On the top of the bed are two sets of surface hardened precision machined slideways; these are parallel, straight and flat. They hold the headstock and tailstock and guide the carriage. The precision of the whole lathe is determined by the precision of these slideways. The motors to drive the lathe may be mounted in the bed or 'headstock'. The headstock holds the gear trains to allow a variety of cutting speeds and feeds to be selected. The headstock also contains the hollow spindle that drives the workpiece holder. The gearing in the headstock transmits drive to the lead-screw and the feed rod. The lead-screw is engaged to the carriage when cutting screw threads. The feed rod rotates and causes movement of the carriage and cross slide during operation of the lathe. The apron contains the components for transmitting the movement from the lead-screw and feed rod to the carriage and cross slide. The carriage, or saddle, can move parallel to the major axis of the lathe and supports the cross slide. The cross slide rests on guides on the carriage and can move at right angles to the major axis; it holds the tool post that carries the cutting tools. When cutting the tool is clamped tightly into the toolpost or toolholder. The tool must be rigid with the minimum of overhang otherwise it will vibrate or 'chatter' and produce a poor surface finish on the component. Both the carriage and the cross slide can be fed by hand or machine drive by engaging or disengaging the transmission.

In operation, the workpiece is attached to the spindle via a number of alternative workpiece holders. For example, if the component being machined is long a centre will be inserted into the spindle and a driving plate attached; the other end of the work will be held in the centre contained in the tailstock. If the work is very long a roller support will be placed opposite the cutting tool to prevent deflection of the work due to cutting forces. More often a chuck will be used as the workholder, see Figure 9.8. These have either three or four jaws. The jaws of the three jaw chuck move in unison when clamping a component; they are therefore useful for holding circular bar or previously turned components. The jaws of four jaw chucks move independently and are therefore used for clamping and centring non-circular components. Collets are quicker to use than chucks and are again used for circular components, see Figure 9.9. They are hollow, therefore they can be used when machining components from long lengths of bar. The bar can be held in a support and passed through the hollow spindle and collet in the headstock. The collet clamps the material and the turning operation is carried out. On completion the component is parted off, the collet then unclamps and the bar is fed through the required length for the next component.

Although the centre lathe serves well to explain the basic turning operations, it is not suitable for high volume production work. One that is suitable is the turret lathe. In this lathe a multi-sided indexable toolholder or 'turret' is mounted on the near side of the cross slide; a toolholder with a parting off tool is often mounted on the opposite side and another multi-sided turret (often hexagonal holding six tools and called the

Figure 9.8 Three-jaw chuck (3 and 4 jaw chucks usually much larger than collets).

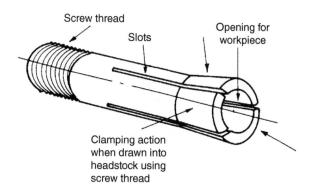

Socket for chuck key

To operate jaws

Workpiece

Jaw movement

Figure 9.9 Round collet for cold rolled or previously machined material (openings may also be square or hexagonal).

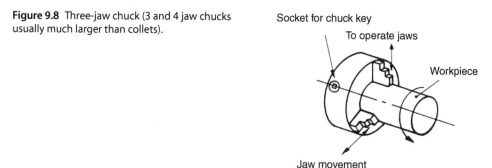

Screw thread

Slots

Opening for workpiece

Clamping action when drawn into headstock using screw thread

main turret) is mounted on a ram in place of the tailstock. These machines often operate automatically and are able to carry out multiple operations, such as turning, drilling, boring, screwcutting, facing and parting off, on the same workpiece very quickly.

Another type of lathe used for very high volume work is the multi-spindle screw machine. Here rather than one spindle a number of hollow spindles are used, each being fed by long rods of material from a rod holder and feeder; there may be four, six or eight spindles. Mounted on an end slide there will be a number of toolholders and tools equal to the number of spindles. This slide does not rotate but feeds the tools axially to the workpieces. There will also be tools mounted radially on cross slides. In this machine, all tools cut simultaneously then withdraw to allow the spindles to index round to the next position. The cutting operations are therefore being carried out in parallel rather than in series. This means that a product can be produced every few seconds, the cycle time being governed by the longest cutting operation.

9.4.2 The Milling Machine

Just as the lathe is the most used machine for producing rotational or cylindrical components, the most used machine for producing non-rotational or prismatic components is the milling machine. Again there are several types, but here the principles of operation can be examined by considering just two – the knee type horizontal and vertical mills, see Figure 9.10.

These machines have the following major components: a base, column, knee, saddle, worktable, overarm, milling head, spindle and cutter. The base supports the other elements and therefore should be made of a material that has strong compressive strength and can absorb vibration, for example, cast iron. The worktable rests on the saddle, which is supported by the knee; this is a cantilever unit guided by slideways on the column. The saddle provides vertical adjustment and the saddle and the table allow mutually perpendicular horizontal axes adjustment. The column contains the motors and gearing to drive the spindle. In the vertical type mill the spindle is vertical and is held in the milling head. In the horizontal machine the spindle is horizontal and drives an arbor that holds the cutting tools; an overarm is used to provide outboard bearing support.

Some milling cutters are shown in Figure 9.11. Figure 9.11a shows a light duty cylindrical, or slab, cutter for surface milling on a horizontal mill; in general, the heavier the duty the greater will be the helix angle relative to the axis and the fewer the number of teeth. The helical form allows each tooth to engage the work gradually and, since more than one tooth is usually in contact with the work at any moment, shock and chatter are reduced, thus producing a smoother surface. The keyway is to allow the cutter to be keyed to the arbor to prevent slippage. Many other types of cutter are used on the horizontal mill, most of them being disc shaped. For example, there are slitting saws for cutting through the material, slotting cutters for producing slots of specified depth and thickness and form cutters for producing any desired profile. Figures 9.11b,c show an end mill and slot drill, respectively; these are used in vertical milling. The end mill has a number of teeth, for example, four, on the circumference and one end. It is used for creating flat horizontal and vertical faces and for profiling. The slot drill usually only has two teeth on the circumference and end. It is designed to cut while sinking into the workpiece or traversing across it and is often used for creating slots and keyways.

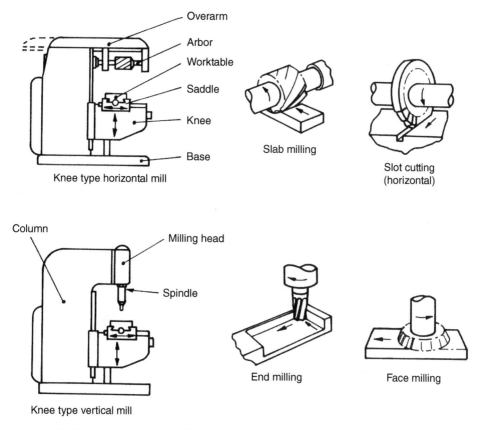

Figure 9.10 Horizontal and vertical milling.

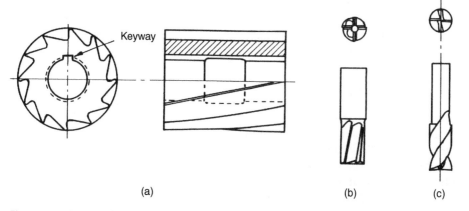

Figure 9.11 Some milling cutters. (a) Cylinder or slab cutter. (b) End mill. (c) Slot drill.

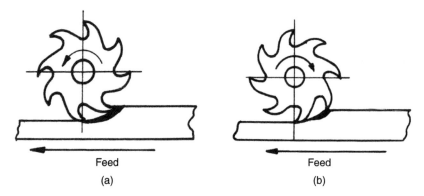

Feed   Feed

(a)   (b)

Figure 9.12 Methods of milling. (a) Conventional or 'up' milling, maximum chip thickness at the end of cut. (b) Climb or 'down' milling, maximum chip thickness at the beginning of cut.

In milling, the direction of cutter rotation relative to the direction of feed is important. The two methods of cutting are shown in Figure 9.12 for horizontal milling. Conventional or 'up' milling is shown in Figure 9.12a. Here the maximum chip thickness occurs at the end of the cut. This means that, provided the cutting teeth are sharp, the tool life is not unnecessarily reduced due to workpiece surface quality and the cut will be smoother. Disadvantages are that there is a disposition towards chatter, the work tends to be pulled upward off the table thus making effective clamping essential and the chips or cuttings are liable to pile up in front of the cutter, thus causing it to constantly have to cut its way through previously cut material. Figure 9.12b shows climb or 'down', milling. Here the maximum chip thickness occurs at the beginning of the cut. It has the advantages of tending to assist the clamping action by pushing the workpiece down onto its location; this makes it suitable for slender or flexible workpieces, also the chips are deposited behind the cutter, making the cutting process more efficient. Disadvantages are that due to the impact made by the cutter as it enters the work the surface should not have any surface scale as this will greatly reduce the tool life; also, since the cutter tends to pull itself into the work, backlash in the system must be reduced to a minimum. This is to avoid the cutter taking unpredictable 'bites' into the workpiece that would create shock loads causing damage to the tool and other parts of the machine.

9.4.3   The Drilling Machine

Drilling is probably the most common operation in machining and although it is relatively simple it can present difficulties. These difficulties usually arise because the cutting tool, that is, the drill, has a large length to diameter ratio and is therefore flexible. Also when drilling deep holes a build-up of cuttings in the hole will cause the drill to jam and even break; provision must therefore be made to allow these cuttings to be removed. Drilling can be carried out on a lathe or vertical mill, but often a 'drilling machine' is used. The basic construction of a hand operated vertical drill press is shown in Figure 9.13; this type is also sometimes called a pillar or column drill. The work being drilled may be held in a 'drill jig', which is hand held and free to move in the horizontal plane, the drill being guided to the appropriate points by passing through drill bushes held in the drill jig. The horizontal compliance is necessary to prevent the drill being twisted due to slight

Figure 9.13 Hand operated vertical drill press.

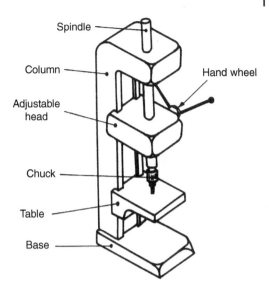

Spindle

Column

Hand wheel

Adjustable head

Chuck

Table

Base

positioning errors. This differs from a 'fixture' used in milling that clamps the workpiece rigidly to the milling table. Other types of drilling machines are: gang drilling machines that have a number of drilling heads in a line, the workpiece being passed from head to head for each required drilling operation; multiple spindle machines that can drill a number of holes in a component in one operation; radial drilling machines that have the drill head mounted on a radial arm, which can rotate around the support column – these are used for large workpieces and, finally, turret drills, which can have a number of drills mounted on an indexable turret – these are often numerically controlled.

The structure of the widely used twist drill is shown in Figure 9.14. It has three basic parts, the conical point, the body and the shank. The shank is held in the drill spindle or chuck and may be tapered or parallel. The body has helical grooves, often two, called flutes along which the cuttings travel upwards and coolant can travel downwards. The remaining surface of the body is termed the 'land'. Immediately behind the cutting edge the full diameter of the drill is maintained for a short distance before the body diameter is reduced to provide the clearance necessary to minimise friction. This short distance may simply be termed the 'land' (British) or given the name 'margin' (US). At the point of the drill the lands terminate and where they meet with the leading edge of the flutes they form a cutting edge or lip. The core of the drill is termed the web; this provides strength to the drill and is really its backbone. Where the web meets the point a chisel edge is produced. When the drill is rotating the speed of the centre of this edge is virtually zero; this of course reduces the cutting efficiency and therefore increases thrust forces, causes deformation and creates unwanted heat. When drilling large diameter holes the problem is removed by first drilling a small diameter pilot hole; otherwise the problem can be reduced only by making slight modifications to the shape of the web at the point.

Various other cutters are used in drilling machines and some of these are shown in Figure 9.15. Combination centre and countersink drills provide starting points for subsequent drilling. They are also used for drilling workpieces that will subsequently be turned between centres in a centre lathe; the countersink provides protection for the edge of the centre hole. Countersinks provide chamfers on the edges of previously drilled holes;

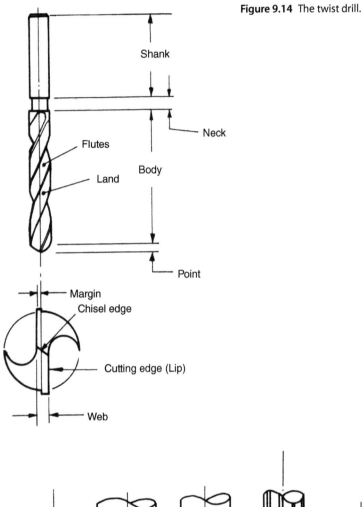

Figure 9.14 The twist drill.

Shank

Neck

Flutes

Land

Body

Point

Margin

Chisel edge

Cutting edge (Lip)

Web

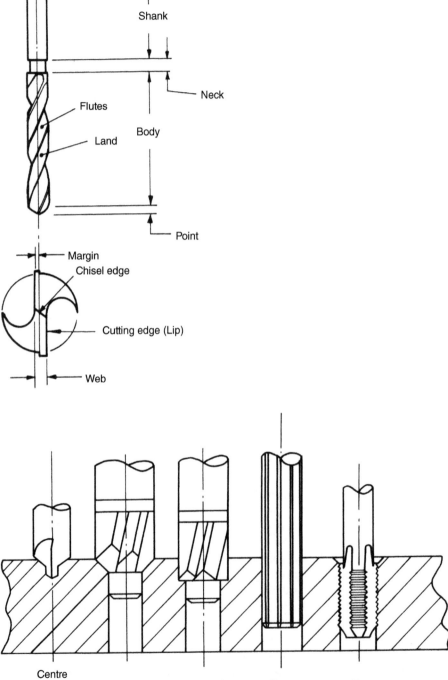

Centre
&
countersink

Countersink

Counterbore

Ream

Tap

Figure 9.15 Some operations that can be carried out on a drilling machine.

this removes the sharp edge and acts as a lead in for subsequent insertion of screws, bolts and so on. They are also used for providing recesses for countersunk head screws. Counterbores are used for increasing the diameter of the previously drilled hole a short distance below the surface; again the hole can be used for recessing bolts and screws. Reamers are used for removing very small amounts of material from a previously drilled hole; they leave a smooth straight surface of high dimensional and geometric precision. Taps are used for producing screw threads in holes; tapping is not a drilling operation but it is often carried out at slow speed in drilling machines.

9.5   Other Cutting Processes

There are many other machine tools, but those mentioned in the previous section serve to illustrate the basic principles involved. This section briefly discusses some of the other cutting processes available.

9.5.1   Thermal Cutting Processes

Processes that use heat to cut material include oxyfuel cutting, plasma arc cutting and laser beam cutting.

Oxyfuel cutting is very widely used for cutting ferrous and non-ferrous metals, the oxygen that supports combustion is mixed with another gas such as acetylene or propane to give the oxyacetylene and oxypropane cutting processes. For non-ferrous metals the material is melted by the flame produced by igniting the mixed gases and blown away to form a cut known as the kerf. With ferrous materials the oxidation process is used. In order for the metal to be raised to a high enough temperature for oxidation, that is, 'burning' to start, a gas-flame torch is used. This takes in oxygen and the fuel gas from separate supplies and mixes them to raise, for example, the temperature of steel to around 870 °C. Once the combustion starts the oxygen required for the reaction continues to be supplied by the torch and theoretically no additional heat need be added. In practice, the fuel gas continues to be fed to compensate for heat lost by conduction, convection and radiation. Handheld cutting torches are commonly used, but for large scale work, for example in cutting steel plate in ship production, numerically controlled machines arc used. These machines have one or more cutting torches mounted on an overhead gantry that traverses the plate being cut. The process is used to cut profiled shapes in sheet and plate, and is also used to prepare edges of plate for subsequent welding. Accuracies of ±1 mm are common, although closer tolerances are possible under carefully controlled conditions. Cutting speeds are normally in the range of 100–2000 mm min^{-1}.

Plasma arc cutting differs from oxygen cutting in that plasma is not dependent on a chemical reaction with the material being cut, as in the oxidation of steel. In the basic process, an electric current in the form of an arc is passed between a tungsten electrode and the workpiece; this is called a 'transferred' arc. In addition to the arc, which provides the highly concentrated heat to melt the metal, a high velocity gas stream removes the metal from the kerf. The gas is supplied through a nozzle around the electrode. Thus whereas the oxygen cutting process is exothermic developing large amounts of thermal energy, the plasma cutting is endothermic, requiring large amounts of energy to be put into the process. This process provides extremely high temperatures, around 33 000°C, and is therefore ideal for cutting metals, in particular the high alloy or stainless steels and non-ferrous metals that are not so easily cut with the oxygen process. Cutting speeds of around $8\,m\,min^{-1}$ are possible in $6\,mm$ aluminium. The combination of high temperatures and the jetlike action of the plasma produces narrow kerfs and very smooth surfaces.

Laser cutting utilises the fact that laser light can be focused to a very small focal point thus permitting very high power densities. The heat created at the point is used to melt or evaporate the material being cut. High power carbon dioxide lasers are now widely used in industrial applications. The laser is useful as a cutting tool only on materials that will absorb the laser light to a large extent. With the carbon dioxide laser wood and plastics cut easily; however, the absorption rate on mild and stainless steels is only 16–20%. For this reason a high pressure oxygen jet is usually used around the laser beam to assist oxidation of the material, the resulting temperature being in the region of $11\,000°C$. Laser cutting produces a high quality cut edge, with a very narrow cut width. The area around the cut is not adversely affected by heat and metal distortion is low.

9.5.2 Water Jet Cutting

In this process, a jet of high pressure clean water from a waterjet gun is directed at the material being cut. The water pressure is very high at around $3500\,bar$ and the jet narrow, between 0.2 and $2.5\,mm$ wide. The material being cut is usually lightweight glass reinforced plastic or carbon fibre a few millimetres thick. On materials such as these cutting speeds between $50\,min\,s^{-1}$ and $125\,mm\,s^{-1}$ are achieved. It is used to trim excess material, or cut internal and external profiles, on three-dimensional components such as crash helmets and car door and body panels. Because the jet is hazardous the process is very appropriate for robotisation. The water jet gun is mounted on the wrist of an industrial robot that will probably have six degrees of freedom. The work is also sometimes mounted on a programmable computer controlled turntable to provide maximum manoeuvrability.

Review Questions

1 State four advantages and four disadvantages of using metal removal (machining) processes to create a product. (8 marks)

2 Name three types of saw and their uses. (6 marks)

3 What are the two essential qualities of a cutting tool? (2 marks)

4 What is the purpose of using materials other than tool steel for making cutting tools? Name two such materials. (6 marks)

5 Explain the two basic machining methods for producing shape. (4 marks)

6 Fully discuss the factors that influence how a material is machined; refer to the desired qualities of the finished component, machinability and the cutting parameters. (10 marks)

7 Describe the conditions that lead to the production of a continuous chip when machining and discuss, with reasons, whether or not you think this type of chip is desirable. (6 marks)

8 Describe the conditions that lead to the production of a discontinuous chip when machining and discuss, with reasons, whether or not you think this type of chip is desirable. (6 marks)

9 Name three types of cutting tool wear that occur when machining and briefly note the measures that can be taken to ensure maximum tool life. (6 marks)

10 With respect to metal cutting what do you understand by the term 'built up edge' (BUE) and what is its significance? (5 marks)

11 Describe five basic components of a centre lathe and their function. (10 marks)

12 Describe three methods of holding a workpiece in a lathe and state when and why each would be used. (6 marks)

13 Name one type of lathe suitable for high volume production and very briefly describe its construction. (5 marks)

14 What are the basic components of a horizontal milling machine? Illustrate your answer with a sketch. (10 marks)

15 Name three types of milling cutter and give examples of their use. (6 marks)

16 What are the differences between 'up' and 'down' milling? (4 marks)

17 Briefly, and by means of a sketch, describe the construction of a twist drill. (5 marks)

18 Apart from drilling, name three operations that can be carried out on a drilling machine? Include brief sketches to illustrate your answer. (6 marks)

19 Describe the oxyfuel cutting process. (4 marks)

20 Describe the plasma arc cutting process. (4 marks)

21 Describe the laser cutting process. (4 marks)

22 Discuss the relative advantages and disadvantages of the three thermal cutting processes, and give examples of where each would be used. (10 marks)

23 Describe the water jet cutting process. (4 marks)

24 Discuss why you think water jet cutting is becoming more popular in manufacturing. (5 marks)

10

Deformation Processes

10.1 Introduction

There are at least three ways in which to classify deformation processes. One is to use the terms 'bulk' deformation and 'sheet' forming. Bulk deformation implies that the ratio of the surface area to the volume is relatively small and that the process will significantly alter the cross sectional area and shape; an example of this is forging. It is this bulk deformation that is considered in this chapter, it is an important process as it can provide fine control over the grain structure of a component and hence impart directional strength properties. Sheet forming involves changing the shape of material without significantly changing the cross sectional area; examples of this are blanking, bending and drawing and these are considered in Chapter 11.

Another classification uses the terms 'primary' and 'secondary' deformation. Primary deformation signifies the changing of shape of a piece of material from its cast ingot form into another shape. Examples of this would be rolling, forging and extrusion to produce standard shapes such as slabs, billets or rods. Secondary forming processes take the output from the primary process and further work the material to produce a finished or semi-finished product. These processes may again be rolling, forging or extrusion, but this time typical products would be foil, bolts or window frames.

A third classification relates to the temperature at which the deformation work is carried out. Thus a deformation process may be termed 'cold', 'warm' or 'hot' working.

Cold working is carried out when the material is worked well below its recrystallisation temperature (recrystallisation was discussed in Chapter 6). If the material is strain hardening then the deformation that occurs when cold working produces improvements in strength, hardness and surface finish when compared to working at higher temperatures. No energy needs to be expended on heating and contamination of the component is minimised. Cold working also allows better dimensional control that in turn facilitates production of interchangeable components. However, higher forces are required for deformation, hence the necessity for heavier and more powerful equipment. Cold working also reduces the ductility of the material; this may cause it to fracture after a number of deformations have taken place unless annealing is carried out between deformations. Imparted residual stresses and directional properties can also be detrimental unless carefully controlled.

Warm working is carried out below the recrystallisation temperature but above normal cold working temperatures. Compared to cold working it has the advantages of

Essential Manufacturing, First Edition. Gordon Mair.
© 2019 John Wiley & Sons Ltd. Published 2019 by John Wiley & Sons Ltd.

reducing the forces required for deformation and compared to hot working it provides better dimensional control. The exact temperature at which it can be carried out depends on the material.

Hot working is carried out above the material recrystallisation temperature. At these elevated temperatures the material strength is reduced and its ductility increased thus making it easier and cheaper to deform. Large changes of shape can be made without the material cracking. The structure is improved as internal pores are welded shut and impurities distort and flow along the grain structure – this allows directional properties to be selectively imparted. Disadvantages are the need to heat the material, poor dimensional control and the oxidation of the material surface. As well as producing a poor surface on the hot worked product the oxidised layer can also be disturbed and forced into the material as it is worked, making the achievement of a good surface finish difficult at subsequent machining operations.

10.2 Rolling

Rolling involves squeezing the material between rollers rather in the way that an old fashioned mangle or wringer was used to squeeze clothes dry. Large reductions in section are made by hot rolling the material. Only when finishing rods, strip, sheet and foil is cold working used since changes in section are slight and high surface finishes are desired.

In the past the raw material for rolling was always a cast ingot; however, with the increasing use of continuous casting in integrated steel plants, more rolling of 'concast' slabs is carried out. In the most general case, rolling steel from the ingot, the process can be described as follows. While still hot, the ingots are placed in gas fired furnaces called soaking pits. There they remain until they have attained a uniform working temperature of about 1100°C throughout; this is well above the recrystallisation temperature but still below the melting point, which for steels is between 1370 and 1530°C. The ingots are then taken to the rolling mill where they are rolled into blooms, billets or slabs. A bloom has a square cross section with a minimum size of approximately 150 mm². A billet is smaller than a bloom and may have any square section from about 40 to 150 mm². Slabs may be rolled from either an ingot or a bloom, they have a rectangular cross section with a minimum width of 250 mm and a minimum thickness of approximately 40 mm. The width is about three or more times the thickness, which can be over 300 mm. The principle of operation of a rolling mill is shown in Figure 10.1.

The mill used for rolling an ingot into a bloom is known as a 'cogging mill' and is generally of the two high reversing type. The heated ingot is placed on a conveyor comprised of powered rollers. These power driven rollers are provided at both sides of the mill, and since they also are reversible the ingot can be passed to and fro between the pressure rollers. The whole process is monitored and controlled from an overhead pulpit that is equipped with instrumentation and closed circuit television screens. From here the power driven manipulators can be controlled that allow the ingot to be turned over, moved laterally and even straightened while on the conveyors. It is important to ensure the metal is maintained at a uniform pressure as this controls metal flow and plasticity. The rolls themselves are cooled by running water over them to ensure that they maintain their own strength and hardness properties. The rollers gradually wear

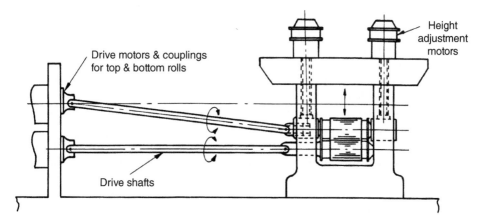

Figure 10.1 Principle of operation of a two high reversing mill for steel rolling.

in use and therefore need to be removed periodically for turning in a lathe to restore their original profile. The rollers for making smooth flat plates would be ground rather than turned (see Chapter 9 for a turning and grinding description).

The mill used for rolling ingots into slabs is known as a 'slabbing' mill. It is very similar in construction to a cogging mill but with a few differences. Since a slab is much wider than a bloom and the width to height ratio of the section is high, the rollers require a long flat profile and the lift of the top rollers must be sufficient to allow the edges of the slab to be rolled when lying on its side. There are also high pressure water sprays incorporated to dislodge the scale from the wide surface of the slab.

Subsequent rolling operations allow blooms to be rolled into large round bars, small slabs and heavy sections such as 'I' beams; many are also rolled into billets of approximately square section with rounded corners. Further rolling of billets converts them into rod, bar and structural sections such as T beams and angle and channel sections. Further rolling of slabs produces plates; these can be again rolled to produce sheet, which could then be rolled to produce foil. These subsequent operations are carried out in rolling mill 'stands' of a variety of configurations, some of which are shown in Figure 10.2.

To produce strip from slabs, for example, a number of stands are arranged in a line. The hot slab enters the first stand where it is reduced in thickness. As it emerges its speed is much faster than when it entered, therefore, the next stand in sequence must be ready to accept it at this higher speed. There will be a number of stands and by the

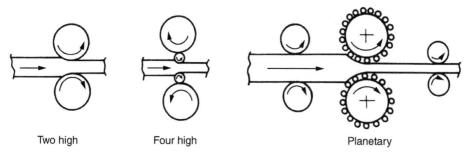

Two high   Four high   Planetary

Figure 10.2 Some rolling mill configurations.

time the material is passing through the last one it is being cold worked and travelling at extremely high speed. It emerges from the stand onto a coiler where the strip is wound ready for dispatch to the customer.

In the rolling stand, the smaller the diameter of the roll used, the shorter will be the length of contact of the roll circumference for a desired reduction in material thickness. This means that a higher pressure is exerted on the material, thus producing a greater deformation. The disadvantage is that the decreased size of the rolls reduces their stiffness. Therefore, to prevent their deflection under load, back-up rolls of thicker section are required. This leads to configurations such as the 'four high' and 'planetary' arrangements shown in Figure 10.2, which allow large reductions in section to be performed in a single pass.

Typical rolled products would be plates for building ships' hulls; structural beams of various sections for the construction industry; hexagonal section bar for making nuts and bolts; steel sheet for making panels for cars, washing machines and cookers; copper alloy strip from which small components such as electrical terminals could be blanked and aluminium foil.

10.3 Forging

As with rolling, forging can be carried out hot, warm or cold, depending on the characteristics required in the finished product. In forging, plastic flow takes place in the material when it is subjected to compressive forces via presses or hammers. Figure 10.3(a) shows a drop hammer used for forging. Forging usually produces products of extremely good mechanical properties; they are much stronger, for example, than components produced by casting. The raw material for forging has often been already hot worked, for example, billets that have been hot rolled. Just as these billets have had their mechanical properties improved from that of the ingot from which they were made, so forging of them into a finished product further improves their properties.

Grain formation in metals and the importance of grain structure were mentioned in Chapter 6, forging provides a good method of controlling this grain structure. Consider the following. If a bar of rolled steel is sectioned longitudinally and the cut surface is smoothed and then deeply etched by a strong acid, then the surface will show closely spaced fine ridges indicating that the structure of the bar consists of fibres running in the direction of rolling. The properties of ductility, impact strength and toughness will be found to be at a maximum when measured in this direction. As the axis of the material approaches 90° to that of the grain flow, the decrease in the ductility and impact strength is considerable (Figure 6.5). The direction of the grain flow can be controlled to a large extent by the sequence of operations in the forging process. For example, in the closed impression die for a conrod shown in Figure 10.3b the first operation is to bend, that is, 'preform', the bar or billet to an approximate shape. The grain flow lines will then follow this shape in the finished forging. The subsequent stages shown in Figure 10.3b allow the billet to be taken to the finished shape progressively; this allows maximum control of the finished grain structure. Space is allowed in the final impression for excess material, flash, to spread out around the finished component. This flash is usually trimmed off separately.

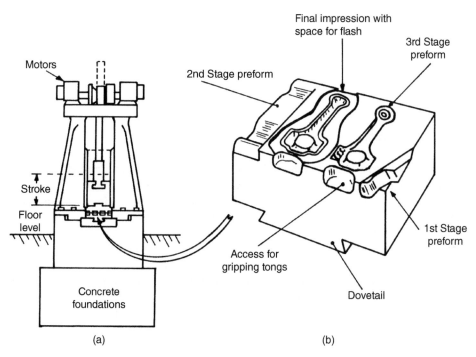

Motors

Stroke

Floor level

Concrete foundations

(a)

Final impression with space for flash

3rd Stage preform

2nd Stage preform

1st Stage preform

Access for gripping tongs

Dovetail

(b)

Figure 10.3 (a) A friction drop hammer forging press and (b) the lower half of a closed impression die for a conrod.

Forging has a number of advantages over casting and machining. As already noted, plastic flow can be controlled to obtain a fibrous structure with maximum strength in the desired direction. The process also produces components with a higher strength to weight ratio than with cast or machined parts of the same material, a fine crystalline structure is obtained, internal pores are closed and physical properties are improved. Compared to sandcasting, smoother and more accurate shapes are obtained. However, there are also some disadvantages: for example, it is not possible for components as intricate as those obtained by casting to be obtained. The size of component is limited by the size of press available, scale inclusions are possible from the oxidised metal surface and, for closed impression forging, dies are expensive, which means the process is not suitable for small production quantities.

Two types of die can be used in forging. In the open face die the skill of the operator is very important. These are used in 'hammer' or 'smith' forging where the heated metal is hammered either with hand tools or between flat dies in a mechanised hammer. This is the oldest type of forging, not very different from that undertaken by a blacksmith, but close accuracies are difficult to obtain and complicated shapes cannot be made. In closed impression die forging the die is made the shape of the desired finished product. Impact or pressure forces the material in its plastic state into the die cavity; this was the type shown in Figure 10.3b.

Closed impression dies are used in drop and press forging. For example, in the gravity type hammer the impact pressure is developed by the force of the falling ram and die as it strikes upon the material placed in the lower fixed die, the force being entirely dependent on the weight of the hammer, the top die and the distance dropped. In press forging a slow squeezing action is employed in contrast to the rapid impact blows of the hammer. The squeezing action thoroughly works the entire section of the material. The presses can either be mechanical or hydraulic, the latter being slower but able to exert higher forces.

10.4 Extrusion

In the extrusion process solid material is placed in a closed container and subjected to pressure which causes the material to flow out through a die. The shape of the die opening determines the shape of the section produced, analogous to squeezing toothpaste out of a tube or icing a cake. The principle has long been utilised in processes ranging from the production of brick, to pipe manufacture, to macaroni production. The principle of extrusion is shown in Figure 10.4. The following is a brief description of the operation of the process, which can be carried out hot or cold depending on the material.

A billet is placed inside an extrusion press container. The exertion of pressure on the billet via a ram causes the material to pass through the die opening. This rapidly creates a long product of the required section. The outer surface of the billet is chilled on contact with the container wall and is therefore less plastic than the core where flow initially takes place. Therefore, during extrusion this outer skin containing scale and debris from the surface of the billet accumulates between the billet and the ram. This unwanted material eventually moves inwards to form a 'piping' defect in the last of the extruded material; this necessitates discarding the 'butt' end, which may be as much as 10% of the total extruded material. Metals such as aluminium, lead and tin can be extruded cold, whereas others such as steel must be extruded 'hot'. The extrusion of plastics is very common and is described in Chapter 12. Moulding trim, tubes, rods, structural shapes and plastic or lead covered cables are typical products of extrusion.

The extrusion process is able to produce a variety of shapes of good strength, accuracy and surface finish at high production speeds. With the exception of casting no other process can provide as great a deformation or change of shape within a given time. The

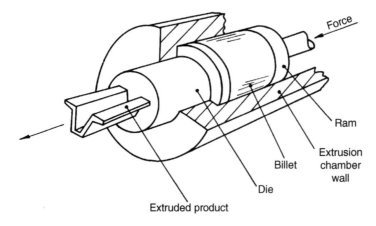

Figure 10.4 The principle of extrusion.

cost of dies is low relative to those required for casting and forging. Almost unlimited lengths of a continuous cross section can be produced from the one die and extrusion machine. For simple sections, production runs as short as 150 m can be justified due to the low die cost. Shapes that would be difficult or impossible to roll, such as hollow bars and sections with re-entrant angles, can be extruded. However, extrusion is about three times as slow as roll forming. The main disadvantage with extrusion is that for economic production the cross section of the product must remain constant over the full length.

There are various types of extrusion. For example, in direct extrusion, shown in Figure 10.4, the flow of metal through the die aperture is in the same direction as the ram movement, the ram being the same diameter as the container. The metal is extruded through the die until only a small amount remains in the chamber; the extrusion is then sawn off close to the die and the remains, containing the contaminants, are discarded. In indirect extrusion, the ram is hollow. A die is mounted over the bore of the ram and when force is applied the metal flows in the opposite direction of the ram movement and through the bore of the ram. Due to the reduction in friction between the billet and the container wall less force is required by this method. Two limitations of the process are that the ram is weakened and adequate support for the extrusion is difficult. Other methods include backward, impact and hydrostatic extrusion.

Review Questions

1   What do you understand by the terms 'bulk', 'sheet', 'primary' and 'secondary' deformation?

2   Discuss what is meant by the terms 'hot', 'warm' and 'cold' working.

3   Outline the advantages and disadvantages of hot and cold working in terms of processing costs and finished product quality.

4   Describe the rolling process used to produce steel slabs.

5   Briefly discuss the use of rolled products.

6   Why does the forging process generally allow the optimum mechanical properties of a material to be realised in a finished product?

7   Briefly discuss the differences between hammer and press forging.

8   With the aid of a sketch, describe the extrusion process.

9   Discuss the advantages and disadvantages of metal extrusion.

10   List five typical metal products that would be made by extrusion.

11

Pressworking

Presswork involves carrying out the processes of shearing, bending and drawing in a press that usually holds a press tool. Presswork is the most economic method of mass producing components from sheet metal. Historically, the use of presses allowed the economic production of products so that they could be sold at affordable prices. Products such as clocks, cash registers and adding machines were totally dependent on their components being produced in press tools. However, although products utilising electronics have largely replaced the old mechanical and electromechanical designs, there is still a demand for components made from pressed steel, aluminium and brass. The ubiquitous motor car would be far too expensive for wide ownership if it was not for the pressed steel panels from which its body is constructed. Frames and panels used for 'white goods' such as domestic washing machines, fridges and cookers are made in by presswork. The chassis, electrical tags and connectors, and other sundry components found within televisions, phones, computers and other electronic goods are made in presses. Also kitchen utensils, hinges, radiator panels and many other products are all made from pressworked components.

Presses are capable of very high production rates since the time to produce one component is simply the time for one up and down cycle of the ram carrying the press tool, and the time it takes to feed into the press a new portion of material. In fact, using tools with multiple dies can allow many smaller components to be produced every second. Most press operations are carried out with the material in its cold state. The press itself consists of a machine frame (see Figure 11.1), supporting a bed and a ram, a mechanism to make the ram move at right angles to the bed and a means of powering the mechanism. The power may come from electric motors or hydraulic rams. For fast operations such as punching, blanking and trimming, a press that uses an electric motor to drive a large flywheel is ideal. The energy from this flywheel is transferred to the ram through mechanisms such as cranks, eccentrics or gears. For slower operations such as those involved in squeezing or drawing hydraulic power is better.

Some of the operations carried out by presswork are shown in Figure 11.2. In cropping, the material is sheared across its whole width. It is important to remember that when designing and manufacturing the tool that in *blanking*, an aperture is created by shearing, *the material that is removed is the required part* and it *takes the dimensions of the die* (there always needs to be clearance between the punch and the die). *Holes* in components are created by *piercing* the *dimension of the hole* taking the *dimensions of the punch*. In bending, two-dimensional deformation takes place, care being required in the tool design to prevent cracking and maintain dimensional accuracy. In drawing, the

Essential Manufacturing, First Edition. Gordon Mair.

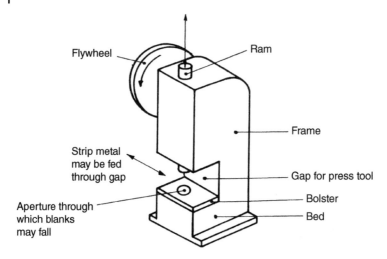

Figure 11.1 The basic configuration of a gap press

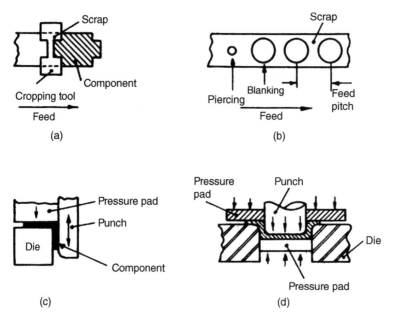

Figure 11.2 Some pressworking operations: (a) cropping, (b), piercing and blanking, (c) bending and (d) drawing.

material is stretched in three dimensions to typically produce a cup shape and careful design of the tooling is again required to prevent wrinkling or rupturing of the metal.

For most products the press is fitted with a 'press tool' that is in two parts, see Figure 11.3 in which a very simple piercing and blanking tool is shown. The top part is fitted to the ram of the press and is therefore free to reciprocate vertically, it is often fitted with the 'punch'. The bottom part is fixed to the bed of the press and usually contains the 'die' it may have a through hole to allow punched blanks or components to fall through the bed of the machine into a bin or conveyor belt. Assuming the press

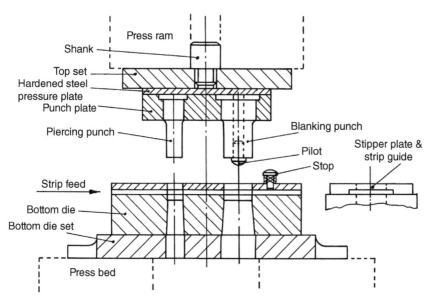

Figure 11.3 A simple piercing and blanking tool.

has adequate capacity it is easy to change from one product to another by simply changing the press tool. The tools themselves can be very expensive depending on their complexity. The top and bottom 'sets' may be made from cast iron and the punch plate from steel. The punches and bottom die will be made from hardened steel as would be the pressure plate which prevents the hardened punches being pushed into the top set. The alignment between top and bottom parts is maintained by pillars set into the bottom die and bushes in the top die.

A strip of cold rolled material is fed through the stripper plate in the direction shown until it is midway between the piercing punch and the blanking punch. In this case the stripper plate, which pulls the material off the punches, is fixed and integral with the strip guides. The material may be fed by hand while the system is immobilised, but usually an automatic feed is used. On the first downward movement of the top tool the piercing punch creates the central hole of the washer. When ram with the top section of the tool has risen the material strip is then fed forward until it reaches the spring loaded stop. In the simple configuration shown this stop may be temporarily pushed down by hand while the press will be immobilised. In the case of hand-fed material the press is immobilised by a failsafe method such as a light guard (see Chapter 28). On the next downward movement of the tool the pilot on the blanking punch locates on the previously pierced hole before the punch comes down and punches out the finished washer. The purpose of the pilot is to ensure concentricity of the inner hole with the outer circumference of the washer. The strip is then fed the pitch length shown between each stroke of the press.

The press tool described here is a two stage tool but far more complex multi-stage or 'progression' tools are commonly used. These have multiple punches and a large number of stages through which the material is fed before the first component is produced. It is apparent that after the material has passed through the final stage each press stroke produces a component and this is why presswork can produce such high volumes of components at such rapid rates and why it remains such a useful process. Other tools,

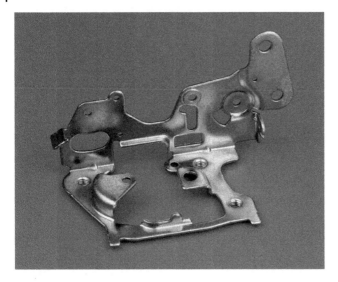

Figure 11.4 A typical small component produced by presswork.

Figure 11.5 A car body panel produced by presswork. Source: Image Courtesy of Audi.

called 'combination' tools, combine a number of operations in the one stage, for example, blanking and drawing. These can be used for producing components that are dish, cup or tube shaped.

The component shown in Figure 11.4 will have been produced in a progression tool where initial operations will have involved the piercing of the holes, the strip will then have been progressed through the press tool for subsequent operations such as drawing and blanking and creating the 90° bend.

In the car industry body panels such as those shown in Figure 11.5 are produced in large transfer presses. In these operations the body panels are blanked from sheet steel or aluminium, they are then lifted and transferred mechanically to the next pressing operation and so on until the panels are completely formed.

The selection of the specific pressing process and the design of the press tools require much skill and experience. Knowledge of the properties of the material that is being worked regarding shear strength and ductility is essential. The actual clearance between the punch and the die is important. It is clear that the punch has to be smaller than the die and the amount by which this is necessary is generally dependant on the shear strength and the thickness of the material being worked. Typically, this will be between 10 and 35% of the material thickness. As the punch first enters the material it produces a slight curvature on the surface of the surrounding material and smooth sides on both the pierced and blanked surfaces. As it passes further downward the material fractures and a slightly wider and rougher surface on the sides is apparent. The width of the part pushed out by the punch will be that of the die and the width of the hole left by the punch will be that of the punch. If the part has not been put through another process, such as tumbling to remove burrs and sharp edges, or a shaving operation where a small amount of material is additionally removed from the sides to improve accuracy, then it is possible to see the smooth and rough witness marks and so understand the direction in which the part was formed.

This chapter is intended to illustrate the principles of presswork and introduce some types and operations. However, just as parts produced by pressworking are ubiquitous and numerous so also are the varieties of presswork available – many more than are mentioned here. They are likely to be the major source of low cost, mass produced, sheet metal part for many decades to come.

Review Questions

1 State reasons for the continuing popularity of presswork as a means of producing components.

2 Name four types of operation that are commonly carried out by presswork.

3 Briefly hand sketch a simple piercing and blanking tool.

12

Plastics Processing

12.1 Introduction

This chapter is concerned with the manufacturing processes used for making plastic components. Plastic materials were discussed in section Chapter 6, where their importance to modern technology was also noted, and the production of plastic was described in Chapter 7.

One of the major advantages of plastics is that, by choosing a suitable moulding or forming process, a complete component can be produced that requires little or no finishing operations. As the number of options available to the engineer when selecting the process is large, the optimum will be found only by considering the advantages and limitations of each process, the design of the component to be made, the tool or mould design and the processing characteristics of the plastic being used. A few of the main thermoplastic moulding and forming processes are now explained.

12.2 Extrusion

This is a continuous process in which one or more Archimedean feeder screws are mounted within a cylinder as shown in Figure 12.1. This cylinder is heated and as the screws rotate they plasticise and convey the plastic along their threads from the feeder hopper to a die. The section of this die determines the cross sectional shape of the extrusion. Extrusion machines come in a range of sizes with screws ranging in diameter from 25 to 200 mm.

Typical products are tubes, rods, sheets and continuous lengths of almost any profile. By modifying the process it is possible to produce products with a combination of materials, for example, by incorporating pressure rolls just after the die the extruded plastic can be bonded to a substrate of another material such as a fabric, paper or metal. By modifications to the die it is possible to include metal sections or wire within the extrusion. In the dual extrusion process plastics of different properties can be combined so that, for example, an extruded strip may be produced, one side of which is rigid and suitable for attachment to (say) a refrigerator door and the other side flexible, thus suitable for forming a seal. Extruded sections are important in providing the plastic in a suitable form for processes such as blow moulding and blown film production. Since extrusion is a relatively fast process, it is suitable for high volume production. To cut the material to length as it emerges from the extruder a flying saw is often used that travels

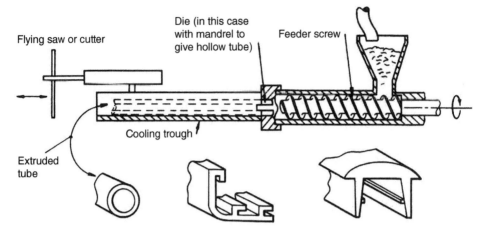

Figure 12.1 The polymer extrusion process and typical polymer sections.

with the material as it cuts. The range of size and sections able to be produced is wide and component surface finish and precision is very good. The machines themselves are expensive, but once the initial investment is made a wide variety of sections can be produced by simply changing the die.

12.3 Blow Moulding

This is used to produce hollow products such as plastic bottles this process utilises short lengths of extruded tube called *parisons* as shown in Figure 12.2. After extrusion, while the plastic is still hot, an appropriate length of the parison is located in the mould. When the mould closes the bottom end of the parison is closed and compressed air is passed in through the open top. The parison inflates and adopts the shape of the mould cavity. By maintaining the air pressure the plastic is held in contact with the water cooled cavity surface until its shape is stable.

Bottles of a multitude of shapes, sizes and transparent, translucent or opaque colours are produced by this method. Soft drinks, detergents, cosmetics, shampoos and so on

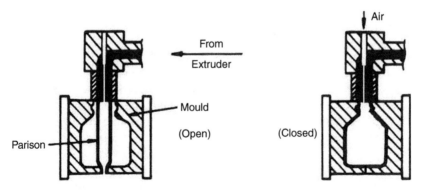

Figure 12.2 The blow moulding process.

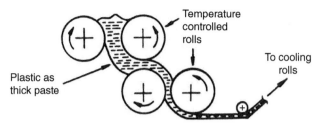

Figure 12.3 Calendering.

are all contained in blow-moulded bottles. Many other lightweight hollow components are produced by this method, although the surface finish is not usually as good as that produced by injection moulding. However, relatively good quality products can be produced at high production rates.

12.4   Calendering

Continuous production of sheet and film plastic is obtained by this process. Beginning as a thick melted paste, the plastic is formed into sheet by passing it between and around a series of nip rollers as shown in Figure 12.3. A calender usually comprises four temperature controlled rolls. A polymer, which may be plastic or synthetic rubber, is fed to the first pair of rolls where the roll separation controls the feed rate. The separation of the following rolls determine the final sheet thickness. Plastic film and plastic and rubber sheets are produced by this process, and a particular advantage is that plastic or rubber can be laminated to other plastic films, sheets of paper, fabrics or metal foil.

12.5   Vacuum Forming

Products of relatively simple form are produced by this process from thin thermoplastic sheets as shown in Figure 12.4. The plastic is clamped in a support frame and exposed to radiant heat until it reaches a pliable state. The frame is then drawn down over a pattern

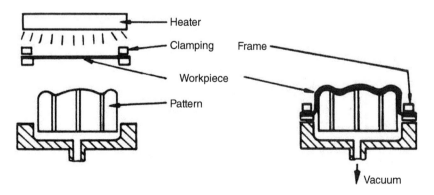

Figure 12.4 Vacuum forming.

and hermetically sealed to the chamber walls, as shown in Figure 12.4. A vacuum pump now withdraws air from between the pattern and the plastic, thus causing atmospheric pressure to force the plastic against the surface of the pattern. The combination of the vacuum and the mechanical action on the plastic as it is pulled over the pattern causes the plastic to adopt the shape of the pattern. The plastic, cooled rapidly by air, becomes rigid again and the pattern is withdrawn.

Products such as margarine tubs, egg containers, disposable drinking cups and larger items are produced by this method. Care has to be taken with pattern design as the stretching of the plastic tends to create thin walls at component corners.

12.6   The Blown Film Process

In this process, a tube is extruded vertically while at the same time it is inflated by the introduction of compressed air (see Figure 12.5). The extrusion is drawn upwards and through nip rolls as shown, these rolls prevent the air escaping immediately. The air flow is controlled to produce a constant bubble size and wall thickness. The film is flattened by the rolls as it is drawn onto a wind up unit. Between the rolls and wind up unit there may be perforating blades, heat welding clamps and handle cutting punches, these being used to produce the rolls of disposable 'plastic bags' widely found in supermarkets. Larger bags and sacks for refuse disposal and holding agricultural products are also produced by the process.

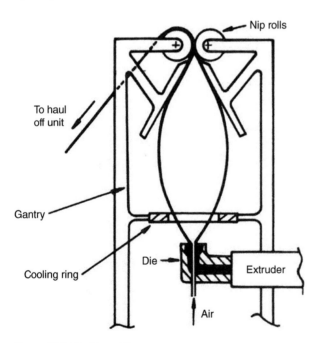

Figure 12.5 The blown film process.

12.7 Injection Moulding

For high volume production, injection moulding is the most widely used and economic of all the thermoplastic processes. Figure 12.6 shows the basic elements of an injection moulding machine, and Figure 12.7 shows the process schematically. The plastic, which is in powder or granular form, is loaded into the feed hopper. This may be done manually, or automatically via tube conveyors from bulk storage bins. The plastic falls from the hopper into the heated machine barrel; this barrel contains a rotating and reciprocating screw, which acts as both a plasticiser and an injection unit. As the plastic passes along the screw flight the root diameter of the screw increases, this compresses the plastic. The shearing action between the particles of plastic generates heat, thus changing the plastic to a semi-fluid state. Heat is also added in a controlled manner via the heater bands around the barrel. Immediately in front of the screw is a chamber for receiving the compressed plastic. As the chamber is filled, the screw is forced backwards until it trips

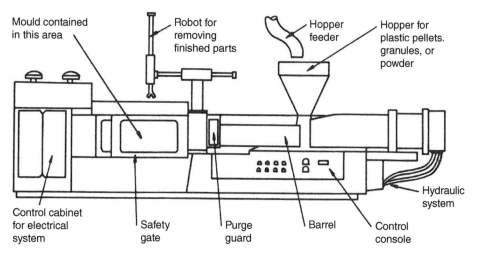

Figure 12.6 Main elements of an injection moulding machine.

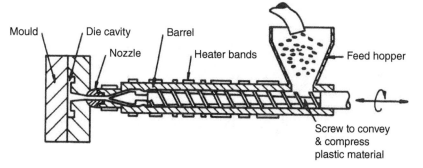

Figure 12.7 A reciprocating screw injection system.

a limit switch that signals the control system to cause the hydraulics to force the screw forward. This causes injection of the plastic into the cavity of the mould, the plastic being prevented from flowing back by a one way valve in the nozzle of the moulding machine.

The process is used to produce very large production quantities of components of widely ranging size. Production rates are also high and the components may have an excellent surface finish. With thermoplastic materials there is no waste since the runners and sprues can be regranulated and reused. Disadvantages are that the moulding tools and machines are expensive, closed containers cannot be made without additional assembly work and complex shapes and re-entrant angles in a design further escalate the tooling costs.

12.7.1 Injection Moulding Tool Design

Due to the importance and widespread use of injection moulding in manufacturing, a little more detail is given here on the design of the moulding tools. The purpose of the tool is first to form the plastic to the desired shape and second to cool the formed product. Figure 12.8 shows a simplified sectional sketch of a two plate injection moulding tool.

The locating ring on the fixed half ensures that the mould is properly located relative to the moulding machine and the guide pins ensure that both mould halves are kept in alignment with each other. The plastic is injected into the fixed half of the mould, the nozzle from the injection moulding machine barrel mating with the tool sprue bush. The other half of the tool is attached to the machine clamping system and can move in the directions shown. When the halves are clamped together the plastic flows into the mould cavities through a central sprue and runner system. The plastic then cools as water flows through internal channels in both mould halves. After the cooling period is completed, the moving half is pulled back. The moulded components, together with sprue and runners, remain with the moving mould half. This is due to the plastic in the sprue having flowed into the notch of the sprue puller shaft and the slight contraction of the cooled plastic also tending to make the components stick to the convex surfaces

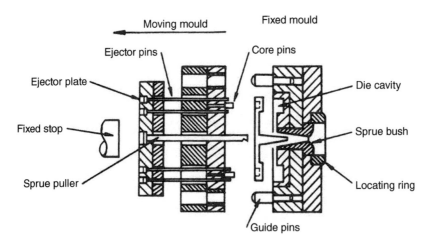

Figure 12.8 A two-plate mould for injection moulding.

of the moving mould. As the mould moves back the ejector plate hits against the fixed stop thus causing the ejector pins to move forward relative to the mould; this pushes the moulding forward and off the mould surface. The moulding may then be allowed to fall down into a bin or conveyor or it may be lifted out of the machine by an industrial robot. These handling robots are sometimes equipped to separate the components from the sprue and runner system.

The example discussed refers to a multiple cavity mould. It is also possible to have single cavity moulds that do not require a runner system. The decision whether to use single or multiple cavity moulds is an economic one. Factors that influence this are: number of components to be produced and rate of production – multiple cavity moulds will be best where these factors are high; the precision required in each component; the type of plastic; the capacity of the moulding machine and the position of the mould parting line. The decision on where to place the parting line depends on a number of factors, including the shape of the article and the number of mould cavities. Components with re-entrant shapes that cannot be released in the normal direction of the mould opening require moulds with more than one parting line. To allow these components to be made facilities such as side or rotating cores may need to be used. Figure 12.9 shows a cam operated side core. As the mould opens right to left the slide rides up the pin shown in the right mould half. This causes the side core to be withdrawn from the moulded component, so allowing a hole to be created at right angles to the normal direction of tool movement.

One method of achieving internal screw threads in a component for high volume production is shown in Figure 12.10. As the mould moves linearly over the screw it causes rotation of gear A, which is internally threaded onto the fixed screw. This causes gear B, which is keyed to a shaft with the threaded core at one end, to rotate. The shaft is also threaded and runs in a screw threaded bore of the same pitch as the core. Thus movement of the mould causes rotation and extraction of the core.

The surface finish of the cavity should be created with great care since every scratch or mark on the surface will be transmitted to the component. Unless a textured surface is desired most mould cavities are highly polished and this is an expensive process.

The designer of the multi-cavity injection moulding tool must also carefully consider the layout of the sprue runner and gating system. The 'gate' is the small aperture where the material from the runner enters the mould cavity; it is also the point where the finished component is broken off from the runner. Figure 12.11 shows an end view of these elements.

Figure 12.9 A method for creating a hole at right angles to the direction of travel of the moving half of a moulding tool.

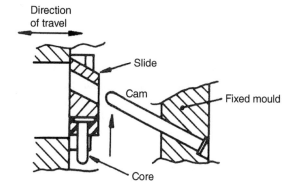

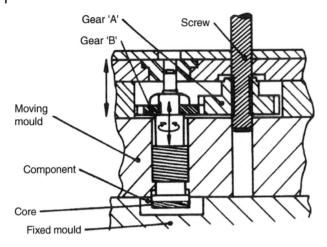

Figure 12.10 A method for creating internal screw threads using a rotating core.

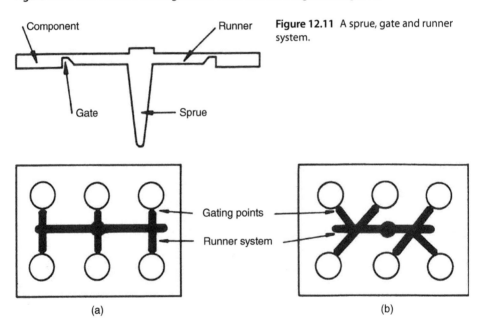

Figure 12.11 A sprue, gate and runner system.

Figure 12.12 A runner system design for optimum flow: (a) poor and (b) better.

The runner system should be arranged in such a manner as to allow each cavity simultaneously to receive an equal amount of pressurised plastic at each injection. Thus the tool design should provide the shortest possible flow route for the plastic between the sprue and each cavity. Figure 12.12a shows a poor layout since the two cavities nearest the sprue will be filled first; this makes it almost impossible to produce mouldings of uniform quality, since these two cavities will receive extra pressure as the injection continues. Figure 12.12b shows a better layout, where each cavity receives an equal share of plastic simultaneously; also the cold slug wells at the end of the runners trap partly cooled polymer. A bad cavity layout should always be avoided since it can cause problems such as stress build up in a component, differences in component sizes and mould release problems. It is also important to give close attention to the position and size of the gates. This is because gates determine the flow pattern of the plastic within the cavity and therefore, since plastics evidence molecular orientation in the direction of flow, their positioning and size will also affect the mechanical properties of the finished component.

12.7.2 Plastic Component Design for Injection Moulding

Figure 12.13 shows the use of metal inserts in a plastic moulding. These are inserted into the component either by placing them in the moulding tool and allowing the plastic to flow around them before cooling or by press fitting them into the previously moulded component. These allow high strength features to be added where necessary, although very large metal inserts should be avoided as differences in thermal expansion create stresses.

Just as with any component design functional requirements such as load bearing, resistance to adverse environmental conditions, dimensional constraints, service life, ergonomic and aesthetic specifications must always be met. The designer of injection moulded components must also be familiar with the constraints and advantages of the polymer material being used, the capabilities and limitations of the injection moulding process and the implications of the component design for the mould tool design. Additionally, the plastic product designer should also consider the following points.

- Large flat surfaces should be avoided as they tend to warp. Corrugated, grooved or curved surfaces are better. If flat surfaces must be used warping can be minimised and the surface strengthened by the use of ribs. Care needs to be taken when designing ribs, however, as they can cause light sink marks on the opposite surface to where they are located. This is due to the rib creating an additional mass of material that contracts more than the surrounding thinner material. Sink marks can be minimised by ensuring that rib thicknesses do not exceed two-thirds the thickness of the main wall and their height does not exceed three times the thickness.
- Corners should be rounded since as well as producing lower stress concentrations than sharp corners they also offer less flow resistance to the injected plastic.
- The component wall thickness should be kept constant as much as possible; where changes in section are necessary, these should be made as gradual as possible.
- Consider that as the plastic in the mould cools it also contracts. Therefore, should the sides of the component be made parallel in the direction of mould opening, it will be

Figure 12.13 Metal inserts in a moulding, the flanges and the knurled head fix or 'key' then inserts into the mould.

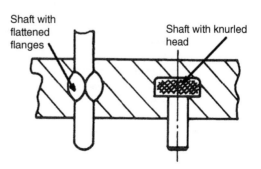

Shaft with flattened flanges

Shaft with knurled head

evident that the moulding will tend to shrink onto the convex portions of the mould and so make removal difficult. For this reason, a slight taper, called 'draft', must always be provided in the component design. This is more apparent on large components as they require greater draft.

- Re-entrants, or undercuts, should also be avoided or at least kept to a minimum, as they will add extra expense to the manufacture of the injection moulding tool. This is obvious from the earlier section where it was shown that re-entrants often demand the use of cam operated side cores or complex rotational cores. Even small re-entrants such as engraved characters should be avoided and replaced with transfers.

Figure 12.14 shows a fictitious component exhibiting poor and improved design. The points mentioned previously are included, along with some additional ones. For example, the rectangular hole in the back wall is expensive to make since it requires a rectangular cam operated side core. A circular opening would be cheaper and changing the opening to a slot completely removes the need for a side core. Large masses of solid plastic, like the pillars in the sketch, cause sink marks – hollows are better. In the component on the right there is a threaded hole that is threaded right up to where the hole meets the surface. This causes a knife edge, which is easily damaged; a recessed thread is better and in this case the thread form has been rounded. In conclusion, as with any design, the simplest component design will be the best one.

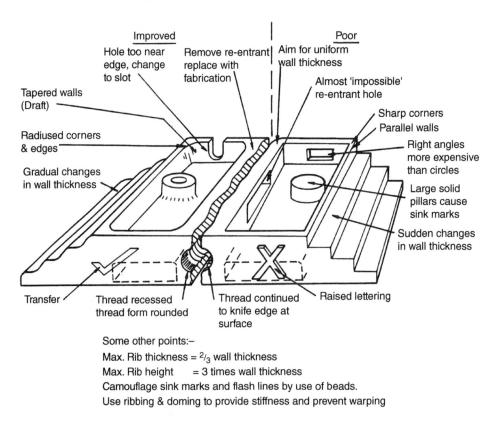

Figure 12.14 Some aspects of plastic component design.

Review Questions

1   What four factors does an engineer need to consider to ensure that the optimum manufacturing process is selected for making a plastic component?

2   Describe the process used to produce continuous lengths of plastic product of constant cross section.

3   Describe the blow moulding process, and the type of product to which it is most suited.

4   What is the calendering process used for?

5   Describe the vacuum forming process and give examples of typical products produced by it.

6   Describe the process used to produce disposable plastic bags and refuse sacks.

7   Why is the injection moulding process best suited to high volume production?

8   Explain how the injection moulding process works (no sketches are necessary).

9   Describe the basic construction of an injection moulding tool.

10   Why is it advisable to eliminate re-entrants in an injection moulding design?

11   Explain why the designer has to pay particular attention to the manner in which the plastic is transferred from the moulding machine nozzle to the mould cavity.

12   Why are ribs necessary in some plastic mouldings and why are their width and height important?

13   What is the purpose of 'draft' in a plastic component?

13

Additive Manufacturing Processes

13.1   Introduction

Compared to some of the other processes noted here, additive manufacturing is very new. For example, a casting of a copper frog has been found that was probably made around 5000 years ago, whereas additive manufacturing only appeared commercially around the late 1980s with the introduction of stereolithography. The reason for the late appearance is that the process was only possible due to the confluence of microprocessor-based technologies, lasers, high precision manufacturing and the ability to directly interface with computer aided design (CAD) systems that led to 'direct digital manufacturing' being a term used to describe the process.

Additive manufacturing equipment takes its fabrication information from a CAD model. This model may be created by practitioners from any number of disciplines such as product design engineers in the aerospace and automotive industries, architects, artists, industrial designers, sculptors, surgeons, dentists or by individual hobbyists using domestic additive manufacturing equipment. The CAD model should be volumetric, that is, a solid model, thus providing full three-dimensional data for the finished object. The data in the model is generated in such a manner that a large number of horizontal slices through the model can be created. Each slice contains the data required for the additive manufacturing equipment to lay down or cut through one layer of fabrication material. The height of these slices contributes to the vertical resolution of the finished fabrication. The file format for this operation should ideally allow dimensions, colours, textures and materials to be represented.

13.2   Advantages of Additive Manufacturing

1) Components can be produced directly from CAD models with no intermediate work, such as the creation of an expensive mould for a casting process. This greatly reduces the lead time for prototyping and also provides the opportunity for producing individualised products for the consumer or even allowing the consumer to produce their own products or replacement components at home using desktop equipment.
2) There are few restrictions on design. Finished complex components can be made in polymers, metals or ceramics that would be difficult or impossible to make in a single operation by machining or any other process.

Essential Manufacturing, First Edition. Gordon Mair.
© 2019 John Wiley & Sons Ltd. Published 2019 by John Wiley & Sons Ltd.

3) Physical models can be made directly from 3D laser scanning of other objects, for example in reverse engineering, or in medicine directly from CT and MRI scans to allow better decision making before surgery. Examples of applications are included at the end of this chapter.

13.3   Disadvantages of Additive Manufacturing

1) It is not economical for large production volumes as only one or a small number of parts can be produced in each cycle and the cycle time is slow.
2) In current commercial systems the resolution and surface finish is relatively low compared to, for example, precision machining. Research is continuing to improve this and in the meantime post manufacture finishing treatments such as additional machining and polishing can ameliorate the problem.
3) The range of materials available for additive manufacturing is relatively limited to traditional manufacturing processes, although as the technology improves this range is becoming wider.

13.4   General Types

There are a number of ways of classifying additive manufacturing technologies. At the time of writing one important and established method is that of the International Organisation for Standardisation (ISO) and the American Society for Testing and Materials (ASTM). This can be found under ISO/ASTM52900-15: Standard Terminology for Additive Manufacturing – General Principles – Terminology. It is divided into Binder Jetting, Directed Energy Deposition, Material Extrusion, Material Jetting, Powder Bed Fusion, Sheet Lamination and Vat Polymerisation. Figure 13.1 below shows these and under each is an example of a process that represents the particular technology. The following sections then briefly explain the processes.

The size of the artefacts created by the processes shown in Figure 13.1 is dependent on the size and configuration of the additive manufacturing equipment. They all could

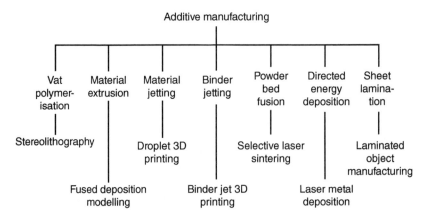

Figure 13.1 Principal additive manufacturing starting materials and examples of processes.

be capable of creating a product of up to one cubic metre, although much smaller sizes are more common, particularly with the liquid and powder bed systems. At the other extreme, much larger items can be created for depositing molten materials, such as in the Fused Deposition Modelling process by using industrial robots or specially designed cartesian machines. Also note that figures relating to layer thickness and resolution or tolerance are approximate as alternative manufacturers or suppliers may offer different specifications. Factors influencing this include the equipment being used, product size and delivery time requested.

13.4.1 Vat Polymerisation – Stereolithography

Stereolithography is a good process for producing models that are to be used for demonstration, prototyping or for ergonomic analysis as they do not require very good mechanical properties. However, it is excellent for producing small intricate parts with fine detail.

In this process, a viscous photosensitive polymer resin is selectively cured using a directional laser beam. Materials such as acrylic and epoxy liquid polymers will cure and solidify when exposed to the light of an ultraviolet laser as shown in Figure 13.2. Following a path determined by the CAD model the laser traces out one layer in the x-y plane of the fabrication, whereupon the build platform moves downward an equivalent distance to the next required horizontal slice and the laser solidifies the next layer. A typical layer thickness is between 0.05 and 0.15 mm. This continues until the fabrication is finished, see Figure 13.3. If any part of the object is temporarily separate from the main body for part of the build, or is otherwise needing support, then additional scaffolding will need to be created during the build process. This will be designed to be easily removed later. Between the laser scan of each layer a recoating wiper blade is drawn across the surface to ensure it is level across the whole work area. When the process is completed the build platform is raised and the resin is drained to allow the fabrication to be removed.

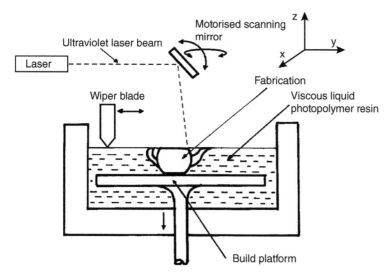

Figure 13.2 Stereolithography (SL).

Figure 13.3 This multipart engine model used stereolithography to achieve a detailed prototype. Source: Image courtesy of Vaupell Rapid Solutions.

The fabrication may be further exposed to UV light if additional curing is required. Finished products are commonly up to $500 \times 500 \times 500$ mm although fabrications over 2 m long can be created. Tolerances for stereolithography in the x–y plane are around $\pm\ 0.13$ mm for the first 25 mm then ± 0.05 mm for every additional 25 mm. Along the z-axis the tolerance is ± 0.25 mm for the first 25 mm then $\pm$ for every additional 25 mm.

Mask Projection Stereolithography reduces the build time of SL artefacts. This is due to replacing the single scanning laser beam with a dynamic mask. This allows an entire cross-sectional area of the fabrication to be exposed to ultraviolet light at the same time rather than progressively. The mask can be altered instantaneously for each layer by the use of a digital micromirror device (DMD). This type of mirror is composed of hundreds of thousands of microscopic mirrors each of only around 13 μm square arranged in a rectangular array. Each of the mirrors can be electronically switched in microseconds to an 'on' or 'off' position thus creating a dynamic mask where each mirror may be considered as one 'pixel'. Layer thicknesses and fabrication size is the same as for laser scanning stereolithography.

The materials used in stereolithography are not thermoplastics but they rather mimic thermoplastic properties. One of the results of this is that if they are exposed to moisture and ultraviolet light in service for long periods they suffer degradation in a change in mechanical properties and appearance. Although they can be protected by painting or plating they are not usually intended for long term use.

13.4.2   Material Extrusion – Fused Deposition Modelling (FDM)

In Fused Deposition Modelling (FDM) a thin filament of a thermoplastic polymer such as ABS or a polycarbonate is initially extruded onto the build platform then progressively added to in order to form the finished fabrication. Wax could also be used in order to create patterns for investment casting. In Figure 13.4 the extrusion assembly moves in the x–y plane depositing the filament in a layer corresponding to a horizontal slice from

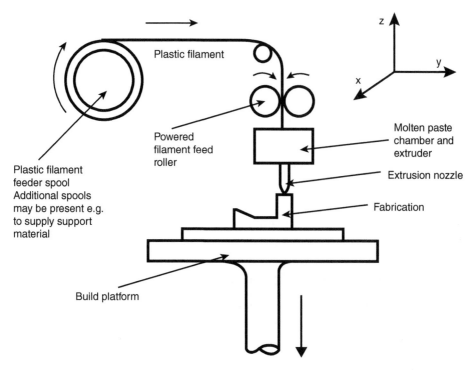

Figure 13.4 Fused deposition modelling (FDM).

the CAD program. The build platform may move down or the extrusion assembly move up the necessary distance for the next layer to be deposited.

Other configurations of the extrusion system are possible, for example, instead of one that is cartesian, a six-axis industrial robot may be used to hold the extrusion nozzle. The extruded material, heated to just 0.5°C above its melting point, is at such a temperature that it fuses seamlessly by cold welding to the previously deposited material. Thus as each subsequent layer is added the finished fabrication takes shape. An additional filament and nozzle supplying easily removed support or scaffolding material may also be used when aspects of the fabrication are fragile. Since the movement along the *x*-, *y*- and *z*-axes is mechanical it is slower than those systems that use a laser in the fabrication process. However, it is very popular and FDM equipment is widely available from high end industrial systems to those that can be used by hobbyists at home. An example part is shown in Figure 13.5.

Parts with very good strength to weight ratios, and thermal and chemical resistance can be created. Also complex shapes both externally and internally can be created that would be difficult, expensive or impossible to achieve with traditional processes. Typical layer thicknesses could be between 0.1 and 0.5 mm with tolerances of around ±0.1 mm depending on size. Part size is determined by the configuration of the extrusion system that could be a gantry type or an industrial robot. The largest machines can produce artefacts of many cubic metres but these are very unusual. A machine capable of 1 cubic

Figure 13.5 A component produced by FDM that would be difficult to make in one operation by conventional manufacturing methods. Source: © 2018 Stratasys Ltd.

metre would normally be classed as large and most machines will be producing much smaller parts.

13.4.3 Material Jetting – Droplet 3D Printing

Material Jetting – Droplet 3D Printing equipment utilises a cartesian configuration in which a printer head moves in an x–y plane over a build platform in a manner similar to an inkjet printer. The build platform will move down in the z-axis a distance of one layer after each horizontal slice is completed. A number of printer nozzles mounted on the gantry allow droplets of different materials and colours to be combined in-process. The molten droplets are deposited on the build table then onto the previously deposited layers until the artefact is completed. The droplets are instantly cured and solidified by ultraviolet light as they seamlessly weld to those that are already laid. After completion the post processing will include removal of any support structures created. Examples of artefacts are shown in Figure 13.6.

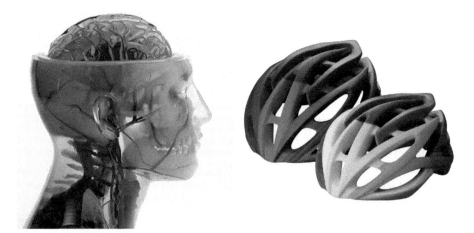

Figure 13.6 A Medical model and bicycle helmets produced by droplet 3D printing. Source: © 2018 Stratasys Ltd.

The process is capable of producing a wide range of attractive and useful models but only in photopolymers. Depending on the polymers used the finished artefact can be hard and rigid, soft and flexible, transparent or opaque. Layer thickness can be as small as 0.013 mm, it is relatively precise and a wide variety of photopolymers in different colours can be used.

13.4.4   Binder Jetting – Binder Jet 3D Printing

Binder Jetting – Binder Jet 3D Printing is a powder bed process so called because a printer head operating in the x–y plane, similar to that in an inkjet printer, ejects adhesive bonding material in the form of droplets onto a layer of powder in the powder bed. This powder may be of a number of materials including metals or ceramics. The powder is held in a chamber under the printer head. As one layer corresponding to one horizontal slice of the CAD model is completed the build platform drops by a distance equivalent to the next layer. Additional powder is then added and the process is completed. Layer thicknesses are between 0.1 and 0.2 mm. The bonded material forms the completed part and the loose powder acts as a support and can be readily removed when the fabrication is completed. When metal powder is used the finished artefacts can be put into an oven to fuse the powder grains together, thus producing a product with similar characteristics to one that is sintered. Artefact sizes of 1 cubic metre would be regarded as large and smaller sizes are more common, see Figure 13.7.

13.4.5   Powder Bed Fusion – Selective Laser Sintering

Selective laser sintering (SLS) allows a wider range of materials than the polymers noted in the first two processes. It can be used to create fabrications from powdered metals, ceramics and polymers.

Figure 13.7 Some components produced by binder jet printing. Source: Image courtesy of ExOne.

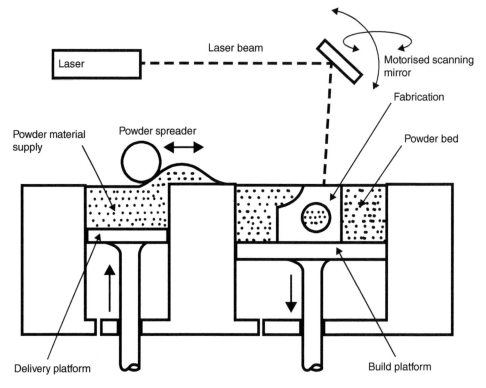

Laser beam

Laser

Motorised scanning mirror

Fabrication

Powder material supply

Powder spreader

Powder bed

Delivery platform

Build platform

Figure 13.8 Selective laser sintering (SLS).

The equipment consists of a powder bed chamber containing the powder from which the artefact is to be made this the powder is supported on the build platform. There will also be one or two chambers containing the supply powder. Although only one supply chamber is shown in Figure 13.8 an additional chamber located to the right of the build chamber can also be present. In this process as the moving laser beam traces out one horizontal slice of the CAD model the powder is fused over the traced cross sectional area. The build platform is then lowered a distance equivalent to the determined layer thickness, which may be between 0.05 and 0.5 mm. Next a powder spreader roller moves across the supply powder and pushes fresh powder across the surface of the powder bed ready for the next scan of the laser. Thus, layer by layer the finished artefact is produced with unfused powder acting as a support where necessary.

Where metal powders are used the whole assembly may be contained within a nitrogen filled cabinet in order to prevent oxidation of the build material. When the fabrication is complete the excess powder can be removed and the finished product extracted. Tolerances in the x–y plane are ±0.13 mm for the first 25 mm then and additional ±0.13 mm for each additional 25 mm and in the z-axis ±0.25 mm for the first 25 mm then an additional ±0.13 mm for every subsequent 25 mm. Maximum sizes would normally be no larger than 1 cubic metre, although it is often smaller. Figure 13.9 parts (a) and (b) show fabrication examples.

(a)

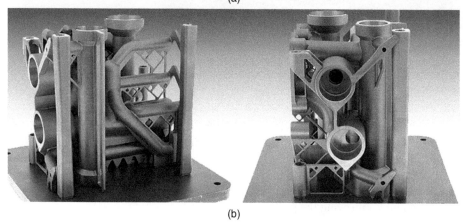

(b)

Figure 13.9 (a) A physical polymer wire-frame model created by SLS. Source: Image courtesy of CIDEAS Inc. www.buildparts.com. (b) A complex metal component created by SLS © Copyright Renishaw plc. Source: All rights reserved. Image reproduced with the permission of Renishaw.

13.4.6 Directed Energy Deposition – Laser Metal Deposition

Directed Energy Deposition – Laser Metal Deposition uses a powerful laser beam that is focused from a nozzle onto a substrate to create a weld pool. Metal powder or wire is then added simultaneously to the weld pool via the nozzle and through an inert gas such as argon to protect the material from oxidation. The metal solidifies rapidly and the fabrication is then built up by horizontal layer corresponding to the slices generated by the CAD model.

Typically, each layer may be between 0.1 and 1.0 mm, however, the weld bead can range from 0.03 mm for micro manufacturing to over 5 mm for coarser work. The substrate may be mounted on a platform with one or more degrees of freedom and the nozzle could be held in an industrial robot having six degrees of freedom. This means that finished parts of complex geometry can be created in a single operation, see Figure 13.10.

The process can also be used to combine different materials such as in cladding, for example, a lightweight but soft base metal can be clad with a thinner but hard-wearing metal as the two materials will be seamlessly fused together during the process. A wide

Figure 13.10 A vane produced by laser metal deposition. Source: Image courtesy of RPM Innovations Inc.

range of metals can be used such as titanium, steel, nickel and aluminium. An advantage of the process is that it is possible to control the grain structure and hence the resulting mechanical properties of the artefact. Part sizes normally range from a few centimetres up to any size the manipulation system can provide although normally no more than a metre cubed. At the time of writing, Directed Energy Deposition is still not widespread and research into improving the process is ongoing.

13.4.7 Sheet Lamination – Laminated Object Manufacturing (LOM)

Here, the starting material is in the form of sheet held on a supply roll as shown in Figure 13.11. The material is unwound and indexed over a platform between support rollers before being wound onto a take-up roll. The sheet material may be paper, cardboard, plastic or metal foil. At each indexed step a laser is used to cut out a profile equivalent to the corresponding horizontal slice from the CAD model. The laser is set such that it will only cut through one layer of the material at each step. The laser also crosshatches the excess material to enable easy removal of the fabrication. The laser may be directed by a mirror system or the whole laser assembly may move in the x–y plane. Alternatively, a sharp blade can be used to cut the material. After each cut the platform is lowered by the thickness of the sheet material. Each layer is fixed to the layer below either by an adhesive backing on the sheet, which can be activated by a heated roller or by adding adhesive between each cutting operation. The height of the fabrication can be monitored in-process to optimise accuracy. When the process is completed the finished fabrication is removed and can be worked further if necessary. If paper or cardboard has been used the fabrication has the appearance of wood and can be worked as such, however, a sealing operation is necessary to prevent the absorption of water.

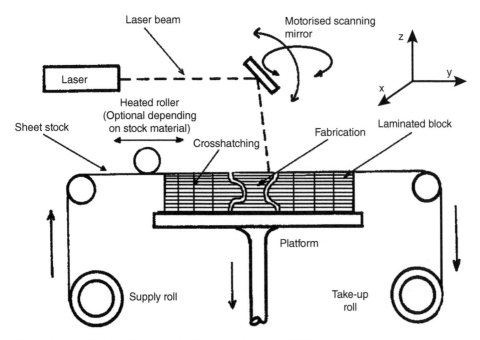

Figure 13.11 Principle of laminated object manufacturing (LOM).

Figure 13.12 Aerofoil section produced from carbon fibre sheets. Source: Image courtesy of Impossible Objects.

Artefacts of 1 cubic metre and above can be created using this method. The vertical resolution is dependent on the sheet and bonding layer thickness and full colour objects can be created directly. However, it is not as accurate as other additive manufacturing process and due to this functional prototypes have not normally been created, nor has it been usual to create parts with complex geometries or internal cavities. As an exception to this recently the use of carbon fibre sheet combined with nylon or other polymers has been used to produce high strength prototypes as shown in Figure 13.12. More common applications include producing patterns for sand casting and, since the materials are inexpensive, relatively low cost conceptual scale models for architects and product designers.

Review Questions

1 Discuss three advantages and three disadvantages of additive manufacturing.

2 Name seven general categories of additive manufacturing processes.

3 With the aid of a rough sketch describe the process of stereolithography and include comments its precision and use.

4 With the aid of a rough sketch describe the FDM process and include comments its precision and use.

5 With the aid of a rough sketch describe the selective laser sintering (SLS) process and include comments its precision and use.

6 With the aid of a rough sketch describe the LOM process and include comments its precision and use.

14

Miscellaneous Metalworking Processes

As well as the processes so far mentioned, there is a multitude of other manufacturing processes widely used in industry today. Some of these are briefly described in this chapter. This chapter is simply intended to provide an introduction to some additional processes of which a good product designer, manufacturing engineer or industrial manager should be aware.

14.1 Electrodischarge Machining

Early in the last century, the introduction of harder, stronger and more heat resistant metals, which were difficult to machine by say conventional turning, drilling or milling, led to the introduction of other processes such as electrodischarge machining (EDM). The process is sometimes termed 'spark erosion', since it relies on the eroding effect of electric sparks crossing between two electrodes, one electrode being the tool and one the workpiece being machined. If both tool material and workpiece material are the same then the greater erosion occurs on the positive electrode; therefore, to minimise tool wear and maximise metal removal rate, the workpiece is made positive. The principle of operation is shown in Figure 14.1. The electrodes are immersed in a dielectric fluid, that is, a liquid that can sustain an electric field and act as an insulator, such as paraffin. The sparking occurs due to heavy electrical discharges across the gap between the electrodes. The interval between the sparks is around 0.0001 s, the sparks releasing their energy in the form of local heat. The local temperature is approximately 12 000°C causing the spark to melt the material and form a small crater. The gap between the tool and the workpiece is maintained accurately by the servomotor. The dielectric fluid is pumped and therefore carries away the eroded metal particles. It is a very useful process since as long as the workpiece is electrically conductive it does not matter how hard, tough, brittle or heat resistant it is. Very delicate work can be carried out as there is virtually no force between the tool and the workpiece. Since the shape of the machined hole conforms exactly to the tool electrode, shapes of almost any cross section can be produced although as the tool itself can be eroded a new tool may be required for finishing. A popular use of this process is the production of dies for presswork.

Essential Manufacturing, First Edition. Gordon Mair.
© 2019 John Wiley & Sons Ltd. Published 2019 by John Wiley & Sons Ltd.

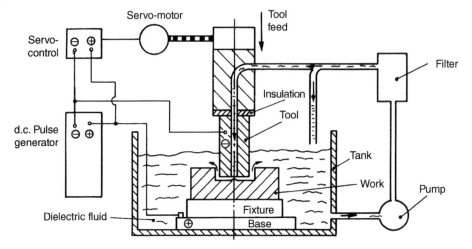

Figure 14.1 Electrodischarge machining.

14.2 Electrochemical Machining

In *electrochemical* machining (ECM), the tool is made the cathode (−ve or negative electrode) and the workpiece is made the anode (+ve or positive electrode). The process involves electrolysis, that is, the workpiece is progressively dissolved by electrochemical action. An electric current is passed between the workpiece and the tool via an electrolyte that is pumped rapidly through or around the tool (this operation is actually the reverse of metal plating). The electrolyte is a conductive solution of an inorganic salt such as sodium nitrate. The tool is progressively fed into the workpiece creating a blind hole or aperture which is the inverse shape of its own. The flowing electrolyte carries away the dissolved particles. Both the tool and workpiece need to he conductors of electricity, the tools often being made of brass, copper or stainless steel. The process is used for mass production of complex shapes in conductive materials. It has the advantage that there is no wear on the tool and stress free machined surfaces are produced. The tooling, however, can be expensive due to development costs.

14.3 Chemical Machining

In chemical machining no external electrical circuit is necessary. The material is 'machined' by chemical processes similar to those found in corrosion, for example, etching is a chemical machining process. In chemical milling, the workpiece is first cleaned, then the areas not to be etched are masked by a material impervious to the chemical reagent. The workpiece is then dipped in or sprayed with the chemical. After an appropriate time the workpiece is cleaned and the masking removed. This process is widely used for the production of integrated circuits, but can also be used for machining much larger workpieces.

14.4   Ultrasonic Machining

One of the mechanical metal removal processes ultrasonic machining uses a tool immersed in a slurry containing abrasive particles. The tool is vibrated at high frequency as it is fed into the workpiece while maintaining a gap. Within this gap, the abrasive is impelled by the vibration against the workpiece material, thus abrading it. The process has the advantage that it can be used to machine almost any material. Other mechanical processes are hydrodynamic jet machining, which uses a high velocity fluid jet for slitting; abrasive jet machining, which uses a gas jet loaded with abrasive particles for a wide variety of machining operations, and abrasive flow machining, which uses a flow of abrasive slurry for edge finishing and polishing.

14.5   High Energy Rate Forming

Included in high energy rate forming (HERF) are a number of processes that form components at a rapid rate by using the application of extremely high pressures at very high velocities. In explosive forming, the high pressures resulting from the detonation of chemical explosives are used for forming sheet metal components. Figure 14.2 shows the principle. A die is made of concrete and lined with an epoxy resin; a facility is included to allow extraction of air from the cavity between the workpiece and the die. The workpiece of sheet or plate metal is clamped across the die so that an airtight seal is formed. The whole assembly is placed underwater and a shaped explosive charge placed in the water above the workpiece. Air is evacuated from under the workpiece, and the charge detonated. The shock wave passes through the water to the metal that is deformed by the rapid change in pressure, thus taking the form of the die. Explosive forming has many advantages. The size of equipment necessary for producing large workpieces is reduced, die costs are relatively low and explosives are also cheap. Materials difficult to shape by other methods can be deformed in this way, for example, some high strength refractory metals too brittle for conventional techniques are found to respond well when subjected to high momentary stresses. Prototypes can be made without expensive tooling. The method is quick and capable of producing parts to tight tolerances. However, it

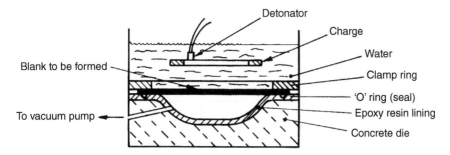

Figure 14.2 Explosive forming a steel bowl-shaped product.

also has some disadvantages, for example, for very large components the dies are heavy and special lifting tackle needs to be on hand. The number of variables inherent in the process can also make development costs high, and due to the use of explosives and the need for safety it is often not possible to integrate the process into the production facility.

14.6 Powdered Metal Processes

These processes, sometimes simply termed 'powder metallurgy', involve pressing fine powders into a desired shape. This is usually carried out in a metal die under high pressure. The compact is then heated, that is, 'sintered', for a specific period; the temperature at which this is carried out must be below the melting point of the main metal involved. The combination of heat and pressure cause the major metallic constituents to weld together at their points of contact. This means that a porous component can be formed for use as (say) a filter, or different metallic or non-metallic constituents can be added to impart desired characteristics to the final product. For example, graphite is added to improve lubricating properties in sintered bearings. The principle of operation is shown in Figure 14.3.

Powdered metal may be obtained by any of the following methods. Machining, which produces coarse particles; crushing, for fine irregular particles; shotting, which involves pouring metal through a sieve and cooling it by allowing the sieved particles to drop into water (this produces spherical or pear shaped particles), and atomisation, in which the metal is sprayed, thus producing particles of irregular shape that can be made to any size required by adjusting the spray. Iron and copper are the two main bases of the metals used, for example, iron and brass are used for small machine parts and bronze is used for porous bearings.

In Chapter 9, cemented tungsten carbide cutting tools were mentioned. It is therefore worthwhile using them here as an example of the sintering process. (i) Tungsten is obtained from the mineral scheelite by chemical and mechanical treatment. (ii) The

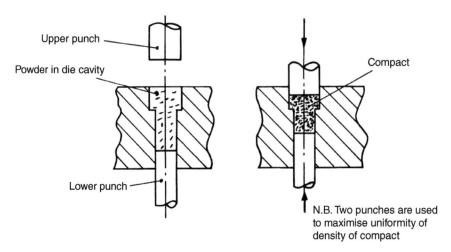

Figure 14.3 The pressing or briquetting process in powder metallurgy, two punches are used to maximise density uniformity of the compact.

tungsten is mixed with carbon powder and heated to form tungsten carbide; these are the hard particles of the tool. (iii) Cobalt is mixed in; this is the binding material or cement; other additions may also be made, such as the carbides of titanium, tantalum and niobium, to improve wear and crater resistance. This mixing is carried out during crushing in large mills. (iv) The powder mixture is pressed into compacts that are presintered at 900°C in a protective atmosphere. (v) The compact is then sintered at 1300–1600°C. At this temperature the binding material has melted and dissolved part of the carbides; 10–50% of these are in a molten state. The cemented carbide shrinks some 20% linearly and 50% by volume and becomes non-porous.

The cemented carbide cutting tool is an example of how powder metallurgy is used to produce items from materials that are difficult to machine. Other types of products are as follows. Complex shapes that would otherwise require much machining, for example, gears that can be produced accurately with a fine finish and can be impregnated with oil.

Products where combined properties of two materials are required, for example, electrical contacts made with copper or silver to provide high conductivity, and tungsten, nickel or molybdenum to provide resistance to fusion. Also electric motor brushes containing copper for high current carrying properties, and graphite for lubrication; or non-porous bearings made from soft tin held within a hard matrix of copper. A different product is the porous bearing made from powdered iron or bronze and containing oil 10–40% by volume, these bearings require no lubrication during their service life. As a final example, porous metallic filters can be made with extremely fine pores constituting around 85% by volume.

Powder metallurgy has the advantages that no scrap is produced, no machining is required, complex shapes can be produced, a variety of materials can be used with combinations of mix and structure, high production rates can be achieved and only semi-skilled or unskilled labour is required for production. Disadvantages are that the cost of powder materials is high, the cost of dies is high, the strength of the component is usually less than a forged, cast or machined component and the design of the component is limited since constant density of the component is often difficult to achieve in components of complex shape.

14.7   Pipe and Tube Manufacture

This section is not concerned with a specific process, but rather a specific product. Pipe or tube manufacturing processes are briefly considered together for convenience. They involve a variety of processes, most of them already mentioned except for welding, which is covered in Chapter 16.

Canes of bamboo joined together after having the internal diaphragms knocked out may have been the earliest form of pipe. The Aztecs and Ancient Greeks used earthenware pipes to transport water and the Romans used lead piping for their water schemes. Today, metal pipe is mostly made from strip metal or solid ingots; thick walled pipe is also made from plate.

Butt welded pipe is the most common; it is used for conveying gas, water and wastes, and for structural applications. In the process, strip steel is heated in a furnace to welding temperature and passed through a series of forming rollers that bend it into a circular section. The edges of the strip are slightly bevelled to allow them to meet accurately when formed. In continuous butt welding the strip is supplied in coils, the ends of the coils being flash welded together to provide a constant unbroken flow of material through the rollers. A 'flying saw', that is, a saw that travels linearly at the same rate of the pipe while cutting, ensures that the process is continuous while still allowing the pipe to be cut to length. Pipe is made by this method in sizes up to 75 mm diameter.

In electric butt welding, once again strip is made continuous by flash butt welding the beginning of one cold rolled coil to the end of the previous one. The strip is passed through rollers to form the tubular shape then finally passed through three centring rollers that hold the formed tube in position while two additional electrode rollers, located on either side of the seam, supply current to generate the heat of welding. Good quality tubes are produced by this process, good bore concentricity, precise control over wall thickness and accuracy of diameter are all obtained. Also the high quality surface reduces the possibility of pockets of corrosion forming when in use, making them suitable for boiler or pressure applications. Pipes up to 400 mm diameter with wall thicknesses of 3–13 mm are made by this process. Larger diameter pipes are usually fabricated by forming plates to shape in special presses, then using the submerged arc welding process to join the seams. Pipes from 400 to 1200 mm diameter and 6–15 mm wall thickness are produced by this process.

Seamless pipe can be produced by centrifugal casting, extrusion, piercing or rotary forging. Centrifugal casting was explained in Chapter 8. The extrusion process was explained in Section 10; for tube production, direct extrusion is used but with a mandrel to shape the inside of the die. A billet is placed in the extrusion chamber, the mandrel is pushed through the centre of the billet and then the ram advances pushing the metal through the die and around the mandrel. Aluminium and plastic are simpler to extrude than steel, which may be extruded up to about 75 mm diameter.

In the piercing process, a solid billet is centre punched then heated in a furnace to hot working temperature. It is then passed between two specially designed and oriented rolls that cause the billet to be squeezed in one direction as it is both rotated and advanced axially. This complex motion causes an aperture to open in its centre as it is forced over a piercing mandrel. While still at working temperature the billet, now in tubular form, is passed through grooved rollers and over a plug to bring it to the desired diameter and wall thickness. Seamless tubes up to 150 mm in diameter can be produced by this method, though further working can be used for pipes up to 610 mm diameter. To produce even higher quality seamless pipe suitable for containing fluids at high pressures pierced billets are passed through the rotary forging process. This involves passing the tube over a mandrel, which maintains the internal diameter while squeezing the material between forging rolls.

14.8 Metal Finishing Processes

After initial production, by whatever process, components usually need some form of finishing or surface treatment. Often the component will have to be cleaned in some way to remove residue created by the previous process, or rough edges or excess material will have to be removed. Then to protect the metal when exposed to atmosphere, or to provide an attractive appearance to the customer, a surface treatment of the metal will be necessary. A few examples of these finishing processes are given here.

Components that have been sand cast often have sand particles stubbornly adhering to their surface. This spoils the appearance and also causes excessive wear on cutting tools if the component has to be machined. The sand is therefore removed by an *abrasive cleaning* process such as shot blasting. In this process, sand or steel shot is impelled against the component at high speed by utilising compressed air.

After components have been hot worked their surfaces oxidise and a black scale is formed. This scale has to be removed before further work is carried out as it can lead to surface defects in the finished component. Scale removal is carried out by the process of *pickling*. This involves dipping the metal parts, which must have been cleaned to remove oil and dirt, in a bath of dilute acid. If controlled properly, the result will be a smooth, clean, scale-free component. Abrasive cleaning can also be used for removing scale.

A process used for removing rough edges, after say machining or blanking and piercing, is *tumbling*. Here, components are placed in a specially designed barrel mounted on trunnions. Sometimes abrasive particles such as pellets of granite are added. As the barrel rotates the components, with which the barrel is almost filled, are drawn upward before sliding down onto those below. This gentle tumbling action causes the component edges to be smoothed and rounded.

Among the finishing processes *electroplating* is very popular. The process may be used to improve appearance, wear and corrosion resistance, or to add material for the purpose of increasing the size of the component. Most metals can be plated and even plastics if first coated with a conductive material. Figure 14.4 shows the principle of the process. The components to be plated are made the cathode and are suspended in a solution of dissolved salts of the metal to be deposited. Also suspended in the solution is a slab of the metal to be deposited, this is made the anode. A direct current voltage is applied and metallic ions migrate through the solution to the surface of the cathodic components. They then lose their charges and are deposited as a metal plating. Steel, aluminium and die cast zinc components are all commonly plated with metals such as tin, chromium, copper and for smaller components such as jewellery, with gold and platinum.

Anodising provides a corrosion resistant surface that is also decorative. It is an oxidation process that converts the component surfaces to a hard, porous oxide layer. In this process, the components are made the anode and they are suspended in an acid bath; chemical adsorption of the oxygen from the bath results. A variety of surface colours

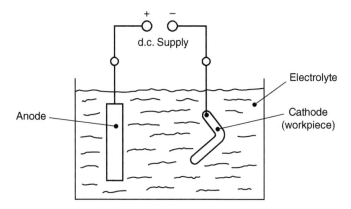

Figure 14.4 The electroplating process.

can be obtained by the use of organic dyes; black, silver, gold, grey, red and bronze being popular. The process can also be used to provide a suitable base for painting. Anodised surfaces are particularly popular on aluminium products.

The familiar process of *painting* is widely used for providing corrosion protection and an attractive appearance on many products. At least two coats are usually applied. The first coats have three tasks, to ensure paint adhesion, to enhance corrosion resistance and to produce a smooth flat surface by filling in small cavities caused by surface imperfections. The final coats are more highly pigmented and provide attractive colours and textures. Three methods of paint application are normally employed. Small parts or components that require their entire surface coated in paint can be dipped. This process involves lowering the component into a vat of paint either manually or automatically while passing along a conveyor. Good surface finishes are difficult to achieve as the paint can leave wave marks as it runs off, and it is not economical where only thin coatings are required. A second method is spray painting. This can be done by hand, which is unhealthy and unpleasant for the painter and often inefficient. Automatic spray painting removes the health hazard, if carried out in an enclosed area. Only a thin coating of paint is usually applied at each coat to ensure a good finish. When using automation, if the components are simply passed through fixed spray heads, much paint can be wasted through overspray; however, industrial robots taught by skilled human painters can greatly improve paint utilisation and quality. Finally, *electrocoating* is used to cover large components such as car bodies. The process involves dipping the product into a vat containing the paint particles in a water solvent. A direct current voltage is applied such that the vat is the cathode and the component the anode. The paint particles become electrostatically charged and they are attracted to the component where they are deposited in a thin uniform coating. This process is an improvement on the ordinary dipping method as it is economic while also providing better coverage of internal surfaces and a better finish that can be thinner and more evenly spread.

Review Questions

1   What process would you expect to be used for producing the following?
   a) High quality seamless pipe 120 mm diameter suitable for carrying chemicals at high pressure.
   b) 10 m lengths of 200 mm diameter steel pipe for low pressure liquid waste disposal.
   c) Two hemispherical pressure vessel end pieces made of steel 30 mm thick and 3 m in diameter.
   d) One octagonal aperture in the die of a press tool; the die is to be of single piece construction.
   e) A decorative matt black finish on the aluminium casing for a microscope.
   f) Porous metallic filters 20 mm diameter by 20 mm long at a rate of 5 00 000 per year.

2   Briefly describe the EDM and ECM processes.

3   State three advantages and three disadvantages of explosive forming.

4 Describe the production of cemented tungsten carbide cutting tools.

5 Discuss the advantages and disadvantages of the powdered metal processes.

6 Briefly describe five different processes used for producing metal pipes and tubes.

7 Describe, with the aid of a sketch, the electroplating process and discuss why it is used.

8 Discuss the differences between the basic paint dipping process and that of electrocoating.

15

Manufacturing Processes in the Electronics Industry

15.1 Introduction

Almost every manufactured product we use today either contains electronic components, or has been designed and made by processes utilising electronics. The wordprocessor being used to write this sentence contains semiconductor based technology, the plastic for the keyboard was made by electronically controlled moulding processes and the assembly of the individual components was done by microprocessor-controlled machines. Motor cars are designed using electronic computers, they are assembled using microprocessor-controlled robots and often they contain such items as electronic ignition, instrumentation and fuel management systems. Our complex and varied telecommunication systems utilising land lines, microwaves and satellites would not operate without electronics. To supply this ubiquitous hardware we rely on the electronics manufacturing industry.

At the beginning of the twentieth century no electronics industry existed. Then in 1904 the thermionic valve was invented and this was followed in 1906 by the triode. These components made possible the construction of many of the electronic devices we know today, albeit in a relatively slow and cumbersome manner. For example, 40 years after the invention of the triode the electronic computer ENIAC required approximately $1400\,\text{m}^2$ of floor space, weighed around $30\,500\,\text{kg}$ and contained $18\,000$ vacuum tubes. However, 1947 saw the invention of the transistor; this semiconductor technology was rapidly developed and by the 1960s many transistors, diodes, resistors and capacitors could be produced on the one 'chip' of silicon material; these were called integrated circuits (ICs). In 1970, an IC that could carry out many of the activities associated with computing was produced; this was the microprocessor. By the middle of the 1980s a million components could be created on a single chip, that is, very large scale integration or VLSI. Today, we have computers that can fit in our pockets whose power would amaze those who worked on ENIAC. At the time of writing there are chips with 15–30 billion transistors using 16, 14 and 10 nm (nanometre) technology. Difficulties arise with the commonly used silicon base as the transistor size approaches 7 nm. This is because as the 7 nm point is reached the electrons can begin to experience quantum tunnelling that normally occurs with barriers of around 3 nm and under. However, work is ongoing to achieve practical sub 7 nm sizes by using nanotechnology and materials other than silicon.

This chapter is concerned with the manufacturing processes used to produce semiconductor electronic components and the printed circuit boards (PCBs) on which they are assembled. The assembly process itself will also be briefly examined. Though not

Essential Manufacturing, First Edition. Gordon Mair.
© 2019 John Wiley & Sons Ltd. Published 2019 by John Wiley & Sons Ltd.

included here, other discrete components are found in electronic systems. For example, resistors may have a wound wire, conducting metal oxide film or carbon powder construction, and capacitors can be made from sheets of metal foil and insulating material wound into a compact canned package. Also to be found may be components such as potentiometers, switches and liquid crystal displays.

15.2   Semiconductor Component Manufacture

Semiconductor materials are so called because they have electrical properties that lie between those of conductors and insulators. However, what makes them useful is that their electrical properties can be altered by adding carefully controlled amounts of impurity atoms, called dopants, to their crystal structure. There are two types of dopant – the 'n' type that makes the material electron rich and the 'p' type that makes the material electron deficient. A component is constructed on a crystal of semiconductor material by selective doping. For example, by creating an n-type element in contact with a p-type element, a p–n junction is produced. This creates a diode junction since the configuration will conduct current only in one direction see Figure 15.1.

Semiconductor components such as transistors and diodes can be manufactured as discrete items for use in a multitude of products such as radios and televisions. They are also used as discrete components in power handling applications where relatively large currents are being manipulated. However, although the basic technology is the same, the most complex conditions occur when large numbers of them are being created on a single chip, that is, the IC mentioned in the previous section. It is this situation, the manufacture of an IC chip, that is considered here.

The chip manufacturing process comprises a number of steps; these are shown in a very simplified form in Figure 15.2.

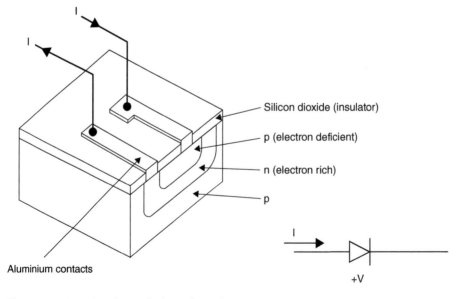

Silicon dioxide (insulator)

p (electron deficient)

n (electron rich)

p

Aluminium contacts

+V

Figure 15.1 A semiconductor diode configuration.

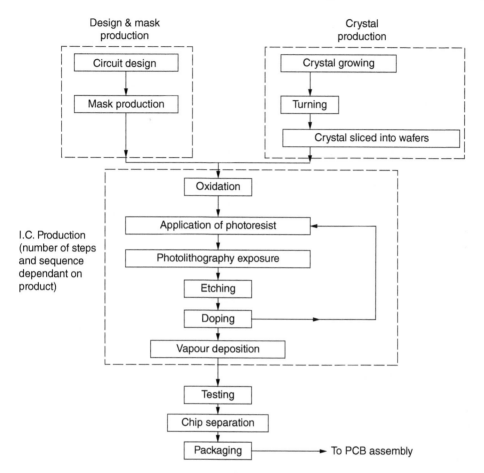

Figure 15.2 An integrated circuit manufacturing sequence.

15.2.1 Crystal Growing

The most widely used material for semiconductor component manufacture is monocrystalline silicon. This is because of its special atomic structure, its oxide (silicon dioxide) that is a very good insulator and can therefore be used for isolating components and component elements on a chip and its abundance in nature. Other materials used are cadmium sulfide for photo-electric cells, lead sulfide for infrared sensors and gallium nitride for LEDs. Another more expensive material is gallium arsenide, ICs fabricated on gallium arsenide operate much faster than silicon based devices and they also have the ability to transmit light; this allows lasers and light emitting diodes to be constructed. However, the fact that the processing of gallium arsenide is relatively complex and that it is expensive means that silicon retains its popularity as a substrate material.

Silicon occurs in quartzite sand, that is, silicon dioxide. It must be purified before it becomes the flawless single crystal needed for chip manufacture. First it is heated with carbon, that is, coal or coke, in an electric arc furnace to produce polycrystalline silicon that is about 97% pure. It is then further processed by heating, fractional distillation as a chloride and vapour deposition to eventually produce polycrystalline silicon that is virtually 100% pure. This is now melted in a quartz crucible; dopant material may also be added at this stage. A seed crystal of silicon is dipped into the melt and then withdrawn very slowly while also being rotated. The melt material solidifies onto the seed crystal so forming a larger crystal, called a 'boule', which is eventually about 0.15 m diameter and 1 m long. This crystal growing is called the 'Czochralski process'. The boule is then inspected for flaws, its electrical properties tested and defective parts removed. Since the crystal growing process does not allow tight control of the diameter size the boule is now turned and ground to produce a precise cylindrical shape between 50 and 150 mm in diameter and 1 m in length. It is then sliced into wafers about 0.5 mm thick using a diamond saw. These are polished to a mirror finish and inspected to ensure that any damage caused by the sawing process has been removed. Crystal growing and wafer production is carried out by companies specialising in this work. Therefore, the final stage is to pack the wafers and despatch them to the component manufacturer's factory.

15.2.2 Integrated Circuit Production

Although the following steps are the basic ones used in IC production, it should be remembered that the actual manufacturing process will involve repetitive implementation of these steps to produce a number of different levels, doped regions and electrical interconnections on the one chip.

15.2.2.1 Photolithography

None of the conventional machining processes are capable of producing the fine definition required to produce billions of transistors on, for example, a $400\,mm^2$ chip of silicon. Resort has to be made to photographic techniques. Even here, visible light cannot be used as its wavelength, from about $0.4\,\mu m$ to just under $0.8\,\mu m$, is too coarse for the definition required in these circuits. For this reason deep ultraviolet light lasers or electron beams with much smaller wavelengths are used to provide very fine resolution.

Photolithography involves printing patterns of the circuits and their elements onto a mask. Since each wafer of silicon is used to produce a large number of ICs the pattern is repeated many times on the one mask. The mask itself is produced photographically on glass from information produced by a CAD system. An image of the patterns is projected from the mask onto the wafer, which will have been previously coated with a photosensitive film. This film, called a photoresist, when exposed to ultraviolet light or electron beams becomes either more ('positive resist') or less ('negative resist') soluble to an etchant solution. Thus, after exposure the wafer can be treated with the etchant to leave the appropriate parts of its surface exposed. Layer fabrication can now take place, these exposed surfaces being subjected to processes such as oxidation, dopant diffusion, vapour deposition or further etching. The design of the IC will determine how often, and in what order, these processes are carried out. Between each step, the photolithography process will be used to produce new patterns.

15.2.2.2 Oxidation

As noted earlier, silicon dioxide is an excellent insulator; it also effectively blocks the dopant diffusion process, adheres strongly to the silicon substrate and is able to be removed by an etchant that does not attack the substrate. It is therefore used to create masks on the chip surface and to separate individual ICs on the wafer; it is also used as an integral part of the circuits and components themselves.

To create an oxide pattern on a silicon substrate the following procedure is carried out. In 'dry oxidation' the silicon substrate is exposed to an oxygen rich atmosphere in a furnace at a temperature of between 900 and 1200°C. This simple process causes oxidation of the silicon surface. When carried out in an atmosphere containing steam the process is called 'wet oxidation'; this produces a higher oxide growth rate but with lower density. Thus a combination of both the dry and wet processes are often used. After cleaning, the oxide surface is coated in a photoresist material that is then pre-baked in an oven at about 100°C. The photolithography process is now implemented as described earlier. Assuming that a positive resist has been used and the exposed resist has been removed by an etchant solution, the remaining resist is toughened by a post-baking operation. This now allows the uncovered oxides to be removed down to the substrate level by another etchant, thus creating the desired pattern. Finally, the remaining photoresist is removed by dipping in a dissolving solution such as acetone.

15.2.2.3 Chemical Vapour Deposition and Dopant Diffusion

This is the process that is used to create the p- and n-type semiconductor regions on a substrate. It is carried out by placing the substrate in an atmosphere of chemical vapours that deposit the dopant. The process is carried out in a furnace at a temperature between 800 and 1200°C. While in the furnace, atoms from the dopant material diffuse into the silicon substrate, thus causing displacement of the silicon atoms.

15.2.2.4 Metal Vapour Deposition

The devices produced by the foregoing processes all need to be interconnected by metal conductors to allow the IC to operate; vapour deposition is the process used to create these conductors. Normally aluminium or aluminium alloy is used, but for higher device densities tungsten may be necessary: again, photolithography and etching are used to create the connector paths. In vacuum deposition the wafer is placed in a vacuum where the metal to be deposited is melted. Under these vacuum conditions the metal boils and deposits itself on adjacent surfaces. The process is therefore so organised that the metal vapour, which propagates out from the melt in a straight line, coats the desired surfaces. Chemical vapour deposition is also used for depositing thin films of non-metallic materials onto a substrate.

After all these processes have been carried out on the wafer a number of times and the ICs have been created, they have to be tested then separated. Testing is carried out automatically by needle probes contacting appropriate points on the circuits. Computer programs run the tests and flawed ICs are automatically marked with an ink dot. The ICs are then separated by either diamond sawing or scribing along lines between the ICs and snapping them off rather like tile cutting. At this stage in the process the ICs are usually called 'dies'.

15.2.3 Packaging of Integrated Circuits

Since the individual dies are fragile and easily damaged they must be given a protective casing. They also need to be provided with connections of a practical size to enable them to be connected into circuits on PCBs and soon. The die may be attached to its protective

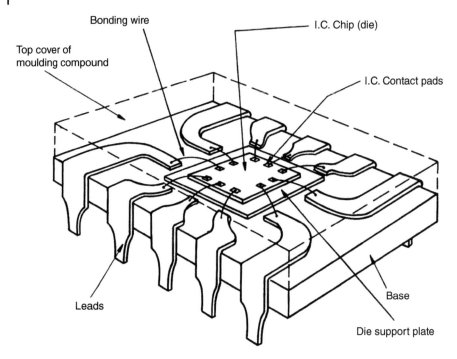

Figure 15.3 An integrated circuit in a dual in-line package.

casing or package by metallising the back then soldering it onto a metal plate in the package, or it may simply be epoxied into place. The package will have metal leads to allow connection to the outside world. The die will have metallised contact pads created by the vapour deposition process mentioned earlier. These contact pads are connected to the package leads by pressure or ultrasonic bonding techniques. The package is then closed and sealed while ensuring that no moisture or contaminants are included, see Figure 15.3.

The package is now given its final tests. Those that must exhibit very high reliability, for example, for aerospace and military use, are given high temperature, temperature cycling, vibration and other tests to ensure that the circuits that would fail early in their service life are identified.

15.3 Clean Rooms

With component dimensions in ICs being much less than 1 μm in size it is apparent that contamination by foreign particles during manufacture could be a serious problem. For example, human hair is between 30 and 100 μm in diameter, bacteria between 0.3 and 30 μm, and tobacco smoke between 0.01 and 1 μm. To avoid contamination by foreign particles special rooms are built for semiconductor manufacture and other fine assembly work; these are called 'clean rooms', and incorporate a number of techniques for ensuring a clean environment.

To prevent contaminated air entering the room, air has to pass through high efficiency filters; these must be almost 100% efficient at removing particles which under the most stringent conditions would be less than 0.1 μm. Charcoal filters may also be used to remove chemical contaminants from the air. To make sure that only cleaned air is present in the room a positive air pressure differential is created between the room and the outside factory. This pressurisation ensures that air will only escape from the room and outside air can enter only through the filtering system.

Although air into the clean room can be filtered, humans inside the clean room present a major source of contaminants. Although eating, drinking and smoking are prohibited, humans still constantly introduce contaminants from breathing, skin particles, hair and clothing. To eliminate this contamination surgical masks, gloves, hair caps and gowns are worn by operators. For ultra clean rooms, the workers may be completely isolated from the environment by wearing one-piece overalls and visored helmets that drape over the shoulders and are connected by tube to an external air supply. A differential air pressure is again created between the inside of the suit and the clean room to ensure that air travels from the room into the suit and not vice versa. Airlocks are used by the personnel when moving in or out of the clean room.

Clean rooms are classified by a number that indicates the number and size of particles that are permitted per m^3 in ISO (International Organisation for Standardisation) standards or per ft^3 in US Federal standards. Ambient city air can contain 35 000 000 particles of 5 μm and larger per m^3. In extreme cases clean rooms can provide atmospheres containing less than 12 particles of 0.3 $\mu m\, m^{-3}$ although most clean rooms will be considerably less stringent than this.

15.4   Printed Circuit Board Manufacture

The individual ICs are seldom used on their own in electronic products. Usually discrete components with operating characteristics that prevent their miniaturisation are also required: for example, components that cannot be integrated such as inductors, large resistors that require extra heat dissipation, large capacitors, power transistors, potentiometers, mechanical switches and other ICs. A means of holding all of these components that will allow them to be connected to each other and to the outside world is necessary. This means is usually provided by a printed circuit board – often abbreviated to PCB. A block diagram of the steps involved in PCB manufacture is shown in Figure 15.4.

PCBs are available in a variety of types and complexities. They are usually rigid, though some are flexible, and they are made from a paper or glass reinforced plastic resin coated with a copper foil. The conductive tracks on the surface of the board are created by screen printing or photolithography and etching. The boards may be single sided, that is, conductive tracks on only one side or double sided. Multi-layered boards are also used. These are equivalent to a number of single or double sided PCBs made into a sandwich, each layer being separated by glass reinforced plastic and interconnected by conductive 'via' holes.

Substrate manufacture. The board substrate is made under clean room conditions since any contaminants on the copper surfaces could prevent proper functioning of the board in service. For single- and double-sided boards the copper foil, usually produced by electroplating, is placed on the highly polished surface of a steel plate resting

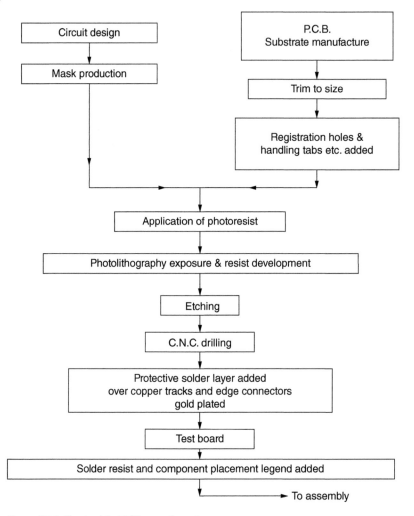

Figure 15.4 Single-sided PCB manufacturing sequence.

in a press. Assuming that a glass reinforced plastic substrate is being created, layers of glass cloth impregnated with an epoxy resin are placed in layers on top of the copper foil until the desired thickness is obtained. Finally, either a release film is added for single sided boards or another sheet of copper foil for double sided boards. Another highly polished steel plate is attached to the press ram. By using high pressures and steam heating of the steel plates the boards are compressed and the resin cured. After cooling, the boards are trimmed to remove excess resin. Usually one press has many plates and can therefore produce many boards at each pressing. Boards are usually produced in standard sizes to allow them to be easily used with automated equipment at later stages. For simplicity, the following stages refer to single sided board manufacture.

Pre-processing. At this stage the boards are prepared for processing by adding certain features. Registration holes have to be created to assist with board location in subsequent processes; these will enable, for example, the board to be located properly while component holes are drilled automatically at appropriate positions on the tracks. Handling tabs may also be added by stamping to facilitate later automated handling. The board may also be given a unique identification at this stage by adding a bar code or alphanumeric characters. The board is then thoroughly cleaned before passing to the next stages.

Photolithography. This process is very similar to that described earlier for IC manufacture. In this case, the photoresist is usually added in the form of a photosensitive polymer. The polymer is held between two dry cover films and supplied in rolls. By removing one of the films the polymer can be applied directly to the board surface. The second film may be removed at this stage or kept in place until later. The board is now exposed to ultraviolet light projected through a mask containing an image of the required track layout. It is then developed by chemical removal of the exposed resist, usually by using a solvent.

Etching. The exposed copper surfaces can now be removed by etching with a solution such as ferric chloride or copper chloride and hydrochloric acid. This is usually done by spraying the board with the etchant, as this provides greater control than total immersion. After all the exposed copper has been etched away, the remaining resist material is removed. This produces the desired copper tracks on the PCB surface.

Drilling. This is carried out to create the holes for component leads and pins, and also for bolting and screwing holes to take larger components such as transformers and heat sinks. The precision with which these drill holes are located is extremely important since automated component insertion methods may be used. A number of boards may be drilled at the same time by creating a stack and drilling through. The drilling will probably be carried out by a CNC, that is, a computer controlled drilling machine. These machines will automatically drill the required pattern of holes based on instructions received from the CAD system used to design the board layout. Drilling is also used to create the 'via' holes in double sided and multi-layer boards. The boards are passed between abrasive rollers to remove the rough edges created around the drilled holes; this is referred to as deburring.

Board finishing. The board now has bare copper tracks and 'lands' (these are the local areas where the track is widened around the lead holes or surface mounted device (SMD) contact points). To prevent the copper oxidising, which would give problems later at component assembly, it is given a coating of solder. This is done by plating or a process known as roller tinning. Edge connectors are usually gold plated for reliability. The board is then tested for continuity of the tracks and inspected for flaws. Finally, the component locations are identified by means of a printed legend and, with the exception of the lands, the board is coated in solder resist.

15.4.1 Assembling Components to Printed Circuit Boards

Having considered electronic component and PCB manufacture, we will now examine how they are assembled together to produce a finished 'populated' PCB. Component configuration and presentation are considered before looking at the final assembly process.

15.4.1.1 Component Configurations

Electronic components can be classified as: (i) leaded, these may have axial, radial, or dual in line (DIP) leads; (ii) surface mounted, these have no leads and (iii) non-standard. Sketches of these configurations are shown in Figure 15.5. Non-standard components may be edge connectors, switches, transformers and so on. Discrete resistors often have an axial lead configuration, transistors have a radial and ICs a DIP. Components produced in leaded configurations can also be produced as surface mounted devices; that is, SMDs. In comparison to leaded components. SMDs allow the production of more densely populated boards, that is boards can be made smaller yet perform the same functions; they have a physical construction ideally suited to automated PCB manufacture, they can be mounted on both sides of the PCB and therefore the only holes needed to be drilled in the board are 'via' holes for electrical connections between each side. However, they may be more expensive than the leaded type and difficult to assemble manually. Because of this and difficulties in making some leaded components in SMD form, PCBs are usually populated by a mixture of leaded, SMD and non-standard components.

Axial and radial leaded components are usually packaged in edge tape bandoliers or sprocket feedable tapes, DIPs in linear stick magazines and SMDs in blister tapes or stick tubes. These are shown in Figure 15.6. In some systems, many bandoliers, each carrying one type of component, are used automatically to create a new bandolier that will have all components sequenced in the correct order for automatic assembly in a special component insertion machine. DIPs can be provided in magazine packages. These can be mounted at the workstation in such a manner as to allow the DIPs to slide out of the magazine onto a rail where they can easily be picked up manually or automatically for assembly. For SMDs, stick tubes are similarly used, or by peeling off a retaining tape, blister packages can be fed past the component pick up points at an assembly station.

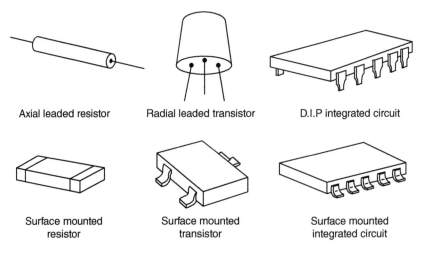

| Axial leaded resistor | Radial leaded transistor | D.I.P integrated circuit |

| Surface mounted resistor | Surface mounted transistor | Surface mounted integrated circuit |

Figure 15.5 Component configurations.

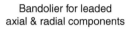
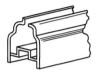
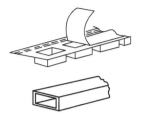

| Bandolier for leaded axial & radial components | Magazine for dual in line packages | Blister tape and stick tube magazines for SMDs |

Figure 15.6 Component delivery packages.

15.4.1.2 Assembly Methods for Leaded Components

Electronic components are assembled to PCBs manually, automatically or by a combination of manual and automatic methods.

Manual assembly methods are used for shorter production runs, where relatively slow production rates of up to say 500 components per hour are to be placed or where the components are difficult to assemble by other means. Leaded components need to have their leads cut to length and pre-formed before insertion, see Figure 15.7a. In manual assembly this can be done by the operator using pliers or by using a special die (in automatic assembly the pre-forming is done automatically using special tooling). The operator lifts the component from a bin or presentation point and places it into the CNC drilled holes in the PCB. These holes will have been marked at the PCB manufacturing stage with identification numbers to assist the operator. A further refinement is to use a machine that presents the components to the operator in a specific sequence; this is synchronised with a projector placed over the board. The projector is supplied with a roll of film that is indexed to allow the position of each component to be illuminated on the board as the operator carries out the insertion. For very small volumes the assembled PCB may have each component individually hand soldered, but it is more likely that the board will move on to an automated soldering process such as wave soldering.

Automated assembly is used for higher volume production. Typically industrial robots will be used where the demand is for around 1800 components per hour to be assembled, since robots can be easily reprogrammed to cope with changing production requirements. For higher production rates, say up to 30 000 per hour, dedicated automated

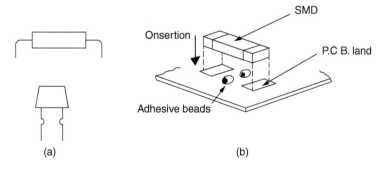

Figure 15.7 (a) Preformed axial and radial leads and (b) SMD onsertion using adhesive.

assembly machines are used. Axial and radial leaded components are automatically removed from their bandoliers or sprocket feedable tapes, their leads are preformed, they are inserted and then their leads are cropped and clinched ready for soldering. DIPs are gravity fed from their stick magazines to their appropriate pickup points.

After insertion the components must be soldered to provide the electrical connections between them and the PCB land areas (lands are the wider parts of the PCB tracks around the component holes). This operation requires the application of a flux to remove metal oxides and contaminants and to distribute the applied heat uniformly across the joint, and a molten solder. For large volumes, production techniques such as wave soldering are used. Here the boards are mounted on conveyor tracks which support them along two edges. The boards then have flux applied to them by one of three methods. They may pass through a bath of flux foam, over a wave of flux created in a bath or they may be sprayed with flux. They then pass through a pre-heating operation that raises their temperature to about 90°C and activates the flux. The board then passes over the surface of a bath of molten solder. A wave of solder is created in the bath as the board passes such that all exposed conducting surfaces are coated in solder.

15.4.1.3 Assembly Methods for SMDs

Since surface mounted devices have no leads, different techniques for holding them in position on the PCB and soldering are needed. Two main methods are used, the first being adhesive and wave soldering. In this method adhesive is applied to the PCB at appropriate points see Figure 15.7b. The adhesive can be applied by a screen printing process or by dispensing from a syringe. The SMD is then placed onto the board, probably using an automated pick and place unit in such a manner as to ensure it is making contact with the PCB lands and the adhesive coated areas. The adhesives used are curable and depending on type the adhesive is now cured by exposing it to ultraviolet light or heat. The components are finally soldered to the board using the flow solder process. This differs from the soldering of the leaded components in that the SMDs themselves have to pass through the solder wave, therefore particular care has to be taken to ensure proper adhesion of the components before soldering begins.

The second method is called the solder paste and reflow solder method. In this, the solder paste acts as both the adhesive and the conductor. The paste is actually a suspension of powdered solder particles in a flux. Just as in the previous technique the paste is applied by either screen printing or syringe. The SMDs are then inserted by automatic pick and place devices. It is apparent that one of the qualities of the solder paste must be that it has a high enough viscosity to ensure that components stay in their positions throughout the process. The boards are then passed through the reflow stage in which the solder is melted and cooled to form the finished conductive joint. Various methods of melting the solder are used, two of which are the infrared and vapour phase. In infrared reflow soldering, the boards are carried on a conveyor past a series of infrared lamps or heaters. It is relatively inexpensive and can handle high quantities, but consistent quality is more difficult to maintain. In the vapour phase method a suitable inert liquid is heated and vaporised in a tank. The PCBs are then placed in the tank where the vapour condenses onto their surfaces. During condensation the latent heat of the vapour is transferred to the solder, causing it to melt. The process can be made continuous by passing the boards through the tank on a conveyor. This method produces good quality joints, but is relatively expensive and slow.

15.5 Conclusion

The manufacture of electronic components and assemblies includes a wide variety of processes that require inputs from many different disciplines in science and engineering. The chapter is intended to provide only an indication of the depth of the subject. For example, one area not covered has been that of testing. Completed assemblies require to be tested in a variety of ways, for example, the operation of the individual components, the continuity of the electric circuit, the positioning and polarity of the components, and the functional performance of the complete assembly are all factors to be considered. Special test rigs, artificial vision systems and chambers where the boards can be subjected to thermal cycling, high humidity conditions, vibration and so on are all used to ensure that products are up to the high quality demanded by international markets.

Review Questions

1 Discuss the significance of the invention of semiconductor technology.

2 What is an IC chip? Sketch a block diagram of the steps involved in its manufacture.

3 Briefly outline the Czochralski process.

4 Describe the process of photolithography and discuss its use in the electronics industry.

5 Why is silicon dioxide an important material in IC manufacture; how is it obtained?

6 Briefly describe the processes of dopant diffusion and vapour deposition, and their use in IC manufacture.

7 Discuss why clean rooms are necessary in IC manufacture.

8 By the use of a block diagram, briefly outline the steps involved in the manufacture of printed circuit boards (PCBs).

9 Describe the differences between leaded components and SMDs, and discuss why the latter are increasing in popularity.

10 Briefly outline the operations involved in PCB assembly.

11 Describe the processes of 'adhesive and wave soldering' and 'solder paste and reflow soldering', and state where they are used.

16

Assembly and Joining

16.1   Introduction

Consider almost any manufactured product: aircraft, motor cars, televisions, computers or ships. All are really composites of a number of different elements. In previous chapters we have seen some of the means whereby the individual items and products can be produced; here, we examine the methods used to join these elements together. The joining process is usually called 'assembly' if mechanical fastening is being used or a number of components and 'sub-assemblies' are being combined. If welding, especially fusion welding, is being used then 'fabrication' is normally the term applied.

This chapter looks at just some of the range of techniques for joining that are available to the manufacturing engineer. The choice of technique will depend on a variety of factors, for example, bolted joints tend to absorb vibrations better than welded joints, but a welded construction would probably be cheaper. Adhesives are becoming a popular alternative to welding and riveting as they are simple to apply, they preserve the appearance of the material and when metals are joined in this way they do not change their metallurgical properties. The choice of the joining process influences the cost of the product, aesthetic qualities and the means of repair and maintenance. Figure 16.1 is a chart showing some examples of the joining processes used in manufacturing industry; it is these that are described in the following sections.

16.2   Mechanical Fastening

These methods may be semi-permanent or permanent; some examples are shown- in Figure 16.2. The semi-permanent methods using screws, spring clips, nuts and bolts and so on allow easy dismantling of the assembly for repair and maintenance. For heavy duty applications bolted assemblies may be selected. For assemblies requiring medium strength and expected to be dismantled by skilled workers, cap screws with hexagonal sockets or hexagonal heads are used. A castellated nut may be used in conjunction with a split pin to prevent rotation in situations where a low tightening torque has been applied to the nut. Self-tapping screws are used in light duty applications, often on plastics, where they make the expense of tapping the screw hole unnecessary. Self-locking plates can be used on thin metal assemblies; they are often seen on domestic goods such as washing machines.

Essential Manufacturing, First Edition. Gordon Mair.

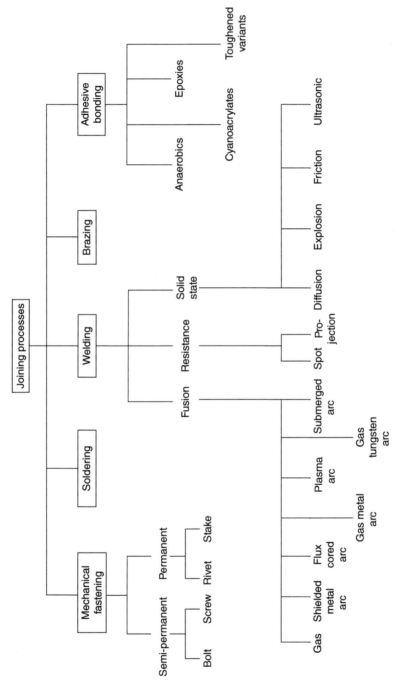

Figure 16.1 A chart showing some examples of the range of joining processes used in manufacturing industry.

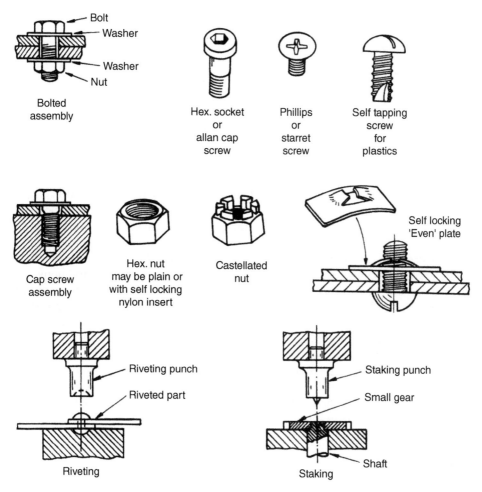

Figure 16.2 Mechanical fastening examples.

The permanent methods of riveting and staking are used only where dismantling of the product is thought unlikely. These methods would be used where maintenance is unnecessary and the product is to be discarded on failure. Riveting involves pushing a headed piece of metal through holes in the two materials to be joined, then spreading out the protruding metal to form the riveted head. The circular rivet hole acts as a crack arrester and therefore riveting is still popular in critical joining processes such as may be found in aerospace product manufacture. Staking is a slightly different process. Here the metal from a shaft protruding into a hole in the part to be joined is spread out against the wall of the hole using a staking punch as shown in the figure. The punch may have a chisel or star shaped point.

16.3   Soldering

Soldering is a process that uses heat, solder and flux to form a joint between two metals as shown in Figure 16.3. There are many types of solder with properties depending on

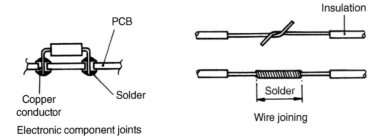

Electronic component joints

Wire joining

Figure 16.3 Soldering examples.

their alloying elements with melting points usually between 180°C and 300°C. There are some with melting points above and below this depending on their application. Traditionally solder has been an alloy of lead and tin, the proportions varying depend on the application, for example, 37% lead and 63% tin for the electronics industry. However, today the use of lead in solder is avoided and even prohibited in some areas due to its toxicity. Lead free solders may contain metals such as tin, copper silver, antimony and others in varying proportions depending on the application. The surfaces to be joined must be cleaned and dried before soldering takes place. Heat and flux is then applied to the joint and solder. The flux further cleans the metal by dissolving oxides on the surface and preventing new oxides forming during heating; it also assists the solder to flow into the joint. The solder may be applied from a coil and an electric soldering iron or gun used to apply heat. For large volumes of electronic components mounted on printed circuit boards wave soldering is commonly used. In all applications solder joints do not provide high strength and should therefore not be used in situations where they may be subjected to heavy mechanical loads.

16.4   Brazing

Brazing is similar to soldering in that the metals to be joined are not heated above their melting temperature. A non-ferrous filler material is used that has a melting temperature above 450°C. Since the parent metal is not melted the filler material must obviously be of a different composition, usually an alloy of copper, silver or aluminium. A flux, or inert atmosphere, is used to help ensure a clean joint. The filler material is drawn into the space in the joints through capillary action, thus necessitating a good close fit between the components to be brazed, see Figure 16.4. Almost all metals can be joined by brazing; less heating is required when compared to welding, therefore

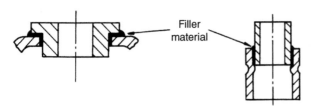

Filler material

Figure 16.4 Typical brazed joints.

less distortion of the workpiece occurs and the process can be mechanised. A major disadvantage, however, is that if heated above the brazing temperature the joint may be destroyed; this limits its application to areas where high temperatures are not expected. It is a stronger joint than a soldered one, but less strong than one that is welded.

16.5   Welding

In welding, two materials are joined together by the use of temperature or pressure, or a combination of both. The materials are usually metal but some welding of plastics is also carried out. The descriptions that follow apply to the welding of metals unless otherwise stated. At the joint the material fuses together to form a solid structure. Figure 16.1 shows the three main weld groupings, that is, fusion, resistance and solid state.

In fusion welding the metals to be joined are brought together and heat is applied to the joint. The edges of the parent metal are often prepared by machining, thus necessitating the use of additional 'filler' metal. The heat melts the parent and filler material at the interface, so allowing a strong homogeneous joint to be formed. In resistance welding the passage of an electric current across the joint interface causes local heating and melting; pressure is applied at the same time and a strong joint results. In solid state welding, pressure is often used in conjunction with appropriate metallurgical conditions; the parent metal is not melted.

16.5.1   Joints and Welds

Four of the basic joints used in welding are shown in Figure 16.5, that is, butt, corner, tee and lap. There are various types of weld used to create these joints and Figure 16.5 shows the fillet and butt welds, which are the most common types. The structure and nomenclature of a fillet weld is shown in Figure 16.6. The edges of the material to be joined require no special preparation. The fusion zone, that is, the area within which the metal is melted, consists mainly of the filler metal with a little of the parent metal. Fillet welds are often used to fill a corner and are very common in structural work, for example, stiffening ribs for a ship's hull may be welded in this way.

A butt weld is shown in Figure 16.7. Where maximum strength is required on thicker materials, some sort of edge preparation of the parent metal is usually necessary, Figure 16.7 shows a 'V' preparation. This preparation is created either by the torch used in originally cutting the plate, or by machining using a single point cutting tool in a planing machine. When welding mild steel, using mild steel for the filler material, it is possible for the welded joint to be stronger than the parent metal.

16.5.2   Gas Welding

A number of welding processes employ the burning of a mixture of fuel gas and oxygen as the heat source for melting the joint. Additional material fed into the weld pool from a filler rod is usually necessary. Although this process can produce good quality welds there is a problem with distortion, and this prevents its widespread use in industry.

Acetylene is a common fuel gas used. An oxyacetylene welding torch is shown in Figure 16.8a. As can be seen in the sketch the oxygen and acetylene, supplied via flexible

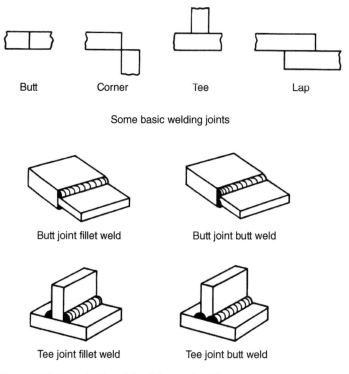

Butt   Corner   Tee   Lap

Some basic welding joints

Butt joint fillet weld   Butt joint butt weld

Tee joint fillet weld   Tee joint butt weld

Figure 16.5 Some basic welding joints and welds.

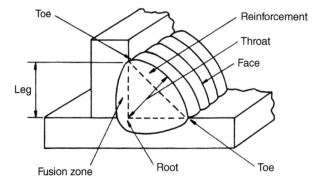

Figure 16.6 Fillet weld terminology.

tubes from pressurised metal bottles, are combined within the handle of the torch. The resulting mixture is ignited at the tip. The ratio of fuel gas to oxygen is controlled using the flow control valves and the three types of flame that can be obtained are shown in Figure 16.8b. Excess fuel gas produces the reducing flame that can be used to weld low carbon and some alloy steels. With equal amounts of fuel gas and oxygen, the neutral flame is produced. This is the most widely used flame as it has the least harmful effect on hot metal. The oxidising flame is produced when the proportion of oxygen is higher. This can be used to weld copper and copper alloys. Gas welding is used for welding thin

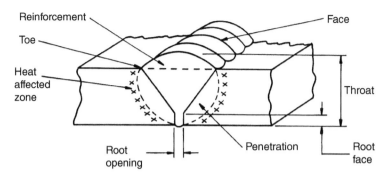

Figure 16.7 Butt weld terminology.

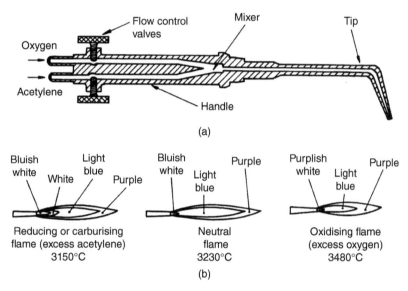

Figure 16.8 (a) Simplified section of an oxyacetylene welding torch. (b) Three types of oxyacetylene flame.

metal sheets and car repairs. It can also be used for welding cast iron. The equipment is portable and relatively inexpensive.

16.5.3 Shielded Metal Arc Welding

Shielded metal arc, manual metal arc or 'stick' welding are all different terms for the same process. Figure 16.9a shows the equipment necessary and Figure 16.9b shows the operation of the process. Stick welding is a very common method of welding because of its portability, flexibility and wide range of applications. In the process the heat created by an electric arc struck between an electrode and the work melts the electrode tip and the parent metal. The current is of low voltage and high amperage. Molten globules of metal from the electrode are carried across the arc into the pool of molten metal at the joint. The pool follows the electrode as it is drawn along the joint, leaving behind solidified seam.

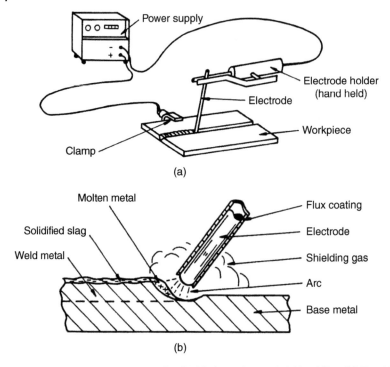

Figure 16.9 (a) Basic equipment for shielded metal arc or 'stick' welding. (b) The shielded metal arc process.

In this process the electrode is coated in a material which provides various functions. The electrode is often called the welding 'rod' or 'stick' and the coating 'flux'. The flux provides; a protective atmosphere that prevents oxidation, arc stabilisation, separation of impurities from the melt, a protective slag that absorbs impurities, reduced weld spatter and alloying elements.

16.5.4 Flux Cored Arc Welding

In 'stick' welding the process has to be interrupted every time the electrode is consumed down to the stub. This causes the welder to lose potential production time as he replenishes his rod in the gripper. The problem could be overcome if the stick could become a length of wire that could be coiled and fed into the weld pool. This is not possible with the normal configuration as the outer coating of flux will crack when deformed. Thus the solution to the problem is to make the electrode a hollow wire containing flux internally in a powdered form. Figure 16.10 shows the principle of flux cored arc welding. It is used in similar applications to stick welding, for example, shipbuilding, general fabrication and construction work.

The process can be self-shielding, where the protective atmosphere comes only from the flux core compounds that generate a gaseous shield during welding, or it can use auxiliary gas shielding, where additional protection is provided by supplying a shielding gas from an external source.

The advantages of flux cored welding are that it provides a relatively high rate of metal deposition with small diameter welding wires; it can be used for a wide range of metal

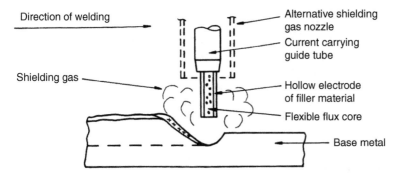

Figure 16.10 Flux cored arc welding.

thicknesses from about 1.5 mm upwards; the process is simple and easy to use and it can be used in any position with the smaller diameter wires, for example, for overhead welding. One disadvantage with the process is that a considerable amount of welding fumes are emitted during operation. This means that additional equipment has to be provided for drawing off the fumes, especially when welding indoors.

16.5.5   Gas Metal Arc Welding

Another name for this is Metal Inert Gas or 'MIG', welding. The process is shown in Figure 16.11. A bare continuous wire electrode is fed through the welding torch from a coil. The torch is water cooled. An inert gas is fed down the annulus of the torch to form a cloud around the arc and weld pool that protects the weld from atmospheric contamination. The gas is often carbon dioxide though other gases can be used. The process provides relatively high deposition rates. The fact that the wire can be fed, and the process monitored, automatically means that it is suitable for use in conjunction with industrial robots. In fact robotic arc welding is now one of the most popular applications for robots.

16.5.6   Gas Tungsten Arc Welding

Tungsten inert gas and 'TIG' welding are alternative names for this process. It is similar to MIG welding except that a non-consumable electrode is used, a hand fed filler rod usually being necessary to supply additional metal to the weld pool. The inert gas in this

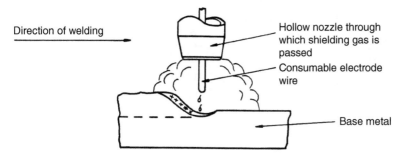

Figure 16.11 Gas metal arc welding.

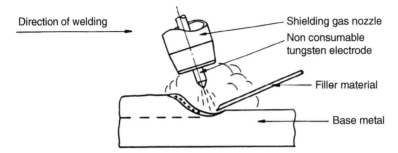

Figure 16.12 Gas tungsten arc welding.

case is usually Argon or Helium. The process is shown in Figure 16.12. It can be used for welding a wide range of often difficult to weld materials such as cast iron, aluminium alloy, magnesium, titanium, stainless steels and so on. The process is ideal for welding thinner sheets of metal under 1 mm thick.

16.5.7 Plasma Arc Welding

Deep and narrow welds in most metals, usually less than 6 mm thick, can be produced by this process. Gas is heated to an extremely high temperature by an electric arc, then forced across to the weld area. The arc temperature can reach 33 000°C and the gas is ionised, that is, it becomes a plasma. The two methods of using this process are shown in Figure 16.13a,b. In Figure 16.13a the arc is transferred across from the electrode to the workpiece, in Figure 16.13b the arc occurs between the electrode and the nozzle, and it is the plasma gas alone that carries the heat across to the joint; this latter method is easier to control though slightly more expensive. Just as in MIG and TIG, an inert shielding gas is also provided. High welding speeds, narrow heat affected zones and reduced distortion are some advantages of this process. A disadvantage is that it can create considerable noise and fumes. This means that measures have to be taken to ensure a safe working environment for the welder and other people in the area.

16.5.8 Submerged Arc Welding

Operating in a similar manner to stick welding this process can be applied to much thicker materials. The joining of steel plates for ships' hulls is a typical example. The main

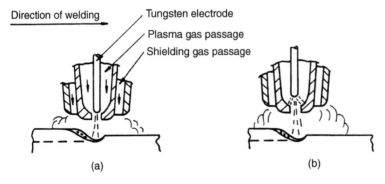

Figure 16.13 Plasma arc welding: (a) transferred and (b) nontransferred.

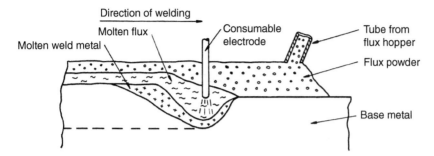

Figure 16.14 Submerged arc welding.

differences in the process are that the electrode is continuous, the flux is supplied in a powder form through a dispensing tube and the path of the arc is within the molten flux so preventing contamination from the atmosphere. Figure 16.14 shows the principle. The process is often carried out automatically; the welding head is suspended from a gantry as it traverses the joint. Rate of welding, feed rate of the electrode wire and flow of flux are all controlled automatically. The welds produced are usually of high quality. A large metal deposition rate is possible but stronger joints are created if a number of smaller welding runs are made rather than a single large one.

16.5.9 Resistance Welding

The previous welding techniques were all fusion types in that the parent metals were melted along the joint. In resistance welding the flow of an electric current across the interface between the two metals to be joined creates the heat necessary for a localised coalescence. The metals to be joined are almost always thin sheet steel.

In resistance *spot* welding, shown in Figure 16.15, the two overlapping pieces of metal to be welded are placed between two water cooled copper electrodes. The system is constructed so that one of the electrodes can be held steady while the other can exert downward pressure onto the joint, as shown in the figure. The pressure at the point of contact is high, as the moving electrode is actuated by a powerful pneumatic or hydraulic cylinder and the surface area of both electrodes is reduced at the point of contact. The combination of carefully controlled pressure and current flow causes the formation of a weld nugget at a spot between the electrodes at the interface of the metals to be joined. The equipment necessary for carrying out this type of welding is not usually very portable. A very common application is in motor car assembly, where the

Figure 16.15 Resistance spot welding.

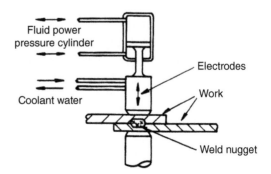

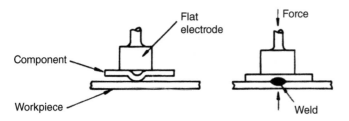

Figure 16.16 Resistance projection welding.

spot welding guns are often wielded by large industrial robots. Almost all joints on the body of a car are spot welded.

In resistance *projection* welding a weld nugget is again created at the interface between the metals to be joined. Again it is made by a combination of pressure and resistance heating from the flow of current across the interface. However, in this case the small contact area across which the current flows is created by forming projections on one of the metal sheets to be joined. The principle is shown in Figure 16.16. The equipment is often mounted on a pedestal similar to some used in spot welding. The electrodes in projection welding are wide and flat, whereas those used in spot welding are narrow, with a small contact area. Strip and sheet metal components and wire mesh products such as supermarket trolleys are welded in this way as the small contact areas at the intersection of wires constitute projections.

16.5.10   Solid State Welding

In solid state welding the material to be joined is not melted directly by an external heat source. For example, in the fusion welding processes only heat was added to form a joint, in the resistance welding methods heat and pressure were used to create a joint. Now, in solid state welding, we consider some of the welds that can be created by using pressure and sometimes heat, but without raising the temperature of the metal above its melting point.

Diffusion welding is used mainly for joining dissimilar or relatively expensive metals. Components for military aircraft using the advanced materials, such as superalloys, can be fabricated in this way. The process involves careful cleaning and preparation of the contacting faces of the parts to be joined. The parts are then placed against each other and pressure is applied for a considerable length of time. Heat is also applied by carrying out the process in an oven; the temperature may be about half of the melting temperature of the material. During the process the weld is formed by the migration of atoms across the boundary at the interface between the two metals. This means that the process is slow and hence relatively expensive.

Explosive welding is used mainly for cladding, that is, joining sheets of metals with one type of property to those of another. For example, armour plating or thin sheets of corrosion resistant material can be bonded to softer and cheaper structural plates of (say) mild steel in this way. The process is shown in Figure 16.17. An explosive charge is placed on top of the metals to be joined. The charge is usually of sheet form and is progressively detonated from one end. On detonation a stress wave is formed that compresses the metal sheets together as it travels along the length of material to be joined.

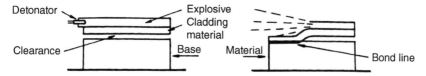

Figure 16.17 Explosive welding.

The kinetic energy in the pressure wave is so intense that mechanical bonding takes place between the metals. The process is of course dangerous and must be carried out under carefully controlled conditions.

Friction welding is used for the rapid joining of two components where at least one has rotational symmetry. It also produces strong joints between dissimilar metals. The process is shown in Figure 16.18. One of the components to be joined is held in a chuck and rotated at high speed. The other component is then brought into contact with the spinning piece and an axial force applied. The friction created generates heat sufficient to cause fusion of the two metals at the area of contact. Typical applications of the process are; joining of motor car transmission shaft elements, joining of bolt heads to bolt shanks and construction of engine valves.

Ultrasonic welding is used for joining thin sheets, foil and wire of similar or dissimilar materials, and it is also used for joining plastics. The process is shown in Figure 16.19. The workpiece is subjected to a downward static force from the mass shown. It is also subjected to an oscillating lateral movement generated by the vibrating transducer that operates at several kHz. The combination of movement and force at the interface of the materials creates a temperature of up to half the material melting point and breaks up any layers of oxides or contaminants. Thus, the combination of pressure and temperature allows the creation of a very strong bond.

There are many other welding processes, but those that have been shown here should be adequate to convey an idea of the variety of applications.

Figure 16.18 Friction welding.

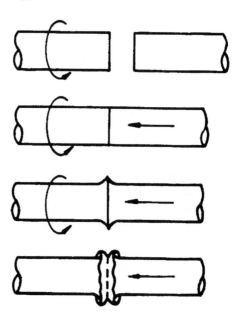

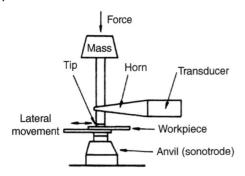

Figure 16.19 Ultrasonic welding.

16.6 Adhesive Bonding

Although adhesives have been in use for decades, recent advances in their design have led to a wide range of adhesive types, and consequently a wide range of applications. Adhesive bonding has many advantages over other joining techniques. For example, the load at the joint interface can be distributed over a large area, reducing stresses in the joint. Adhesives are lightweight. Completely dissimilar materials can be joined together. The structure of the materials being joined is not disturbed in any way by having holes drilled for mechanical fastening or being melted during welding. Thin, fragile or porous materials can be easily joined. As the temperatures at which bonding takes place are relatively low, no distortion or, if metal, damaging metallurgical effects happen to the material. Adhesives are inexpensive when compared to other joining methods. Products joined by adhesives generally look better than those that have been welded or mechanically fastened. The disadvantages are that the adhesive will fail if subjected to too high a temperature; surface preparation of the materials to be joined is critical, therefore this part of the process can be time consuming; the time to cure the joint may also be long and finally, some adhesives are unpleasant or dangerous to work with in an enclosed area.

16.6.1 Joint Loading and Design

Figure 16.20a shows the loading conditions that may be experienced by an adhesive joint and in Figure 16.20b examples of different types of joint. As can be seen, adhesive joints that are subjected to compression loads have the best strength, those in tension and shear are acceptable, but those in cleavage or peel have poor strength. Therefore, when designing a product adhesive joints should be positioned so that they experience compressive forces and tension or shear if necessary, but cleavage and peel conditions should be avoided. The joint design itself should be such that as large a surface area of adhesive as possible is employed. Mechanical strengthening also helps, for example, the use of a glued dowel adds considerable strength.

16.6.2 Adhesive Types

A wide variety of adhesive types exist, each one designed to be suitable for certain applications and materials. Four types are briefly noted here as typical examples.

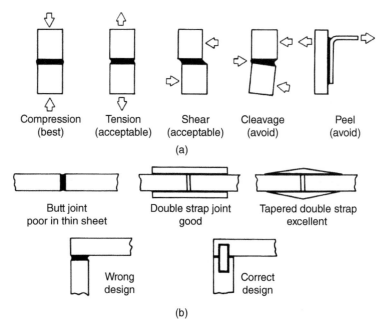

Figure 16.20 Adhesive bonding. (a) Adhesive joint loading conditions and (b) designing adhesive joints.

Anaerobic adhesives are used for making gaskets, thread locking on bolted or screwed components, pipe sealing and so on. They are widely known as 'sealants' or 'locking compounds'. They increase the strength of mechanical joints, which should be close fitting as the adhesive cures in the presence of metal and the absence of air.

Cyanoacrylates are used for the assembly of plastic and electronic components. They cure through reaction with moisture present on the surfaces of the materials being joined. Close fitting joints are necessary, but curing speed may be measured in seconds and a high strength joint is obtainable.

Epoxies are used for large joints and in tough applications such as the bonding of carbide tool tips. They are versatile and give strong hard joints with most materials. They are often in two parts, that is, an epoxy resin and a hardener, which when mixed together produce the adhesive.

Toughened variants are used in critical applications such as in aircraft construction. The anaerobics and cyanoacrylates are acrylic based, the epoxies are obviously epoxy based and the toughened variants are acrylic and epoxy based. A rubber material is incorporated in the toughened variant material to provide high peel strength and shock resistance. They also have a fast cure rate.

This chapter has shown the most popular methods of joining materials together to form sub-assemblies, assemblies or complete products. The choice of the most suitable method for a particular task should take into account the aesthetics of the product, its functional performance and the manufacturing processes available. This again illustrates the need for a team approach to decision making, the selection of the joining method possibly relying on input from, for example, an industrial designer, a mechanical engineer and a manufacturing engineer.

Review Questions

1 Briefly list the five main categories of joining processes used in manufacturing, and state the general type of application to which each is suited.

2 What is 'welding' and what conditions are necessary for it to occur?

3 What is the difference between a fillet weld and a butt weld, and under what conditions would each be used?

4 Describe the welding process that is most likely to be used for making minor repairs to the sheet metal body of a car.

5 Why is 'stick' welding so widely used?

6 What advantage is gained by putting the flux in the core of the electrode in shielded metal arc welding?

7 What is gas metal arc welding and why has it become so popular for automated welding installations?

8 Describe an arc welding process suitable for welding cast iron or aluminium alloy.

9 What are the advantages of the plasma arc welding process? Name one disadvantage.

10 Describe the process you would choose for welding together thick steel plates such as those used for ships' hulls.

11 Describe the welding process widely used in the car industry for fabricating car bodies.

12 What do diffusion, explosive, friction and ultrasonic welding have in common? Name one application for each.

13 Why has adhesive bonding become popular as a joining process?

14 Name one type of adhesive and a typical application.

15 Are joining processes necessary? What can be done to eliminate them?

17

Material and Process Selection

The importance of an integrated approach to all aspects of manufacturing is emphasised here at the process selection stage. To select an appropriate process for a product, due consideration must be given to the material used. Conversely, the design engineer when considering the material for his product must also consider the processes that will be used to form the product. We will therefore, at this interface between product design and product manufacture, briefly consider material selection as well as process selection.

Figure 17.1 shows the factors to be considered when carrying out the material selection and process selection activities; it also indicates their close interrelationship. When selecting a material for a product the designer will try to juggle the material attributes shown in Figure 17.1 to achieve an optimum solution. For example, the material will require specific mechanical properties, it will need to possess certain minimum shear, compressive and tensile strengths, or it may have to exhibit high elasticity, toughness or wear resistance. Physical properties may also be important, for example, thermal and electrical conductivity, optical characteristics or magnetic properties. The designer's choice of material will also be influenced by the expected lifetime of the product, the implications of product failure and the type of environments in which the product will be used. The 'manufacturability' of the material is obviously of extreme importance and will possibly cause the designer to modify his specifications in other areas; for example, if a material possesses particularly high compressive strength and is hard, then it may be too difficult or expensive to produce the product by a manufacturing processes such as forging or extrusion. Thus terms may be coined such as 'weldability', 'castability', 'formability' and 'machinability' to describe how easily the material can be used in a particular manufacturing process.

Considerations common to both the material and process selection decisions are the number of components to be made, their size, their weight, the precision required, their surface finish and their appearance. The possibility of making the product out of one part as opposed to a number of parts requiring joining is also a combined material and process selection problem; for example, selecting a thermoplastic to make a high volume product may allow it to be made in one piece using the injection moulding process. It follows that the type of material selected has a strong influence on the manufacturing process to be used. Table 17.1 shows a sample of a number of processes with their capabilities. The manufacturing engineer's decision on which process to use will be based on the previously mentioned factors, plus some further economic considerations. As well as the production volume, the rate at which the components are to be produced will also influence the process; the costs of labour and equipment must also be considered.

Essential Manufacturing, First Edition. Gordon Mair.
© 2019 John Wiley & Sons Ltd. Published 2019 by John Wiley & Sons Ltd.

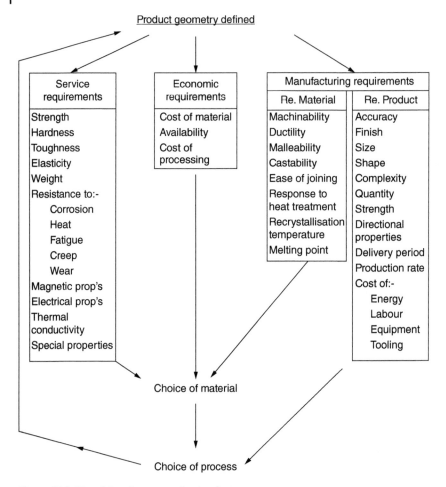

Figure 17.1 Material and process selection factors.

Usually, processes suited to high production volumes are also suitable for high rates of production; they also often have high equipment costs and relatively low labour costs. Conversely, if lower volumes are required the equipment does not need to be so specialised but workers of a higher, and hence more expensive, skill level are required. Table 17.1 provides an indication of typical tolerances, surface finishes, component sizes, materials and batch sizes for some processes. For high volume production, tooling is generally relatively expensive, for example, injection moulding, die casting and closed impression die forging tools.

Note that Table 17.1 is indicative only. As the size of the part increases so often does the tolerance value, for example, in additive manufacturing it may increase by around 0.05 mm per increase of 25 mm in size. Types of a particular process can produce different results, for example, diamond turning can produce optical quality surface finishes of

Table 17.1 Examples of the attributes of some processes and associated materials (Chapter 27 provides information on tolerance and surface texture Ra and µmRa values).

Process examples	Typical tolerance (mm) (size dependent)	Surface texture (µmRa)	Competitive component size (kg or mm)	Typical material	Economic batch size
Casting					
Sand	±1.5	5.0–25.0	Any	Any	Any
Die	±0.05	1.0–2.0	>0.01 kg	Zn, Al	>1000
Investment	±0.05	0.3–3.0	>0.01 kg	Any	>100
Hot Forging					
Open die	±5.0	1.0–25.0	Any	Steel	1–100
Impression	±0.5	1.0–25.0	0.01–10 kg	Steel	>100
Extrusion					
Hot	±0.5	1.0–25.0	1–500 kg	Most	>100 m
Cold	±0.1	0.4–4.0	0.1–50 kg	Most	>100 m
Rolling					
Hot	±0.4	4.0–25.0	>10 kg	Most	>5000 m
Cold	±0.1	0.5–4.0	>10 kg	Most	>5000 m
Presswork					
Blanking	±0.02	0.1–6.0	<10 mm thick	Steel, Cu or	>1000
Bending	±0.2	0.2–0.8	<100 mm thick	Al and other	>1000
Drawing	±0.1	0.2–0.8	<10 mm thick	alloys	>1000
Machining					
Drilling	±0.05	0.8–6.0	<100 mm diameter	Most	All machining
Turning	±0.03	0.3–12.0	<3000 mm dia.	un-hardened	suitable from
Milling	±0.1	0.3–12.0	<1000 mm^2	materials	single units to
Grinding	±0.008	0.2–3.0	<1000 mm^2	excluding	any number of
Lapping	±0.005	0.03–0.2	<100 mm^2	ceramics	parts
Powder Metallurgy	±0.3	0.4–1.5	<1 kg	Any	>1000
E.D.M.	±0.02	0.2–6.0	<300 mm diameter	Any conductive	1–100
Injection Moulding	±0.1	Any, (mould dependent)	<10 kg	Most polymers	>1000
Additive Manufacture					
Stereolithography	±0.05	Related to process resolution, see Chapter 13 for more detail on this aspect	<0.75 m^3	Polymer like material	Single items, e.g. prototypes, bespoke designer goods, medical models, artworks etc.
LOM					
FDM					
SLS	±0.25		<0.5 m^3	Nylon	
DMLS	±0.1		<0.25 m^3	Steel, Al, Ti	

0.001 µmRa (Ra values are discussed in Chapter 27). Also, although some metals such as aluminium can be cold extruded, stronger materials such as steel are more difficult. In practice, by using preliminary information such as that mentioned before the design or manufacturing, the engineer should contact process equipment manufacturers and provide the precise specifications of the component in order to select the most appropriate manufacturing method.

Part IV

Manufacturing Automation

18

Manufacturing Automation – Introduction

The topic of manufacturing automation covers a broad spectrum and this short chapter focuses on the 'what' and 'why' of automation. In the following three chapters, the individual building blocks of automation are initially considered, then industrial robots as an exemplar of the topic and in Chapter 21 the operation of machine vision systems is examined as these are commonly found within many areas of industry.

The word 'automatic', was derived from the Greek *automaton* meaning 'acting independently'. Although it had been used earlier, the term 'automation' came into common usage in the car industry in the early to mid-twentieth century and it implies the concept of mechanisation with the added feature of automatic control. The term 'mechanisation' was a prominent aspect of the Industrial Revolution that began in the eighteenth century and it refers to the application of machinery to do work previously done by humans and horses and so on. Here, we will use the term 'manufacturing automation' to encompass the equipment and techniques used not only to eliminate the need for human work, as with industrial robots, but also to assist the manufacturing process more generally.

18.1 Types of Automation

Dedicated automation is used for making high volumes of the same product at high production rates for long periods of time. The type of equipment used is sometimes called 'hard' automation as it is purpose built to make a specific product and the physical structure and its control are not easily changed to produce different products.

Reprogrammable, or 'soft' automation refers to automatic processes or equipment that can be easily reprogrammed to cope with changes in products or tasks. This type is popular due to the short life cycles of many modern products and the need to satisfy a wide range of different customer demands. Reprogrammable automation fully exploits microprocessor-based control techniques and manufacturing equipment such as industrial robots and numerical controlled machine tools come into this category.

Considering manufacturing automation in its broadest sense the concept can include the whole manufacturing system. The application of computer controlled machines in factories is generally termed Computer Aided Manufacture (CAM) and similarly the use of computers to aid the design process is called Computer Aided Design (CAD). The integration of CAD, CAM and other systems such as a computer-based

Essential Manufacturing, First Edition. Gordon Mair.
© 2019 John Wiley & Sons Ltd. Published 2019 by John Wiley & Sons Ltd.

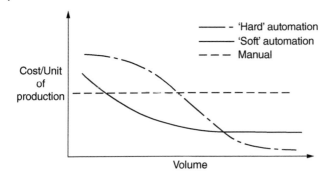

Figure 18.1 The interrelationship between volume and cost per unit for manual methods and hard and soft automation.

Management Information Systems (MIS) is called Computer Integrated Manufacturing (CIM). This in turn can be a contributing part of an overarching Enterprise Resource Management (ERP) system that provides real-time information on many aspects of a company by integrating information from sub-systems such as those from finance, sales, orders, inventory and manufacturing. The term 'autonomous manufacturing' is sometimes used to convey the concept of including intelligent automation in a manner that permeates the entire supply chain with the manufacturing facility as one part of a fully integrated system.

Manufacturing factories are almost never completely automated and indeed labour rates in different countries often determine the ratio of manual workers to automated equipment in any specific location. A graph indicating the interrelationship between volume of product produced and the cost per unit is shown in Figure 18.1 for manual work and hard and soft automation. When using manual labour, as production volume increases, the cost per unit remains the same since proportionately more people have to be employed. With dedicated hard automation a relatively high capital investment has to be made so that the initial cost per unit is high but then decreases significantly as the number of units produced increases. Reprogrammable soft automation comes between manual labour and hard automation. The initial cost is not normally as high as that of hard automation and the cost per unit initially drops quicker than hard automation but eventually as volumes increase then hard automation can be seen to produce units at a much lower cost.

18.2 The Advantages of Automation

Why use automation? It is generally accepted that unemployment, or at least the absence of the opportunity to earn money by working, is a bad thing. Why, then, use a system that eliminates the need for human work? There are many reasons and a number of them are noted next.

Automation, when applied properly, will increase the *profitability* of a company, and therefore in many cases make the jobs of its workforce more secure and better paid. Also the suppliers of, and those who use, the automated equipment will require engineers, technicians and craftsmen to build, operate and maintain the equipment. Often retraining will be provided to improve knowledge and skills. Thus increasing automation does not necessarily mean increasing unemployment. In fact, the opposite can be the case; the countries using the most automation and modern manufacturing techniques often have the lowest unemployment rates.

The use of automation improves *labour productivity*, that is, the cost of labour is reduced while the value of goods produced either remains constant or increases. However, if improvements in total productivity are to be gained then the investment in the automation must be made wisely and with realistic expectations of the equipment's capabilities. The car industry has the most notable examples of productivity improvements through automation and comparisons of the number of cars produced per worker are regularly made between manufacturers in different countries.

Quality is improved by removing the human element. People become tired while working and in repetitive jobs get bored, therefore prone to mistakes. Products produced by manual work also vary in quality due to differences in personal skills. Once automatic equipment is set up to produce good parts it will continue to produce good parts. Any tendency to deviate from the standards set, for example, if a random fault develops or there is progressive deterioration in the process such as caused by tool wear, can be monitored using statistical quality control (SQC) techniques (these are explained in Chapter 27). The process can then be modified as required. With automated equipment, in-process automatic inspection can allow 100% inspection of the product as it is being made, thus aiding the achievement of 'zero defects'. This consistency in operation also results in less wasted material or scrap.

Production rates are increased over those possible by manual methods. Where the work is repetitive and can be broken down into a sequence of simple movements automatic production rates can be many times those of human workers, for example, consider the speed of operation of an automatic bottling plant, newspaper printing press or a special-purpose machine for the population of printed circuit boards. Automated systems can operate 24 hours per day and 7 days per week, with interruptions only for scheduled maintenance. This can be contrasted with the use of humans who require rest, refreshments and lunch breaks, and if 24-hour operation is required then three 8-hour working shifts would be necessary, thus tripling the wage bill.

Labour costs usually increase in line with the cost-of-living whereas an investment made in an automated system is a single expenditure with only running costs subsequently required.

Working conditions are improved where automation displaces human workers from tasks that are dangerous, hazardous to health, unpleasant or tedious. For example, the handling of radioactive materials, spray painting, situations where heat and humidity conditions are high and monotonous assembly work. In each of these areas it is expensive to employ humans as health and safety legislation has to be satisfied through protective clothing, special ventilation, extra rest breaks and so on. Also, if the work is unpleasant or boring, high labour turnover will mean constant hiring and training of new workers to replace those that leave. Thus, as well as being ethical and humane, automation often saves money in these applications.

The predictability and consistency of automation means that the flow of work through the factory can be more easily monitored and controlled. Consequently, this means faster throughput times and reductions of work in progress; this makes the factory more competitive by reducing costs and the time between order receipt and delivery of goods.

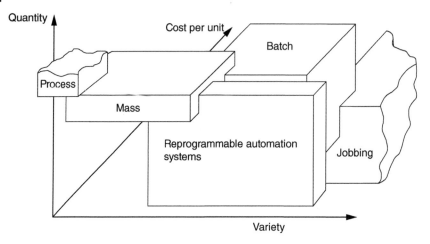

Figure 18.2 Area of application and effect of reprogrammable automation systems on cost per unit of production.

Specifically in relation to soft or reprogrammable automation, we have the ability to achieve the *cost benefits available to mass production* at batch production quantities. One of the reasons it can do this is the flexible nature of the equipment. This means that, unlike hard automation, reprogrammable machines such as industrial robots are able to respond rapidly to product design changes and can be relatively quickly reprogrammed to suit new products as market demand alters. In many situations it takes longer than hard automation to become obsolete thus allowing the cost of investment to be recouped over a longer period. Reprogrammable automation also enables a company to tackle the manufacture of relatively low quantity specialised designs to suit particular market niches, with an attempt to achieve the cost per unit benefits of mass production. This is shown in Figure 18.2 where the various types of production are shown against their relative production volumes, costs per unit of production and variety of products produced.

Finally, we have the ability to *integrate* better the equipment with the overall manufacturing system. This allows much better control within the factory thus reducing lead times and both raw materials and finished goods in stock. This results in increased productivity, quicker response times to customer requests and punctual delivery of orders.

These, then, are some of the advantages of automation and the reasons for its increasing use. It does have some disadvantages, however. For example, its implementation usually requires significant capital expenditures by the company; these will be justified using investment appraisal techniques. Also, if 'hard' automation is used, the flexibility of the company to cope with rapid changes in product demand or design may be limited.

18.3 Typical Examples of Manufacturing Automation

Considering again Figure 18.1, we can see that there are cost advantages for using labour, hard automation and soft automation at different production quantity levels. Additionally, the necessary assembly dexterity, local labour rates and the required production rate are all also important factors in deciding what type of automation to use, or if automation is even required.

Take, for instance, a motor car manufacturing plant. Here, we can see examples of manual labour, hard automation and soft automation all integrated into one mass production system. Manual labour can be found doing jobs that require significant dexterity such as fitting pre-assembled dashboards, wiring looms or wheel nuts. The actual volume here is not the determining factor in what process to use, rather, it is the skill and flexibility of the human operator that is important. Hard automation can be found in the conveyor systems that transport the cars throughout the production process, this is appropriate as the conveyor will require only minor alterations as new products come on-stream. Soft automation can be found in the reprogrammable industrial robots spot welding the cars as they move along the conveyor, this is because in the car industry a number of different models of a car may be going through the production system at the same time. For example, an industrial robot will have to change its pattern of spot welds when an estate version of the car body is followed by a hatchback version (see Figure 18.3). However Figure 18.4 shows an application where human labour is more appropriate than automation – visual qualitative inspection of the car surface finish.

In other industries, such as in the bottling industry, hard automation is almost exclusively used. This is because the product and the bottle design is unlikely to change regularly and production rates are so high that only dedicated special-purpose equipment could keep up with the flow rate. One type of programmable automation that is prevalent in this industry though, is machine vision. This is necessary to ensure high confidence levels in detecting flaws in the glass bottles, correct labelling and making sure that the bottle caps are properly closed. The consequences of allowing damaged bottles, or particles of glass inside a bottle, to be sent out to customers are so serious that a large number of these vision systems are normally deployed.

In the food and pharmaceutical industries we also find automation deployed. This is not only to improve productivity but also to improve hygiene and product consistency. The equipment and materials used must are controlled by regulations. For example,

Figure 18.3 Industrial robots spot welding car bodies. Source: Image courtesy of Nissan.

Figure 18.4 Human labour being used for qualitative inspection. Source: Image courtesy of Nissan.

stainless steel is used as any corrosion in the machine parts could create pockets where bacteria could grow, lubricants cannot be allowed to escape onto the product so special seals are required on moving parts.

The electronics industry uses automation for the manufacture and population of printed circuit boards. Both special-purpose and reprogrammable automated equipment are used. However, for assembly of high volume products whose style and components have very short life cycles, for example, mobile phones, large numbers of people are still involved in manual assembly. This is often carried out in countries with relatively low labour rates and the workforce can easily rapidly switch to new product designs.

This concludes the introductory chapter on manufacturing automation. The following three chapters now consider some important features in more detail.

Review Questions

1 Discuss the differences between dedicated and reprogrammable automation and provide typical examples of the application of each.

2 Describe the typical situations where human labour, dedicated automation and reprogrammable automation would each most appropriately be used.

3 With the aid of a sketch describe the relationship of cost per unit of production to production volume, for human (manual) labour, dedicated (hard) automation and reprogrammable (soft) automation.

4 Fully discuss five advantages of automation over human labour.

19

The Building Blocks of Automated Systems

In this chapter, we will consider some of the components commonly found in automated manufacturing systems. These examples show how actuation of these systems can be obtained, how control of movements can be effected and how information on the system can be fed back to the controllers.

19.1 Cams

A purely mechanical component cams are probably one of the earliest control devices; these are used to control the movements of cam followers, which are in turn connected to linkages. These linkages are used to move cutting tools, pick and place devices, inspection probes and so on. Three types of cam are shown in Figure 19.1. The cam profile is machined around the perimeter or into the surface of the cam, which is usually made of steel, and it is the profile of the rotating cam that determines the movement of the follower. Cams may still be found in hard automation such as screwcutting machines where the product is produced in high volumes and is unlikely to change over a period of years. However, where product volume is not so high and product changes are more frequent, then it is easier and cheaper to control movements by microprocessor-based programmable systems.

19.2 Geneva Mechanism

Another mechanical device this is used to produce a rotary indexing movement in, for example, a worktable or toolhead of a turret lathe in a hard automation system. Its principle of operation is shown in Figure 19.2. The slotted Geneva plate has equispaced radial slots. An indexing plate has a fixed pin which engages with the slots as it rotates. Since the Geneva plate shown has six slots (other numbers of slots could be used), one revolution of the indexing plate will cause the Geneva plate to rotate one sixth of a revolution, that is, 60°. After each index the locking plunger will hold the mechanism in position until the rotating pin engages again. The number of steps per revolution is therefore determined by the number of slots in the Geneva plate.

Essential Manufacturing, First Edition. Gordon Mair.

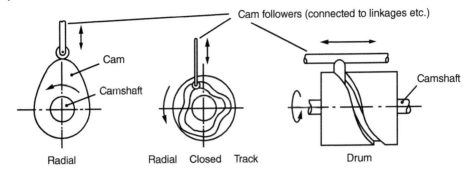

Figure 19.1 Three types of cam used in dedicated 'hard' automation.

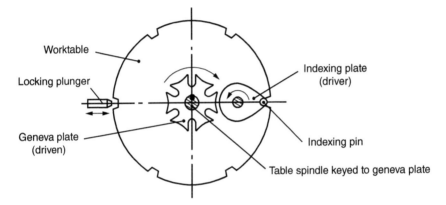

Figure 19.2 A Geneva mechanism.

19.3 Transfer Systems

Transfer of parts between workstations in an automated system is effected by rotary or linear transfer machines and linear transfer lines. The workstations themselves may involve machining, handling, assembly or inspection operations. Linear systems can be used for any number of workstations, but rotary machines are limited to applications where only a small number of operations are involved. The heads of these systems are constructed from basic elements, for example, pneumatic and hydraulic actuators with grippers or inspection probes attached, or cutting tools driven by electric motors. The rotary or linear movement of the parts can be effected by a variety of means such as the Geneva indexing mechanism for rotary machines or the pawl type linear transfer system. A pawl system is used to produce a linear indexing movement and the principle of operation is shown in Figure 19.3. Reciprocation of the transfer bar can be effected by a pneumatic or hydraulic piston or via an electric drive. Since the movement is equal to the spacing of the workheads, the work carriers will be indexed the desired distance along the line with reciprocation. This type of transfer system may be found in various applications, one example being the transfer of large sheet steel car components through presswork operations in automobile manufacture.

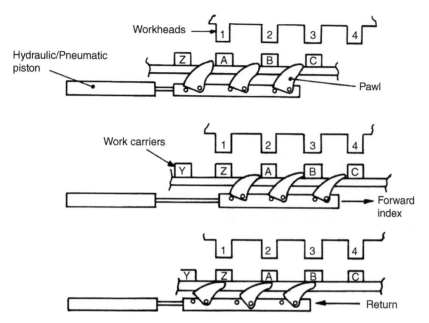

Figure 19.3 A linear transfer system.

These configurations are often found in hard automation systems, where each work-station has been specially designed to carry out one operation before the parts are passed on to the next station in 'line'. They are therefore 'dedicated' machines, since they would need to be completely dismantled and rebuilt for a change in product.

Linear transfer lines, such as those used for car assembly, are not as inflexible as a purpose built linear transfer machine used for, say, machining engine blocks. These lines have workstations located along their length that use a variety of dedicated equipment such as surface treatment plant, reprogrammable equipment such as spot welding robots and human labour used for work that is impractical to automate.

19.4   Conveyors

These come in many forms. For example, roller conveyors are popular for transporting individual units such as boxes or components; the rollers may be powered by electric motors or free running. For bulk material, small components and foodstuffs, powered belt conveyors are often used. A roller conveyor system is shown in Figure 19.4.

19.5   Limit Switches

These are simple electromechanical devices found in almost all automatic systems. They are made in various configurations and three are shown in Figure 19.5. They incorporate

Figure 19.4 Conveyor system.
Source: Reproduced with permission of Pixabay.

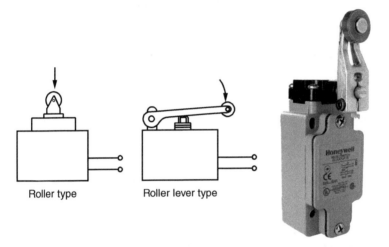

Roller type   Roller lever type

Figure 19.5 Some limit switch configurations. Source: Image courtesy of Honeywell.

a small micro-switch operated by a mechanical plunger. This plunger is caused to move by a variety of means depending on the application. When the switch is operated it sends an electrical signal to the system controller to initiate some action. These switches are used in many applications, for example to limit the travel of a handling device, sense the presence of components on a conveyor line or detect the opening of a safety barrier.

19.6 Fluid Power Components

Pneumatic and hydraulic systems use the fluids of air and hydraulic 'oil', respectively. Pneumatic devices that use compressed air at a pressure of between 1 and 7 kilopascals (kPa) and hydraulic devices that use fluid pressurised to between 70 and 170 kPa are used to drive many automatic systems; for example, hydraulic or pneumatic actuators would be used to operate the pawl mechanism mentioned earlier. Hydraulics are useful where very high loads have to be moved or sudden shock loads are experienced, but due to the problems associated with the high pressures and the possibility of leakages hydraulic machines are now less common in manufacturing systems. Pneumatics, however, are very popular. These devices are cheap, fast, clean, safe and easy to work with. Almost all factories have an air compressor and compressed air system installed (they are also found outside industry for powering 'animatronic' tableaus of characters and dinosaurs etc.).

Pneumatic and hydraulic devices work on the same principle, a pressurised fluid is forced through connecting tubes to an actuator that does the work. The fluid is pressurised by a compressor in the case of air and a pump in the case of hydraulics. The flow of the fluid is controlled in both cases by valves. In automatic systems these valves incorporate electromagnets called solenoids that allow the valves to be opened or closed by electrical signals from a controller. With hydraulic systems much more complex control of speeds and forces can be obtained by using servo-valves that regulate the flow of fluid based on instructions from an electronic control system. The reason hydraulics allow better control is that hydraulic fluid is incompressible while air is not; this means that pneumatic actuators are used for simple movements where the load or actuator can be pushed against a fixed stop. A sketch of some pneumatic components is shown in Figure 19.6 and a simple system is shown in Figure 19.7. The control valve in Figure 19.7 is shown according to a convention that indicates symbolically the two possible positions of the valve that allows air to flow into or out of the pneumatic cylinder.

19.7   Electric Motors for Actuation

For complex control of automated machines, such as numerically controlled machine tools or industrial robots, special electric motors are used. These are called servomotors because they incorporate some sort of 'feedback' device that sends a signal back to the controller indicating the actual response of the motor to a command signal. Two types of servomotors are common, that is, the permanent magnet (PM) direct current (DC) motor and the brushless motor. Electric motors operate on the principle of passing an electric current through conductors in a magnetic field and so producing a torque. In servo-motors the magnetic field is produced by the permanent magnets and the conductors are created by the motor coils; schematic sections of them are shown in Figure 19.8.

The PM DC motor has been the most common due to relatively simple control; however, brushless motors are now popular as the cost of their control has decreased. Brushless motors do not have carbon brushes to wear and so replacement costs and downtime due to maintenance are reduced; they also dissipate heat better and are simpler in construction. These motors are usually controlled by a method called 'pulse width modulation'; this allows control of the current passed to the motor coils thus controlling the motor torque and hence the speed of rotation.

For moving very light loads non-servoed stepper motors are used; these do not need feedback devices. They 'step' round one increment, say 1.8°, for every pulse received from the motor controller. Thus if the controller counts the number of pulses sent to the motor the angle of the motor shaft will be known at any point in time; also by monitoring the rate at which the pulses are sent it can calculate the speed of the motor.

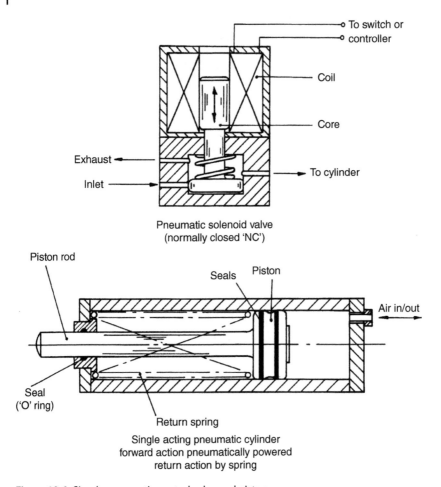

Pneumatic solenoid valve
(normally closed 'NC')

Single acting pneumatic cylinder
forward action pneumatically powered
return action by spring

Figure 19.6 Simple pneumatic control valve and piston.

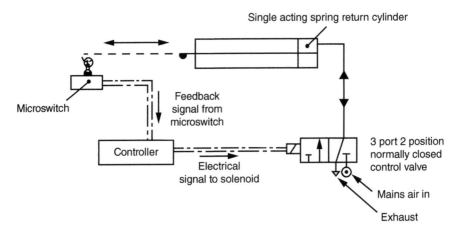

Figure 19.7 Simple pneumatic circuit.

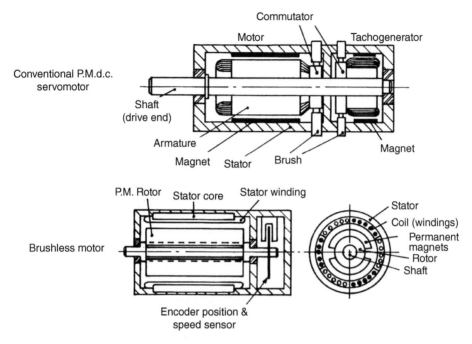

Figure 19.8 Typical electric servomotors used in manufacturing automation.

19.8   Feedback Devices

These devices, used widely in soft automation systems, provide feedback information on displacement and speed to the controller. There is a wide range of types; we will consider just three here: the potentiometer and the tachogenerator, both analogue devices, and the digital optical shaft encoder. Analogue devices produce an infinitely variable signal proportionate to the quantity being measured, whereas digital devices produce a series of discrete pulses whose rate or pattern provides the necessary control information.

A potentiometer is used to provide information on angular displacement. It is essentially a calibrated rotary variable resistor, its principle of operation being shown in Figure 19.9. A direct current supply of volts (V) is applied across a resistance R. The output voltage is measured by tapping over a distance, r. The distance, r, and hence the output voltage, v, will vary depending on the angular displacement of the wiper arm as it rotates about the central shaft thus the voltage, v, indicates the angle of rotation. The shaft is usually attached to the part of the automatic device whose movement is being measured.

A tachogenerator operates as a motor in reverse, that is, it generates a current when its coils are rotated in a magnetic field when made integral with a motor by mounting it on the same shaft. The location of the tachogenerator on a motor was shown in Figure 19.8, it produces a voltage signal proportionate to the speed of rotation, see Figure 19.10. One disadvantage of the tachogenerator is the small dead spot in the middle of the graph where there is a poor response to low speeds.

The optical shaft encoder is a very popular means of obtaining feedback information on angular displacement. Its principle of operation is shown in Figure 19.11. It is composed of a glass disc that may be transparent with opaque sections or opaque

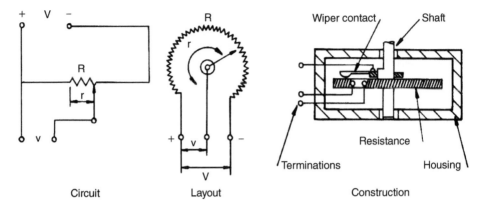

Figure 19.9 The potentiometer, an analogue device for displacement measurement.

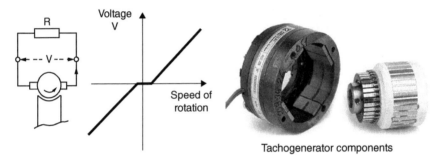

Figure 19.10 The tachogenerator, an analogue device for measuring rotational speed.

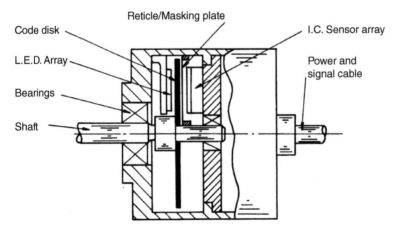

Figure 19.11 The optical shaft encoder, a digital device for displacement measurement.

with transparent apertures. The disc is mounted on a shaft, for example, the shaft of an electric brushless motor or a joint on an industrial robot. An array of LEDs (light emitting diodes) is located on one side of the disc while an array of light sensitive sensors is located on the other. These photosensors produce a voltage when exposed to light. As the sections or apertures pass the light sources so the sensors register the pulses of light by creating corresponding voltage pulses that are sent to the controller. Since the pattern of sections or apertures is known, for example, say 360 apertures per revolution, by counting the number of pulses the angle of the shaft will be known. By noting the number of pulses per second, the speed of the shaft will be known and by noting the change in speed with respect to time, the acceleration or deceleration of the shaft can be determined.

A microprocessor-based controller is usually used with these devices. Shaft encoders are of two types, the incremental encoder and the absolute encoder. The incremental type simply sends a train of pulses to the controller; this means that if power is removed from the system, even for a short time, then the encoder will not know where it is in absolute terms when power is restored. Therefore, when they are used for example on industrial robots, the robot has to be taken to a 'home' or reference point every time power is restored to it. Absolute encoders do not have this problem; they have a coded pattern of sections or apertures that uniquely defines the angular displacement of the disc. Thus if power is lost then restored, the unique pattern of signals from the photocell array will inform the controller of the exact position of the shaft.

19.9   The Vibratory Bowl Feeder

This is the most popular part holding and feeding device for automated systems, being found in almost every factory engaged in automated assembly work. Its construction is shown in Figure 19.12. It basically comprises a cylindrical bowl with a slightly convex base and a helical track running from the base of the bowl up around the internal wall to the rim, where it meets a delivery chute. The bowl is supported on three leaf springs inclined at an angle; these springs are fixed to a heavy base. Between the bowl and the base is a powerful electromagnet. In operation, the bowl is partially filled with the parts that are to be fed. When power is supplied the electromagnet is switched on and off at high frequency. During each instant that the magnet is on it pulls the bowl downwards in a twisting action caused by the inclined springs. Thus as the bowl vibrates at high frequency the parts in the bowl are shaken down the convex base to the bowl wall, where some will rest on the helical track. The high frequency vibratory twisting action causes the parts on the track to be left momentarily suspended in air; they then drop down onto the surface of the track at a point slightly ahead of where they left it. In this way, the parts move up the track to the delivery chute. Once they reach the chute they slide down to the work station under the force of gravity. Careful design of the bowl track can ensure that only parts at the desired orientation are fed to the delivery chute, see Figure 19.13 for an example.

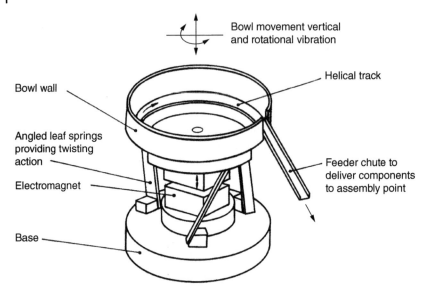

Figure 19.12 A vibratory bowl feeder.

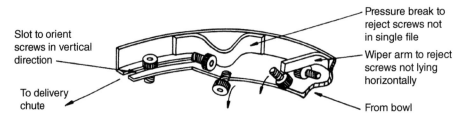

Figure 19.13 An example of bowl feeder track design.

19.10 Programmable Logic Controllers (PLCs)

Programmable logic controllers (PLCs) are microprocessor-based and are used across a wide range of applications both within and without industry. They can have digital and analogue inputs and outputs and can carry out sequencing, timing, logic and calculations. In the manufacturing context they are used to control and monitor equipment and processes and are designed for ease of interfacing and to withstand the rigours of the industrial environment. Both dedicated and reprogrammable automated systems are often connected to PLCs to enable interfacing with sensors and safety equipment. Figure 19.14 shows the basic elements of a PLC and a typical physical configuration although their form can vary widely. The Central Processing Unit or CPU is capable of carrying out all of the decision making, logic and mathematical functions, and also supervision of all the input and output signal. The memory contains the instructions as to how the PLC is to control the external equipment. The low voltage power supply provides power to the controller and for the output signals. A program input device is also required, for example, a keyboard or touchpad.

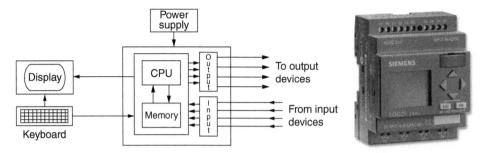

Figure 19.14 A programmable logic controller (PLC) elements are shown on the left and a basic PLC is shown on the right. Source: Image courtesy of Siemens.

19.11 Control of Automated Machines

Finally, a brief discussion on how automatic control is obtained on an automated machine or system. The factors that usually need to be controlled are: the sequence of movements, the speed of movement and, in some cases, the path followed by the mechanism. The control of the sequence of movements can be obtained by sending trigger signals from a microprocessor system, such as a computer or a PLC, to the control valves or motors of the machine. However, speed and path control is more complex and usually involves control of electric current to a motor for electrically powered systems or to a valve that subsequently regulates the flow of fluid in systems that are fluid powered. Terms that are used in describing control systems include: *open loop*, *closed loop*, *feedback* and *servo-control*.

Open-loop control is the type of control found, for example, on cam operated machines or machines under computer control that are subjected to small loads. In this type of control, the machine is expected to operate exactly as commanded by the system controller and no automatic verification is made. This can be used only where light loads are involved and there is a high degree of confidence that the desired action will take place.

Closed-loop control is the type used in almost all computer or microprocessor-controlled machines. It is found in numerical controlled machine tools, industrial robots and automated guided vehicles; outside the factory it is found in such things as advanced SLR cameras, car fuel management systems and aircraft autopilots. All of these artefacts have a common attribute – they are 'mechatronic' devices. A mechatronic device is one that is composed of mechanical, electrical and electronic elements under the control of a microprocessor.

Implicit in the term 'closed-loop control' is the concept of *feedback* this signifies that a signal is sent back from the device being controlled to the controller. This signal contains information on how the device is responding to the controller's commands; the controller can then use this information to modify its commands to ensure that the device performs as required. For example, assume that the controller sends a signal to an amplifier to send a specific current to an electric motor that should cause the motor to rotate at a desired speed. The feedback device on the motor, for example, a shaft encoder, sends a signal back to the controller that the motor is moving too slowly.

The controller will then modify its output signal to cause the current to the motor to be increased.

This type of control is often called *servo-control*. Essentially, servo-control implies a system that involves amplification of a control signal, for example, the braking system in a car where the manual force applied to the brake pedal is amplified by the brake servo-system to produce a larger force at the brake pads. In the types of machines we are concerned with, this type of control is always present since the low power command signal is usually a low voltage signal from the controller. However, in industrial applications servo-control is usually synonymous with feedback and closed-loop control.

Review Questions

1 Describe with the aid of a sketch the operation of purely mechanical automation device, the purpose of the device should also be stated.

2 What is a 'limit switch' and where might it be found?

3 Provide a sketch of a simple pneumatic circuit.

4 What is a servomotor and what advantages does the brushless type have over the more traditional brushed motor?

5 Describe an analogue device for measuring displacement and an analogue device for measuring speed.

6 Describe a digital device for measuring rotational displacement and state how this can be used to also provide speed and acceleration information.

7 State the purpose of a vibratory bowl feeder and provide a brief description of its operation.

8 Briefly discuss the elements of a PLC and give an example of its application.

9 Explain the terms open-loop and closed-loop control.

20

Reprogrammable Automation

This chapter looks in further detail at the reprogrammable automation mentioned in Chapter 18. There are many types of reprogrammable machines and systems used in industry and these include industrial robots, computer numerical controlled (CNC) machine tools and automated guided vehicles (AGVs). These will be described here beginning with the exemplar of industrial robots as they are widely used in manufacturing and they embody many of the aspects of reprogrammable factory automation.

20.1 Industrial Robots

Industrial robots are often found where a variety of products may be made with relatively short life cycles and regular product design changes, such as in the car and electronics industries. They can also be found in such diverse industries as aerospace, pharmaceutical and food production. The robots themselves come in a wide range of configurations and sizes. The International Standards office defines an industrial robot in ISO 8373 as: 'An automatically controlled, reprogrammable, multipurpose manipulator programmable in three or more axes, which may be either, fixed in place or mobile for use in industrial automation applications'. Thus, the robot, as well as being able to operate on its own according to previously given instructions, must also be able to have those instructions changed easily. In summary therefore, an industrial robot can be considered as a *reprogrammable* device used for the *manipulation* of components, materials or tools. The methods used to program them and the ways in which manipulation is achieved vary considerably between robots, but the concept of *reprogrammable manipulation* remains constant. The manipulation aspect therefore differentiates them from CNC machine tools and AGVs that, although reprogrammable, do not manipulate parts, tools or materials.

In addition to the general hard and soft automation benefits outlined in Chapter 18, industrial robots have the following advantages over human labour. In comparison to humans they can be much stronger and are therefore able to lift heavier weights and exert larger forces. Many robots can be very precise in their movements thus making them suitable for delicate tasks. As well as advantages over humans, robots also offer advantages over the specialised, dedicated equipment designed for mass production systems. Dedicated equipment normally becomes obsolete when the product it has been designed to produce reaches the end of its original application. This means that the cost of the design and manufacture of the equipment has to be recouped in a short time thus

Essential Manufacturing, First Edition. Gordon Mair.

reducing profit margins. Industrial robots, however, can be reprogrammed for a new task after their original application has ceased. They can also be more easily adapted than dedicated equipment to accommodate small design changes as the product evolves.

20.1.1   Configurations of Industrial Robots

Robot arms are constructed of links that move in relation to each other on rotational and/or linear joints. The arrangement and combination of these joints define the geometric configuration of the robot. A large variety of configurations is therefore theoretically possible although in practice a relatively small number of permutations are used. The volume of space created within the virtual surfaces swept by the robot arm at maximum and minimum reach is termed the robot work envelope and this, in conjunction with the configuration, classifies the physical appearance of the robot. The most common configurations are shown in Figure 20.1. As is evident in the illustrations the maximum theoretical work envelope is restricted in each case due to the physical construction and constraints of the robot. Each robot usually has a wrist at the end of the arm with additional axes to allow orientation as well as positioning of a tool about a point. Other legacy configurations may still be found but newer industrial robots are mostly confined to the type shown in the figure. Additionally, two arm articulated robots for working alongside humans and multi-jointed snake like robots for the nuclear industry are also available but are not common.

Articulated robots. These robots normally have a rotational vertical 'waist' with upper and lower arms connected by rotational horizontal 'shoulder' and 'elbow' joints. The rotation of the base coupled with the movements of the other two major axes produces three degrees of freedom, thus providing a hollow spherical work envelope. With an additional three degrees of freedom, and a wrist capable of roll, pitch and yaw rotations, they are adaptable to a very large variety of tasks. The base of the robot takes up very little area in comparison to the work volume that can be encompassed by the arm. Straight line motion in any direction requires coordination of all the joint axes movements. Their versatility allows them to be used for such diverse tasks as paint spraying, seam welding and spot welding, adhesive application, assembly and heavy materials handling.

Delta robots. This configuration is comprised of three or four parallelogram arms suspended from an overhead body containing a drive motor for each arm. The arms have universal joints at the top and bottom. The three arm version can have a central shaft that can provide rotation about the vertical axis. The parallel configuration provides very good stiffness and allows a work envelope that is similar to a squat cylinder with a truncated cone attached. Because the arms do not have to support a significant load they can be made of lightweight composite material that results in extremely high speed capability. Over 200 picking operations per minute are possible. This makes the robot ideal for fast, lightweight, pick and place applications such as in the food, electronics, pharmaceuticals and packaging industries.

SCARA robots. The SCARA design (Selective Compliance Assembly Robot Arm) allows a certain amount of compliance or movement in the horizontal plane when attempting to assemble objects in a vertical direction. This is particularly important when inserting one component into another as the necessary precision of the arm

Configuration	Work envelope	Example
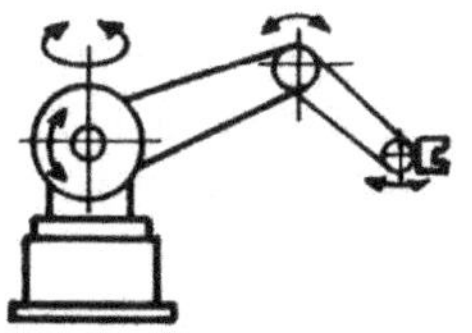 Articulated robot	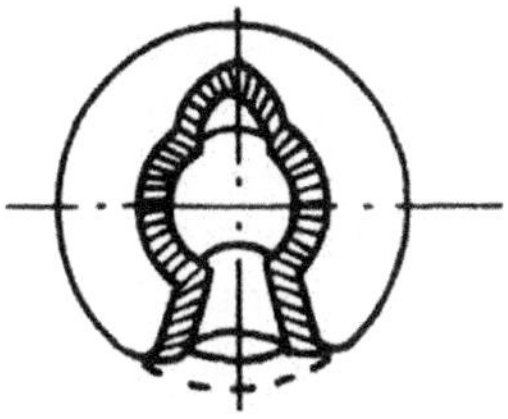 Partial spherical	 FANUC Articulated robot
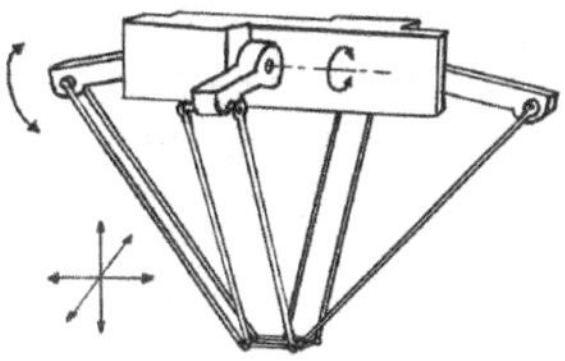 Delta robot	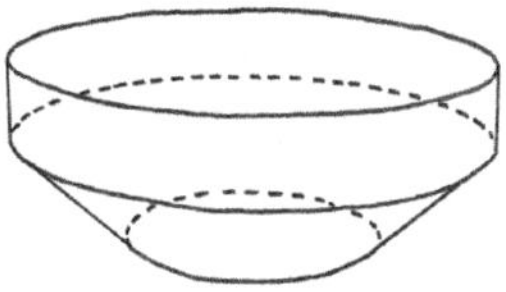 Partial cylindrical	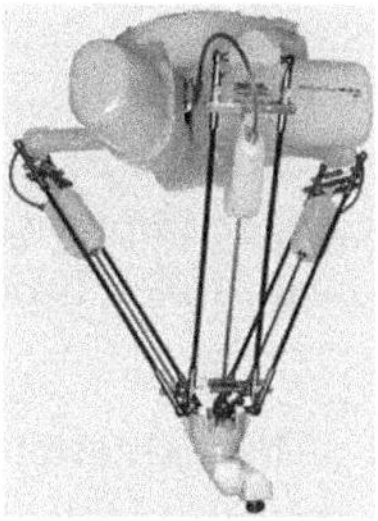 FANUC Delta robot
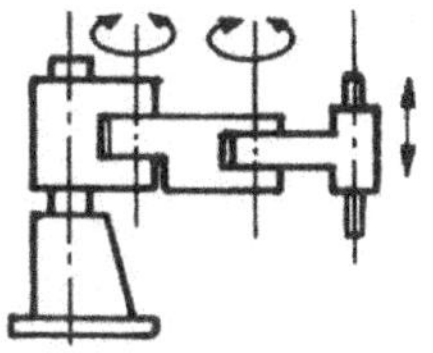 SCARA robot	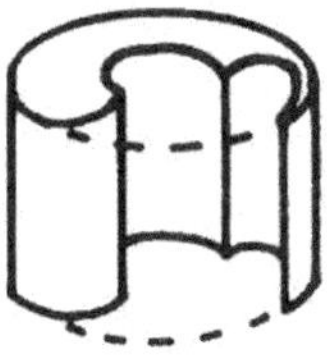 Partial cylindrical	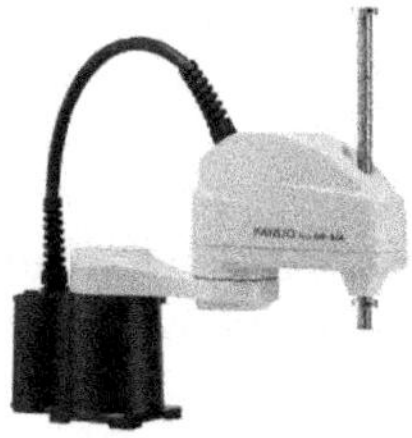 FANUC SCARA Robot
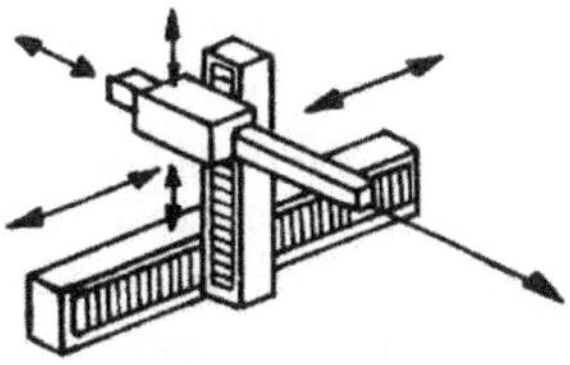 Cartesian robot	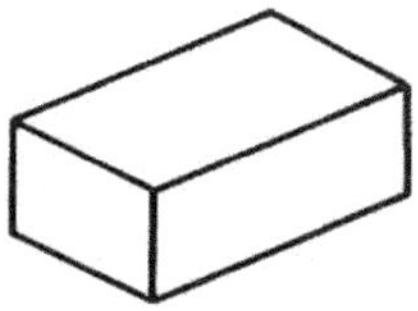 Rectangular	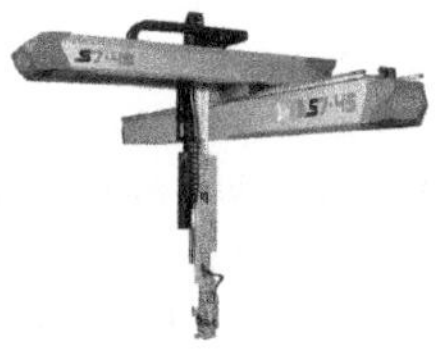 Sepro cartesian robot

Figure 20.1 Common industrial robot configurations. Articulated, Delta, and SCARA configuration images courtesy of FANUC Europe Corporation. Source: Cartesian configuration image courtesy of Sepro Robotique Ltd.

and/or component location is reduced providing there is an adequate lead-in or chamfer on one of the components. These assembly robots, being normally only used for lightweight tasks, are usually of simple construction and do not always require full servo control on every axis. Applications are mostly found in the electronics industry for printed circuit board (PCB) population. Very occasionally larger versions of the configuration can be found in heavy duty applications in areas such as forges, foundries and other places where heavy workloads are met. On all of these robots a partial cylindrical work envelope is produced.

Cartesian robots. This configuration provides a rectangular work envelope. The arm's three major axes are rectilinear and provide movements that are simple to control. Two varieties of this configuration are common: the gantry type robot in which the arm is suspended from an overhead support, and the slideway mounted arm that may have its base and one linear axis mounted on the floor or along the side of a work area. Due to the ease with which a linear axis can be extended these robots can cover a large workspace. They can be used for pick and place operations in the electronics industry and larger types can be used for palletising and machine tool loading. The simplest forms can be used for injection moulding and die-casting machine unloading. Some very large complex examples are used in a gantry form in association with integrated servo controlled turntable fixtures for seam welding.

20.1.2 The Elements of an Industrial Robot

The physical structure and composition of one type of industrial robot is now briefly considered in the next section. The robot shown in Figure 20.2 is a Panasonic articulated arc-welding industrial robot and is one example of the various robot designs available commercially. Although specifically designed for arc welding this type of configuration can be applied to a wide range of tasks and it illustrates most of the basic elements.

20.1.2.1 The Arm

This is the active manipulator that carries out the physical work, it is shown on the left-hand side of Figure 20.2. In this configuration the arm comprises a fixed base, above which is a joint, or waist, connecting to the body of the robot, this allows the robot to rotate about a vertical axis. At the top of the body is another joint connecting the lower arm. This horizontal shoulder axis provides a second degree of freedom to the arm allowing it to sweep out a hemispherical shell. At the outer end of the lower arm is another horizontal joint, or elbow, attached to the upper arm. This provides a further degree of freedom allowing the arm to move *within* the hemispherical work envelope. This robot has another three axes that constitute the 'wrist' movements of the arm. One joint has its axis lying along the centre line of the forearm, the joint itself lying approximately midway along the arm. A further two axes lie at the extremity of the arm at right angles to each other. This robot, by utilising all six axes, has the freedom to position a gripper or tool, frequently called the end effector, at any point in space and in any orientation; in this case the end effector is a welding torch.

To power the robot a drive system is required. Typically, an electric robot like those discussed previously will use DC servo or AC brushless servo motors as the prime movers. The two other drive systems are hydraulic, which is useful for heavy duty work due to the incompressibility of the fluid, and pneumatic, which is low cost, fast, but

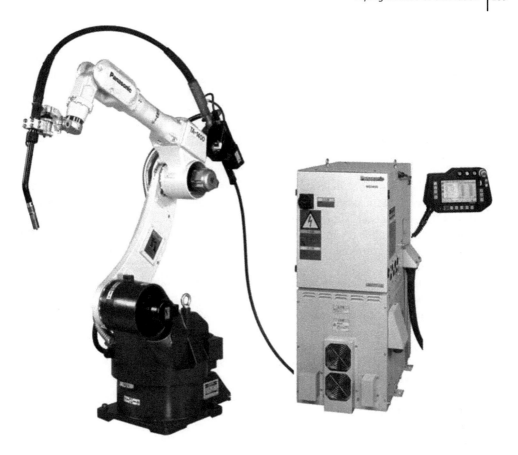

Figure 20.2 The Panasonic TA1400 articulated robot designed specifically for arc welding. *Source:* Courtesy of Panasonic.

usually limited to simple cartesian movements and light loads due to the compressibility of air. In the case of the more commonly used electric robots the torque from the motors is transmitted to the required point through a system of gears, rods, screws, pulleys or belts. A few electric robot designs have their motors mounted coaxially with the joints themselves thus removing the necessity for linkages, although this adds weight to the arm.

Unless the robot is a very simple type, further components will be necessary to provide feedback information on the location, speed and direction of movement of the end effector at any point in time. These will take the form of encoders, tachogenerators and potentiometers as described in Chapter 19. Had the robot been a fluid powered machine there would also have been a need for control devices such as servo valves to be included.

All of these components for drive, transmission and feedback add mass to the arm and thus make it important that their weight is kept to a minimum. The arm framework that supports them must also be as light as possible and yet contain enough rigidity for accurate control.

20.1.2.2 The Controls

On the right-hand side of Figure 20.2 there are two main elements shown; a small detachable teach pendant with a display and a large control cabinet. These items are related to the control function. The cabinet contains the computer and the means for interfacing the information therein to the robot arm and the human operator. The teach pendant can be hand-held by the operator and used to instruct the robot in a step by step manner how to carry out a particular task. Additionally, a PC may be used for two-way communication between the operator and the robot or for preparing complex programs during off-line programming.

Within the control computer, a large number of tasks need to be carried out and these are shown in block diagram form in Figure 20.3. The top and bottom of the figure shows the communication links between the controller and the outside world. The top shows the instructional links from the programmers and also the links that may be necessary if the robot is working within a work-cell with and collaborating with other reprogrammable equipment, for example, CNC machine tools, conveyor systems or other robots.

The controller also effects the arm movement. For example, it may be controlled point-to-point as in assembly or PCB population, point-to-point with coordinated path as when spot welding car bodies on a moving conveyor line or continuous path as in spray painting.

The bottom of the figure shows the links to the robot actuators and from feedback components. It also shows the input from other sensors such as vision systems or signals from limit switches connected to safety equipment. The central block shows all of the functions that need to be carried out by the system in order to allow the robot to move in the required manner.

20.1.2.3 Programming

Industrial robot programming can be carried out on-line or off-line. In on-line programming the programmer is in relatively close physical proximity to the robot and moves it through the desired paths by using a hand-held teach pendant, or by leading it through the required movements by grasping the end of the robot wrist or using a lightweight master arm of the same geometry as the full robot. In off-line programming the programmer sits remote from the physical robot and writes the program with the aid of graphical simulation and a virtual model of the robot.

Some of the tasks that need to be carried out in the control unit relate to programming and the geometric structure of the robot, these are called the forward and inverse kinematic problems. Thus, given the joint coordinates of the robot, to translate these into the world coordinates of the robot end effector is the Forward Kinematic Problem and it occurs during on-line programming. And, given the required world coordinates for the end effector, to translate these into the required joint coordinates is the Inverse Kinematic Problem and occurs during off-line programming. There are advantages and disadvantages to both methods and these are summarised next as bullet points.

On-Line Advantages:

- Direct programming
- Can be observed
- Position data can be read direct from the arm

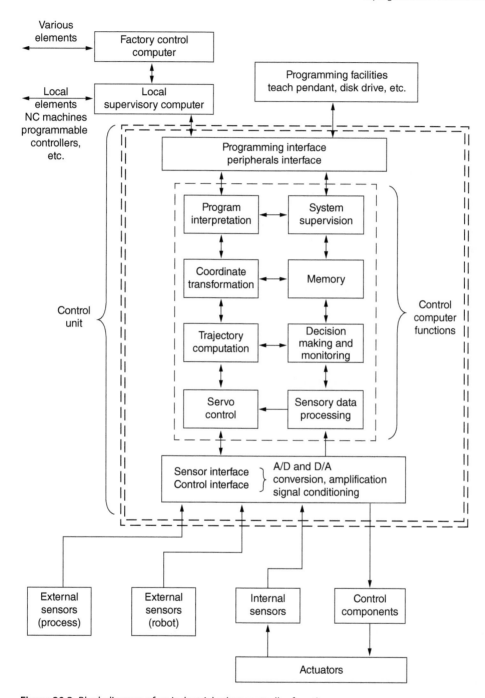

Figure 20.3 Block diagram of an industrial robot controller functions.

- Easy to learn
- Quick implementation
- Operator's skill and experience transferred efficiently to the robot

On-Line Disadvantages:

- Tedious and time consuming for long and complex jobs
- Restricts the full potential of the robot

Off-Line Programming Advantages:

- Robot can continue working productively while programming is taking place
- Potential for full CAD CAM integration
- Simulation can be used to anticipate collisions
- Programs can be written to cope with unusual or unexpected situations and to handle errors in the work situation
- Variations in the work can be more easily accommodated
- The advantages of adaptive and intelligent control can be more fully exploited making the fullest use of sensory abilities

Off-Line Programming Disadvantages:

- Real world contact is lost
- Arm transducers cannot be used to provide position data
- Greater programming skill required

Benefits of Off-Line Programming with Simulation

- Frees the physical robot to continue working so increasing production time
- Can reduce programming time
- Potential collisions can be detected
- Facilitates integration into full CAD CAM system
- Programmer obtains clear picture of planned operation
- Can be used as an aid in robot design by reducing the need for physical prototypes and allowing simulated trials to be carried out
- Assists in selecting an appropriate robot before making a purchasing decision. Work envelopes, estimated cycle times, joint movement limitations and so on can all be examined
- Robot work-cells can be simulated and modified to achieve the most efficient layout
- Simulation is an efficient method for education and training in robotics

20.2   Reprogrammable Equipment Precision

All reprogrammable equipment, such as industrial robots, CNC machine tools and AGVs, have a certain 'precision', that is, how well the equipment can be relied upon to go to a taught position and return to that position when required. Precision is comprised of three elements. *Resolution* is the smallest increment of movement that can be controlled or registered by the system hardware such as shaft encoders or stepper motors and software, for example, number of bits per word to define a location. *Repeatability* is the statistical capability of the system to return to a previously taught point. If the system has high repeatability then it will form a very tight cluster of points about the target point. *Accuracy* refers to the system's ability to reach a point designated in software, for example, as experienced in off-line programming (Figure 20.4a–d). This is shown graphically for the positioning of an industrial robot end effector in Figure 20.4e.

20.3   Computer Numerical Controlled (CNC) Machine Tools

The principles of cutting and the basic cutting operations and machines have been already covered in Chapter 9. There follow some notes on widely used CNC machine tools for cutting material, usually metal. These automated cutting machines may be CNC lathes, or CNC machining centres that provide multi-axis milling, or CNC multi-tasking machines (MTMs) that incorporate the functions of turning, milling, drilling, tapping and other cutting operations all in the one integrated system, see Figure 20.5a,b. MTMs can provide very high efficiencies by allowing parts to be completely machined in the one location without the need for transporting material around the factory for different operations, reducing work in progress and improving the ability to achieve close tolerance on the workpieces.

An NC machine tool is microprocessor controlled and programmed using its own programming language, or 'part program'. In CNC, the machine has its own dedicated microcomputer. This performs many of the basic NC functions with its own stored program and provides flexibility with a powerful computational ability. With DNC (direct numerical control) the machine tools are controlled directly by a supervisory computer that downloads part programs to individual machine controllers. Thus a central computer can hold programs for many machine tools. The communication between the machines and the computer is two-way and the system has to operate in real time, that is, the response time must be so short as to appear instantaneous for each machine.

As with industrial robots, programming can be carried out on-line or off-line. A machine operator may construct a program at the machine then leave it to run and produce a batch of components. Alternatively, programs can be written off-line then downloaded to the CNC machine as required. Raw material can be loaded and finished components unloaded either manually or by integrating the machine with an industrial robot. Thus an autonomous machining cell can created where the machine tool is serviced by the robot. Additionally, the supply of raw material and the removal of finished items can be effected by an AGV.

20.4   Automated Guided Vehicles (AGVs)

Consider the need to automatically transport parts and materials within the factory. Transport and material handling is required within the raw material and bought-in components stores, between these stores and the work areas, between the carriers that take the work to the work areas and the points of work and then between different work

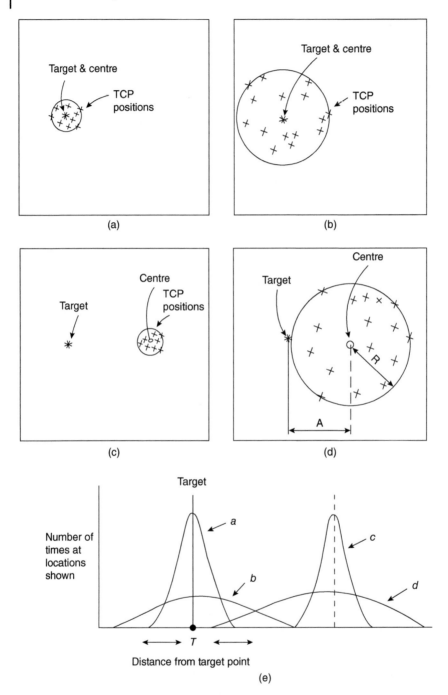

Figure 20.4 Concepts of accuracy and repeatability. (a) High accuracy, high repeatability. (b) High accuracy, low repeatability. (c) Low accuracy, high repeatability. (d) Low accuracy, low repeatability. (e) Graphical representation of the positioning of an industrial robot end effector (TCP refers to the Tool Centre Point).

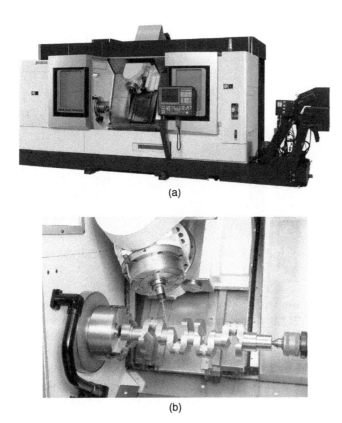

(a)

(b)

Figure 20.5 (a) Multitasking CNC machine. Source: Okuma America. (b) Typical MTM tooling. Source: Images courtesy of Okuma America.

areas and back to the finished goods store or dispatch area. For full automation, all of this handling equipment should have a control system capable of being interfaced with a supervisory computer and be programmed for automatic routing. If the work involves machine tool loading there will probably be a primary handling device, such as a transporter of some kind, to take material from the store to the machine, and a secondary handling device, such as a robot, to transfer the work from the primary carrier into the machine tool. This principle will also apply to assembly tasks. Work holders such as pallets, trays, boxes or racks may be used. There is a large variety of handling devices used to carry these work holders, but the two primary handling devices are conveyors and AGVs. Conveyors have already been discussed in the Chapter 19 and illustrated in Figure 19.4, with a larger system also shown in Figure 18.5. AGVs are used for transporting raw material from stores to machines, components in process between machines and assembly areas and finished assemblies back to stores when complete. Figure 20.6 parts a and b show AGVs transporting partly completed products within a factory. They are essentially mobile pallet carrying platforms and not restricted to the factory as they are also used in automated distribution centres and warehouses.

(a) (b)

Figure 20.6 (a) Very large AGV transporting part of an aircraft fuselage. (b) AGV transporting a completed engine assembly. Source: Images courtesy of Ab Solving Oy.

Some types may use inductive sensors to follow hidden cables embedded in the factory floor. However, for full flexibility, free ranging AGVs provide the most attractive option. These vehicles require no hidden underfloor cables or tracks but instead are able to use their own on-board guidance systems to make their way through the factory. There is a variety of methods whereby AGVs can do this: (i) By internal sensor monitoring where the wheel movement, speed and direction are continuously monitored. This allows the AGV's position to be known at any point in time assuming a 'home' position was first fed into memory as a reference point, this is probably the simplest to implement but is not 'absolute' in the sense that errors can occur due to wheels slipping or skidding. (ii) Inertial guidance utilises an inertial gyroscope that is set up parallel to the required direction of movement. When the AGV deviates from the desired path an acceleration is created at right angles to the direction of motion and this is detected by the gyroscope. This acceleration is used to determine the deviation. The information is then fed into the control system to modify the AGV position accordingly. (iii) Position reference beacons can be set up at specific positions and used by the AGV as it moves around the factory. Various methods may be adopted by the vehicle to pick up its bearing from each beacon as it passes. For example, laser beams may be used to scan for the beacons that could have bar coded information on them. The vehicle will have on-board sensors to pick up the reflected signals and so take appropriate action to ensure the proper route is being followed. (iv) Stereo vision may be used to allow the AGV to 'see' its way along its desired route. The vehicle will hold a map of the factory in its memory and will constantly compare the information from its sensors with that map and associated instructions. (v) LiDAR (Light Detection and Ranging) uses time of flight of laser light to determine the distance to objects along the route of the AGV. Often a combination of some of these methods may be used and much of this guidance technology has parallels with autonomous vehicle guidance systems for public highways. For safety reasons within industry, AGVs have a flexible touch sensitive 'bumper' at front and rear that if touched, say by contact with a human, will remove power, thus stopping the vehicle.

20.5 Reprogrammable Automation and Industrial Robot Safety

Safety in the manufacturing environment is covered more fully in Chapter 28. However, some aspects of reprogrammable automation safety are additionally noted here.

The law must always be adhered to, for example, in the UK there are the Factories Act, the Health and Safety at Work Act and there are various other Acts directly enforced by the Health and Safety Executive (HSE). Over 200 sets of regulations that apply directly to health and safety are constantly emerging or being updated, for example, Control of Asbestos Regulations 2012, Control of Noise at Work Regulations 2005 and The Supply of Machinery (Safety) Regulations 2008. These relate to the UK but every country will have their own equivalents.

With respect to industrial robots there are guidelines and standards that should be used. For example, there are the International Standardisation Organisation (ISO) Industrial Robot Safety Standards ISO 10218-1:2011 and ISO 10218-2:2011. In the UK there is the HSE publication HSG43 'Industrial Robot Safety'. In the USA and Canada there are also standards, that is, ANSI/RIA R15.06 and CAN/CSA Z434, and at the time of writing these are being harmonised with the ISO standards noted before to provide a combined document.

The science fiction writer Isaac Asimov presciently introduced the 'Three Laws of Robotics' in a story titled *Runaround* in 1942. These are relevant today in the real world, they are;

> 1) a robot may not injure a human being or through inaction allow a human being to come to harm; 2) a robot must obey the orders given to it by human beings except where such orders would conflict with the First Law; 3) a robot must protect its own existence as long as such protection does not conflict with the First or Second Law.

These laws are obviously applicable not only to the industrial robots considered here but also to the development of domestic robots.

Three physical regions around an industrial robot can be identified. These are as follows. The perimeter, this is where a 2 m high fence with interlocked gates, guardrail or photo-electric sensors including light curtains can be used, see Figures 20.7 and 20.8. The volume of space between perimeter and robot mean that here pressure sensitive safety mats would be appropriate, see Figure 20.9. The robot arm contact edge could be provided with simple touch sensors such as ribbon switches that would switch the robot off if contact was made with another object or a human.

Industrial robots have some unique safety considerations: the robot arm and end effector move within a work envelope volume very much larger than the robot's own base (except for gantry type robots); as a robot responds to commands from the microprocessor control it can appear to the casual observer to move both spontaneously and unpredictably; the integrity of the control system hardware and software is extremely important; heavy duty robots are often massive and move at high speed producing a hazardous situation and the end effector or material being handled present fast

Figure 20.7 An industrial robot working within a 2 m high safety cage.

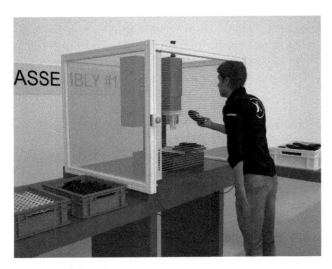

Figure 20.8 Principle of light curtain, a vertical linear array of photo sensors and LEDs. Source: Image courtesy of Contrinex.

moving hazards following paths difficult for the general observer to anticipate. It is also important to note that proper training is necessary for programmers, operators and maintenance personnel, and only approved and trained personnel must use the robot.

Emergency stop buttons should protrude, be easily hit, be easily accessible at all times and should be hardwired; that is, they should not rely on software to stop the system but

Figure 20.9 A pressure sensitive mat. Source: Image courtesy of Rockford Systems LLC.

Figure 20.10 An emergency stop button; note the mushroom shape for rapid operation. Source: Image courtesy of www.tlc-direct.co.uk.

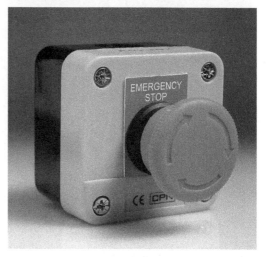

should directly cut off of the power supply, see Figure 20.10. Any start button should be close to a stop button, see Figure 20.11. There should only be one active start button for the system and it should be recessed or shrouded so that only a deliberate act will depress it. Teach pendants should require two-handed operation. Electrics may have to be intrinsically safe if operating in a hazardous atmosphere. Brushless motors may have to be used to avoid sparking. Power cables and enclosures should be protected and flame proof and signal carrying wires must be adequately screened. All safety devices should be designed to fail to safety, that is, 'fail-safe'; for example, electromechanical brakes that hold the robot arm in position if an unanticipated power failure occurs.

Finally, some industrial robots have been designed to work alongside humans without the need for safety cages. For example, the Kawada Industries Nextage robot has a

Figure 20.11 Recessed start button in close proximity to the protruding stop button. Source: Image courtesy of CEF Ltd.

Figure 20.12 The humanoid configuration Nextage robot by Kawada Robotics at the Glory Saitama factory. Source: Image courtesy of Kawada Robotics Corporation.

humanoid configuration. It uses very low power motors and has sensors that allow the robot to stop moving if a human is detected in close proximity, see Figure 20.12.

Review Questions

1 Define the term 'industrial robot'. You may use your own words or you can use the ISO definition.

2 Describe four configurations of industrial robots, include rough sketches of their work envelopes and an example of an application suitable for each configuration.

3 Draw a block diagram that illustrates the tasks required to be undertaken by the control system of an industrial robot.

4 Discuss the advantages and disadvantages of on-line and off-line programming of industrial robots.

5 Explain the three main elements that contribute to the overall precision of reprogrammable equipment and illustrate your answer with a sketch.

6 What do you understand by the terms CNC and MTM in relation to machine tools?

7 Within the context of manufacturing describe the purpose of an AGV and briefly note how 'free ranging' AGVs are be guided.

8 Discuss whether or not you believe Asimov's three laws of robotics are universally applicable today.

9 With respect to safety, what are the three physical regions around an industrial robot and how can they be safeguarded?

10 With particular reference to industrial robots describe five unique safety considerations that are relevant for reprogrammable automation.

21

Machine Vision

Machine vision systems are found in a wide range of applications within industry and they are an important element wherever automation is implemented. This chapter considers the application and operation of machine vision within manufacturing.

21.1 Areas of Application of Artificial Vision

21.1.1 Qualitative and Quantitative Inspection

This large area of application of vision system has many subdivisions. It is an attractive area for investment as, without the use of automation, visual inspection can account for around 10% of the total labour costs of manufactured products. *Qualitative inspection* involves using the vision system for checking a product on an accept or reject basis by examining its attributes. For example, a flaw in a glass bottle, a crack in a casting or a discoloured fruit on a conveyor would be attributes that would cause these items to be rejected. Other aspects of this would be the detection of foreign bodies such as mice or insects on a conveyor in a food processing plant, verification of a completed task in an assembly sequence or checking the degree of swarf build up on a drill bit. *Quantitative inspection* involves using the system to measure the dimensional features of a product or component. Examples of this are measuring the width of steel strip coming from a steel mill, gauging the diameter of a shaft during or after turning or inspecting a completed component dimensionally.

21.1.2 Identification

Rather than check an attribute or measure a feature, the system is used here to identify and classify components or products. The vision process may involve identifying an individual item in a bin of jumbled components, reading characters from a label on a component travelling along a conveyor belt or identifying a component by its shape and other characteristics among other components in a scene. Following on from the identification of a part, the vision system may use this information as an input to a subsequent system, such as a robot, to allow the parts to be sorted and manipulated for machine loading, assembly or routing to a store (see Figure 21.1).

Essential Manufacturing, First Edition. Gordon Mair.
© 2019 John Wiley & Sons Ltd. Published 2019 by John Wiley & Sons Ltd.

Figure 21.1 Vision systems checking bottle fill levels and label integrity. Source: Images courtesy of Omron Industrial Automation.

21.1.3 Providing Information for Decision Making, Guidance and Control

The visual data is used in this case to help the robot or automated system decide what to do next in the current situation. For example, laser vision is used in seam tracking in arc welding, here the visual data give immediate guidance to the robot controller to direct the welding torch thus enabling optimum quality welds. In more sophisticated applications, vision can be used in conjunction with artificial intelligence techniques to determine the assembly order of components, or to enable a mobile robot or automated guided vehicle (AGV) to find its way through an unstructured environment.

21.2 Vision System Components

A complete manufacturing automation machine vision system normally requires six different elements, some of which may be integrated into a single unit:

1) Camera
2) Camera interface and initial image processing equipment
3) Frame store or Field Programmable Gate Array (FPGA)
4) Control processor (note that elements 2, 3 and 4 may be physically integrated)
5) Appropriate scene illumination
6) Interfacing equipment to link the vision system to the external world, for example, to provide information to a quality control system, or to an industrial robot or other manipulating system to take appropriate action depending on the image analysis

The relationship between these elements is shown in Figure 21.2 for an integrated vision and industrial robot system.

The system elements are integrated in the following manner. First, the camera contains the imaging sensor comprised of thousands or millions of light sensitive photocells. The light from the illuminated scene is focused onto the sensor within the camera. This sensor then produces a voltage signal for each photocell that varies proportionately to the light intensity falling on each one. The signals are transferred to the initial processing stage where they are transformed into a format suitable for further processing. As the data are being handled at a very high rate at this stage it is often necessary to use a storage device to hold the data before further analysis. This is carried out by the frame or picture

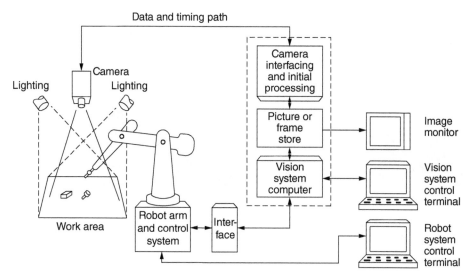

Figure 21.2 Basic elements of a vision–industrial robot system.

store or within a FPGA. At this stage a monitor can be used to view the scene observed by the camera. From here the scene is transferred to the system control processor for analysis. The vision control terminal can be used at this stage for setting the various parameters required. To enable a physical reaction based on the information obtained by this system the vision system needs to be linked to, for example, the control computer for the robot. This is done through an interface that can handle the communication between the vision system controller and the robot controller. Thus the basic process is: (i) acquire the image by means of the camera; (ii) process this image into a format suitable for further analysis; (iii) analyse the image to be able to determine for example the identity and orientation of the articles viewed and (iv) make a decision based on this information. To provide further detail, some of the vision system components are now examined more closely.

21.2.1 The Camera

Cameras can be area array or line-scan, see Figure 21.3. The area array cameras have photocells in the form of a two-dimensional matrix, whereas the line-scan camera has a sensor comprised of a single line of a large number of photocells. At the time of writing a typical industrial vision camera area array may have a matrix of around 2000×2000 photocell although they can be much higher than this. Similarly, in the line-scan camera there can be up to 12 000 photocells but since high resolutions lower the line-scan rate, a resolution of around 3000 may be more typical. The line-scan camera is used to capture data from very fast moving objects that may be travelling many metres per second.

Two main types of sensor are the CCD (charge coupled device) and CMOS (complementary metal-oxide semiconductor). CCD sensors and cameras produce good quality images and are used in applications such as astronomy, low-light and low contrast imaging and surveillance and military imaging. However, CCD cameras can suffer from 'blooming', this occurs when a very bright area in the scene saturates the relevant cells in

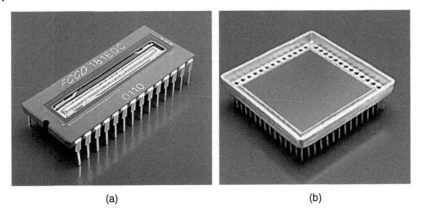

(a)   (b)

Figure 21.3 Machine vision photocell arrays. (a) Linear photocell array of 2592 × 1 CCD elements 10 μm × 10 μm size × 10 μm pitch. (b) Area photocell array of 2028 × 2048 CCD elements 15 μm × 15 μm in size.

the sensor and activates their surrounding cells so that the bright area tends to spread or 'bloom'. CMOS sensors approach a comparable level of performance as CCD and have a much lower cost since standard manufacturing processes can be used for fabrication. Since CMOS sensors also have lower power consumption than CCD sensors they are popular in industrial vision systems, mobile phone cameras, tablet cameras and professional and amateur photography cameras. In area array cameras, CCD sensors are good for grabbing single images of fast moving objects with no distortion as they usually have what is called a global shutter operation, whereas CMOS cameras, although they can have a faster frame rate than CCD, generally have a rolling shutter effect that can create distortions in the fast moving image due to the manner in which the sensor pixels are scanned.

21.2.2   Camera Interfacing and Initial Processing

The next stage in the vision system requires that the video signal from the camera be converted into a form suitable for manipulation by a computer. For this stage there has been dedicated hardware designed that enables the necessary fast processing times to be achieved. Speed is essential as in the factory environment production must be able to continue at optimum rate unimpeded by the processing time of the vision system.

The light falling on each pixel produces an analogue voltage proportional to the light intensity falling on it. Thus the first stage of the process is called quantisation and it involves examining the light intensity at each pixel and allocating it to a specific grey level, in effect digitising the analogue signal proportional to the light intensity at that pixel. This is possible by creating a grey scale with, say, a value of 0 relating to black, 255 relating to white and the shades of grey in between being allocated an appropriate value. Thus, 8 bits of information will be required to define the brightness of each pixel in a 256 level grey scale. To reduce the processing time and complexity, it is possible to reduce the number of levels or shades of grey down to 128 or 64. Taking this principle to the limit, the scene can be reduced to a binary image, that is, black or white, by representing all grey values above a specified level as white and all below this level as black. Only one

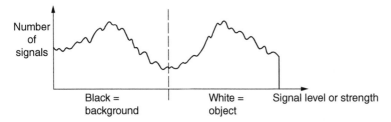

Figure 21.4 Setting a threshold value.

bit of information is then needed to register the light intensity at each pixel, although much more attention needs to be given to lighting to enable a high enough contrast between background and object to be obtained.

In grey scale vision, therefore, a very comprehensive picture of a scene can be built up. This means that a great deal of information needs to be subsequently processed to be able to analyse the scene adequately. For example, to be able to extract from the picture the edges of a component from a cluttered background, a comparison of various contrast changes over the whole scene has to be done. This is possible by comparing the grey levels of each pixel with those above, below and on either side.

The binary vision technique is much simpler, faster and usually cheaper, although for it to work reliably the objects to be viewed must be in relatively sharp contrast to their background and preferably not be overlapping or tangled in each other. The binary image is obtained by using a technique called 'thresholding' that operates as follows. The objects viewed by the camera are reduced to their silhouettes by setting all the light intensities in the scene background to black and all the object light intensities to white. This could, of course, be reversed depending on whether it was, for example, bright steel objects being viewed on a dull surface or objects placed on top of a light-table thus showing up as black on a white background. The threshold value is now set by the vision system operator, see Figure 21.4 This value determines which intensities should be set to white and which to black during silhouette formation. The picture is now available in a digitised form, either in grey scale or binary and at a certain resolution.

Although the discussion here used binary and grey scale vision as examples, colour vision is widely used. This is particularly the case in the food industry where fruit may be being inspected for ripeness and quality as it passes along a conveyor belt, to check the colour of baked cakes or to read labels and check fill levels in a bottling plant. Three-dimensional vision may also be used in some applications where depth information is important.

21.2.3   The Picture (or Frame) Store or FPGA

The frame store is a facility to allow the picture obtained at the previous stage to be held awaiting further processing by the computer. This is not always necessary but is usually part of the system due to the extremely high information transmission rate from the video processor. The computer has to handle large amounts of these data and cannot always operate at a compatible speed with the video, the frame store therefore acts as a buffer between initial picture processing and analysis.

21.2.4 The Vision System Processor

The information now coming to the computer is in a digitised form suitable for further analysis. The first task is to extract or 'trace' the profile or edge of the object from the scene. In some vision systems this can be implemented by dedicated hardware and may in fact be part of the pre-processing stage. In binary vision this edge extraction is carried out by systematically scanning the digitised image for the edges, probably starting at the top left and moving left to right down the frame. A change in signal governed by the threshold limit previously set will signify a change from either background to object or object to background. When the computer recognises one of these edge points it checks the neighbouring points on the line above. This is how it builds up a picture and determines if it is the start of a new component or a continuation of an already scanned one. When the frame scan is fully complete the computer has an 'image' of the edges of the object and features such as holes in its memory.

In grey scale vision the principle is similar but many more computations are required. This is because in grey level vision it is not an absolute value determined by the threshold limit that delineates an edge point. Here edges of objects are found by looking for a sharp rate of change of contrast. This means that each pixel point is examined in relation to perhaps four or eight of its neighbours before a value can be attached to it. This examination can be complex for the system may be used, for example, to find small flaws in a casting or for picking individual objects out from a jumbled heap.

Once the image has been stored in the computer memory analysis of the data can begin. A common approach is to examine the object and compare this image with 'templates' or 'models' held on file. These reference models will have to be taught to the vision system by showing samples to the camera and programming the computer to remember them by measuring and recording various attributes. Thus when the vision system examines a scene later it can compare the attributes of the component it views with those of the component held on file. It can then, depending upon the system being used, recognise components, perceive deviations from the standard models such as flaws, identify the location and orientation of components within the scene or carry out measurement of the object.

Consider, then, that the vision system now has in its memory a record of the scene in the form of an area array of pixels. For this discussion, it is assumed for simplicity that it is a binary system and therefore each pixel is represented by a '0' or a '1' in each memory location. The system can then assess the attributes of the scene or components in view. The following parameters are examples of some of these attributes and how the vision computer can recognise them.

Area. When it is considered that the binary image is made up of black and white pixels with, say, white pixels representing the object, then the sum of the number of white pixels in the image is a measure of the area of the object. This operation can be carried out by the processor as the image is being scanned thus giving the area measurement immediately the scan is complete. Should a specific measurement, rather than a relative value, be required then the actual size and spacing of the pixels together with lens magnification and object distance would have to be considered.

Perimeter. This operation can again be carried out as the picture scan proceeds. Where each scan line changes from black to white or white to black this is an indication of the component edge. By totalling the number of changes an indication of the perimeter length is achieved. The changes that occur due to holes within the object can be eliminated by software if necessary. If a measurement of the perimeter is required then once more pixel pitch and so on must be taken into account.

Perimeter squared/area. This is an indication of the compactness of the object. This feature does not change as the magnification or the scale of the image changes.

Centroid. The centroid is the position of the centre of area of the object. It is the position given when the first moment of area about both the x- and y-axes is calculated and then divided by the object area thus giving the x and y coordinates of the centre of area.

Number of holes. If the binary image is being considered as consisting of a black object on a white background, then every white area surrounded by black must be a hole. The number, position and size of these holes can therefore be calculated, stored and used for identification or inspection purposes.

Maximum and minimum radii. These are the maximum and minimum distances from the centroid to the farthest and nearest points on the perimeter.

Minimum enclosing rectangle. This is obtained by observing the maximum and minimum points on the x- and y-axes of the perimeter of the object.

The relationship of some of these parameters can also be used to determine the orientation of the component. For example, the angular relationship of the maximum radius to the x-axis at a point about the centroid can be used, as can the relationship of radius vectors from the centroid to the centres of any holes in the object. Thus, the very asymmetry that necessitates the additional work of orienting the object is used to provide data that can be used to determine its existing orientation.

Figure 21.5 illustrates a rectangular component and some of the attributes mentioned here; the image on the right should be considered a binary one with the rectangle as white and the background as black.

21.3 Lighting

It is lighting that gives the scene the necessary contrast to enable the component, or component elements, to be differentiated from the background and care has to be taken in selecting the type and arrangement of the lighting system. In binary vision, this requires that the background and object transmit to the vision sensor suitably different light intensity values, for example, the background a white conveyor or illuminated table on which the object shows up as a silhouette or the background could be a matt black surface on which metallic components rest and reflect back to the sensor relatively high light levels. This would certainly be a very artificial situation in many factories where dust and grime are often present covering component and table or conveyor indiscriminately. This is where grey scale and colour imaging can be advantageous as the varying contrast or colour between surfaces can be used to derive the information required to determine the edges or other features of the component. Under the proper lighting conditions it can be used to identify flaws such as scratches and cracks in plastic goods or metal castings. Whatever lighting is used it should be strong enough to prevent interference by ambient light levels surrounding the work area.

Generally, as in photography, manipulation and organisation of light sources can be used to emphasise or reduce shades and highlights. This will facilitate identification of

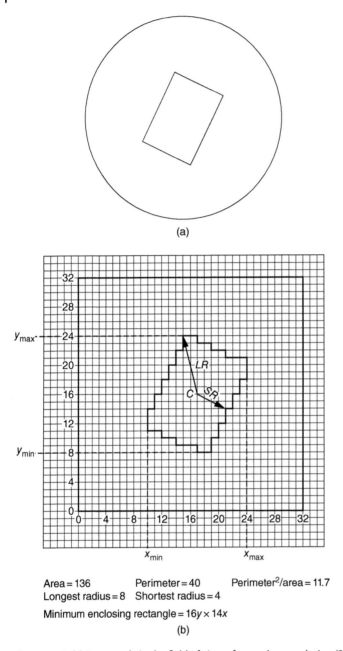

Area = 136   Perimeter = 40   Perimeter^2/area = 11.7
Longest radius = 8   Shortest radius = 4

Minimum enclosing rectangle = 16y × 14x

(b)

Figure 21.5 (a) A rectangle in the field of view of a very low resolution (32 × 32 area array) vision system. (b) The image as it appears to the system in a digital form.

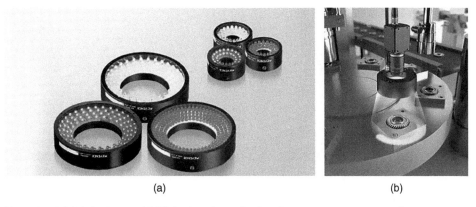

(a)   (b)

Figure 21.6 (a) Light rings and (b) light rings in application. Source: Images courtesy of Keyence UK Ltd.

critical features such as component edges or flaws. The use of filtering may also be applied to create contrast between components and background. This may be done by, for example, using a red filter on the lens to enhance red components on a dark background. An example of circular lighting that can be used to provide an even and shadowless image is shown below in Figure 21.6.

21.4   Some Further Application Examples

Figure 21.7 shows how a monotone system can be used to detect flaws in bricks in a robotic brick palletising application.

In Figure 21.8, a colour vision system is being used to check the quality of fruit on a conveyor system.

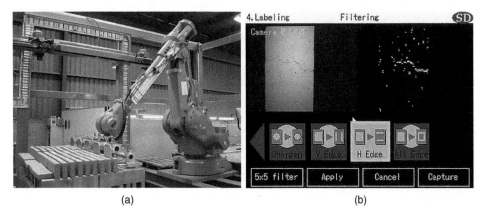

(a)   (b)

Figure 21.7 (a) An industrial robot is shown palletising bricks with a vision system inspecting the bricks before stacking. (b) The image from the programmer's monitor; the grey scale image shows a crack in a brick and the binary image shows how this vision system perceives this as a series of pixels forming a recognisable crack pattern.

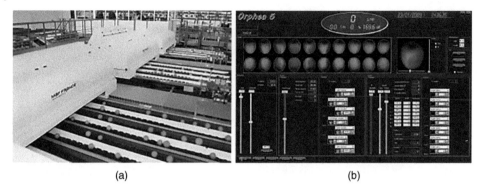

(a)   (b)

Figure 21.8 (a) A fruit conveyor system is shown with a vision system inspecting the fruit before packing. (b) The image from the programmer's monitor. The images will be in colour on the screen and show the quality of the fruit passing along the conveyors. This is monitored by the system that decides automatically whether or not the fruit is damaged or over ripe depending on the colour. Flawed fruit can then be ejected before packing.

21.5   Conclusion

Automation has many facets and components that cannot be addressed in the space available in this text, however, the essentials have been presented here in a form that should convey both the context of manufacturing automation and useful examples of components and applications.

Review Questions

1   Discuss fully three areas of application for machine vision in industry.

2   What are the basic components of a machine vision system?

3   Within a vision system the sensor photocells may be arranged in a linear array or area array configuration, briefly note typical applications of each.

4   The sensor used in the machine vision system may be CMOS or CCD, briefly describe the differences between them.

5   Describe a typical sequence of operations that occurs within a vision system in order for it to recognise a component and to determine if that component has a flaw.

6   Why is the choice of lighting important for machine vision systems?

Part V

Manufacturing Operations Management

22

Production Planning

22.1 Introduction

In the previous chapters, we examined the processes used to create products. But the creation of a product, especially a modern complex one comprising many elements, also necessitates careful planning and control of the production system. This section of the book considers how the resources of materials, machines, money and people, or manpower to retain the mnemonic, are managed to produce competitive products for the world market.

Improvements in management techniques to increase efficiency are constantly appearing. As technology, markets and society change, so also the methods employed by management change. Concepts and techniques such as Simultaneous (or Concurrent) Engineering, Design For Manufacture (DFM), Manufacturing Resource Planning (MRPII), Just In Time manufacture (JIT), Optimised Production Technology (OPT) and Total Quality Management (TQM) are examples. Most of these are referred to by their acronyms. Simultaneous Engineering and DFM were discussed in earlier chapters; some of the others will be examined in Chapters 23 and 26.

We begin this chapter on planning by considering the geographic location of a manufacturing plant, then examining how the internal layout of the plant may be optimised for maximum efficiency. We next look at how to select the most suitable process to make a particular product, before going on to consider project and process planning. In the following sections, we discuss some of the previously mentioned concepts and techniques used to help organise the flow of the work through the manufacturing system, ensure quality and ensure that the manufacturing operation remains economically viable.

22.2 Plant Location

Determining where to locate a manufacturing facility involves the consideration of a wide range of criteria. The relative importance attached to each criterion is dependent on the company involved. For example, a shipbuilding company will require proximity to a sea or river and possibly a steel plant for supply of steel plate and so on. However, an electronics manufacturing company will he more concerned with the proximity to an inexpensive labour source for manual assembly tasks and skilled engineers for equipment design and maintenance. For larger companies, the decision will be made on a

Essential Manufacturing, First Edition. Gordon Mair.
© 2019 John Wiley & Sons Ltd. Published 2019 by John Wiley & Sons Ltd.

global basis. Criteria such as government incentives to attract foreign investment are important, for example, many USA and Japanese companies have factories in the UK due to factors such as low rates for the use of land, special grants and skilled local work-forces. Also much foreign investment is made in the developing countries where labour rates are still relatively low. Conversely, companies may decide to set up a manufacturing facility in an industrialised country simply to avoid trade barriers or the cost of exporting the complete product to it; for example, cars. For smaller scale activities, proximity to the end user is no longer of such importance and proximity to natural energy resources has also become of less importance. Companies now view the world as their market place and take advantage of modern methods of transportation and communication.

It is therefore apparent that the plant location problem is by no means trivial, and that the interrelationship of many criteria must be considered. The decision will be based on a combination of financial incentives offered by each country to invest in a specific locality, the land space available and building costs and the infrastructure of the area, for

| Criterion | Weight | Possible factory location | | | | | | | | | |
		A		B		C		D		E	
Skilled labour	7	2	14	3	21	0	0	1	7	4	28
Large pool of unskilled labour	8	5	40	2	16	0	0	4	32	2	16
Motorway	7	3	21	2	14	1	7	3	21	4	28
Airport	4	1	4	3	12	4	16	2	8	2	8
Sea/River	0	2	0	5	0	5	0	2	0	1	0
Housing	5	4	20	3	15	0	0	3	15	4	20
Amenities	5	3	15	2	10	0	0	2	10	3	15
Potential for expansion	7	2	14	1	7	5	35	3	21	2	14
Availability of grants/Incentives	8	1	8	2	16	5	40	1	8	3	24
Safety	2	3	6	2	4	5	10	2	4	2	4
Planning constraints	5	2	10	3	15	5	25	4	20	2	10
Environmental impact	4	3	12	2	8	4	16	1	4	2	8
Total		164		138		149		150		175	

Proximity to : applies to rows Skilled labour through Amenities.

Figure 22.1 Decision matrix for plant location: site E is most desirable.

example, communication and transport links, skill level and cost of local labour, proximity to essential services and resources and availability of local sub-contractors.

Figure 22.1 shows an example of a decision matrix that could be used to select a location for a new factory. *A, B, C, D* and *E* are possible sites, The criteria important to the company are listed down the left-hand side of the matrix. Each criterion is allocated a weight between 0 and 10, proportionate to its desirability to the company. For example, this company needs a location close to a large pool of unskilled labour, the availability of skilled workers, the presence of financial incentives and the space for eventual expansion. The boxes created by the rows and columns are split into two as shown. In the top left-hand corner of each box, a number between 0 and 5 is inserted signifying the ability of the site heading the column to satisfy the relevant criterion. After all the corners have been filled, the multiple of each number and the criterion weight is entered in the bottom right hand of the box. The sum of the multiples for each column is now calculated. The site with the highest score at its column base is the one most likely to best satisfy all the needs and wants of the company.

22.3 Plant Layout

First, we will consider the classic patterns of plant layout that are still commonly found and are effective when used appropriately. We will then consider the use of component classification and coding that can be utilised to help construct what are called 'cellular layouts'. This classification and coding of parts has implications for the whole manufacturing system, the overall concept being called group technology or GT. Group technology improves the overall efficiency of large manufacturing companies by identifying similarities in parts and grouping them into families. This produces benefits from the optimisation of equipment and tooling usage and the possibility of realising many of the benefits of standardisation.

Proper layout of a manufacturing facility is essential since it will determine the work and material flow throughout the factory. Some of the benefits of good layout design will be to: maximise productivity, ensure best possible use of floor space, simplify handling of work and materials, improve equipment and labour utilisation, reduce throughput time, minimise product damage, improve safety and working conditions, reduce travel times for personnel and materials and simplify production control.

22.3.1 Classic Plant Layout Patterns

The problem of laying out a work area occurs in many situations, for example, a garage, a hospital, a laboratory and even the domestic kitchen. Our concern is with the manufacturing area of a factory. Typical layouts can be identified as follows.

1) *Random layout.* Likely to be very inefficient; this type of layout may be found in small factory units where the production volume and production variety has gradually grown. This may be seen in the initial stages of the evolution of a 'start up' manufacturing company.
2) *Functional layout.* This is a common but relatively inefficient type of layout. It is often used where the production of a large variety of products in batch volumes is required,

for example, the components for large water pumps for ships or desalination plants and components for sub-assemblies for the aerospace industry such as radar equipment or jet engines. Similar processes are grouped together, thus creating physical areas such as drilling, turning, milling, injection moulding and casting departments (see Figure 22.2a). The reasons the method is inefficient are that the material transport routes are long since material is transferred from department to department; work in progress is high and in a large plant the tracking and control of individual orders can be difficult. The advantages are that specialist supervision and labour can be employed, and there is flexibility in the processing of the work as it can be transferred from department to department any number of times and in any sequence. The reasons for grouping together common specialised equipment such as injection moulding machines or foundry equipment are obvious; however, the advantages of grouping together the machining equipment such as lathes or milling machines are not so obvious, and today the concept of cellular systems, mentioned in point 6, is more appropriate in many situations.

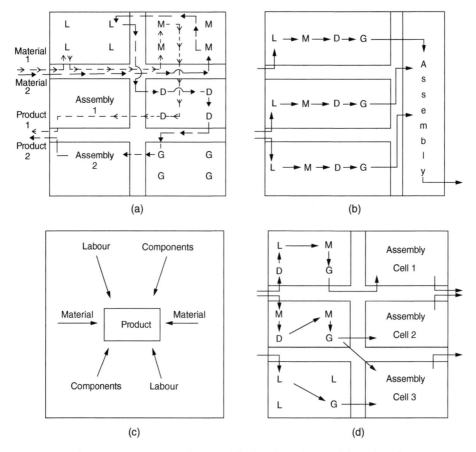

Figure 22.2 Plant layout patterns: (a) 'functional' for batch production, (b) 'product' for mass production, (c) 'fixed-position' for jobbing/project production and (d) 'cellular layout' for batch production.

3) *Product layout.* Used for the mass production of goods where a small variety is involved and production volumes are very high, for example, motor cars, television sets, electric motors and refrigeration compressors. It may also be called a flow or line layout. It is much more efficient but less flexible than the functional layout. Work in progress is minimised and jobs are easily tracked through the system. A reliable and steady demand is essential due to the high investment in capital equipment, much of which will be special purpose and ideally only one product will be involved. Because of its serial nature (as shown in Figure 22.2b), the layout is very sensitive to machine breakdown or disruption to the material supply. It is therefore important to have rapid attention to breakdowns and reliable material deliveries.

4) *Fixed position layout.* This is used for the manufacture of large, high cost, single artefacts, for example, ships, offshore oil platforms, spacecraft and communication satellites. It differs from the previous three layouts in that the product remains static while the workers, tools and equipment come to the work rather than the reverse (see Figure 22.2c).

5) *Process layout.* This applies in the process industries such as plastics or steel manufacture. Here, the technology of the process determines the layout, for example, the location of fractionating columns and pipework or blast furnaces and continuous casting equipment.

6) *Cellular layout.* This requires firstly an explanation of *workpiece classification, coding* and *GT*.

This topic could have been appropriately introduced at a number of points throughout the text since it is a method of describing a manufactured part. Its use has implications for component design, process planning, plant layout and purchasing. When properly implemented in conjunction with computerised manufacturing systems, it is also an aid to improving the overall efficiency of the operation. This process, which allows grouping the components used and produced by a company into families, is also the first step in the implementation of GT. The concept of GT attempts to maximise the advantages to be gained from similarities in the design and processing requirements of products; these advantages are particularly important in large companies producing and using many thousands of parts.

Both design and manufacturing attributes can be used as a basis for the classification. Similarities in features of the component design such as the shape, dimensions, tolerances, surface finish, holes or re-entrant angles and length to width or length to diameter ratio, are used as classification criteria. Similarities in manufacturing features such as the type of process used, the sequence of operations, the equipment employed and the production volumes and rates required can all be used to identify families of components. It should be noted that many of the design and manufacturing criteria are interrelated, for example, the tolerance and surface finish obtainable is dependent on the process used.

The actual code used to define the part may be devised by the user company themselves to suit their own particular circumstances, or an 'off the shelf' industrial coding system may be purchased and tailored to suit. Each alphanumeric character in the code defines an attribute of the part and, although the code should be as simple and easily understood as possible, the number of characters can be large. The 'MultiClass' system, for example, uses up to 30 digits. However, once allocated, the part code can be stored in a computer along with the codes for all other company parts. The computer is then used to sort the parts into families according to their codes.

Once the part families have been determined, it should be possible to manufacture all the parts from one family by using a group of common processes or machines. A grouping of machines like this is called a manufacturing 'cell'.

A cellular manufacturing cell is the means of creating the products identified by the classification and coding process as belonging to a single family. The cell is an autonomous manufacturing unit that can produce a finished part and it will contain one or more machines. This aspect of group technology is most commonly applied to machined parts, the cells usually containing numerically controlled machine tools, which are described in Chapter 21, such as machining centres, lathes and milling machines (see Figure 22.2d). Thus a company that has employed GT throughout its operations will often have a large number of these cells within its factory. The cells may be manned with one operator tending a number of machines, but they are often computer controlled and utilise robots for material handling. Consequently, productivity and quality are maximised, and throughput times and work in progress can be kept to a minimum. Due to its flexibility the cell layout system is particularly suited to the manufacture of products in batches and where design changes often occur.

22.3.2 Other Considerations

Once the general type of layout has been decided, more detailed planning needs to be done. Space requirements for equipment, materials and personnel must be allocated. Material handling methods will be considered, for example, whether to use bulk transfer means such as conveyors or to use discrete load transporters such as fork lift trucks or automated guided vehicles. Safety is obviously very important and aisles of adequate size must be available to provide safe access to all work areas; aisles must be clearly marked, usually by white painted lines, and kept clear at all times. Provision must be made for services such as electricity for lighting and machine power, data transfer links, compressed air, water and possibly automatic removal of swarf from machining installations. Storage space for materials and tools will be made but it should be remembered that in a modern factory work in progress must be kept to a minimum; this means that large storage areas on the shop floor for it should not be necessary.

The detailed layout will also be determined by carrying out studies to ensure minimisation of material flow, personnel movement and work handling. Many techniques exist to aid this process, for example, the use of scale drawings, three dimensional models, string diagrams and computer simulation, are all common. Their description and use cannot be covered fully here, but some of the techniques are also used in method study and these are described in Chapter 24 on Work Study.

22.4 Project Planning

As a scenario for this section we will assume that a small manufacturing company has decided on a location for its plant and has in fact selected a suitable building on an industrial estate. We will further assume that the manufacturing processes to be used have been selected and that the plant layout is being designed. Concurrent with the layout design, the necessary machines and equipment will be ordered and their installation planned. This is the project planning stage and certain techniques have been developed to assist the planner carry out the task efficiently and allow progress to be monitored. Here, we briefly consider three aids for the project manager, that is, the Gantt chart, Critical Path Analysis (CPA) and the Work Schedule chart. These can be used for any type of work where a large number of items or activities have to be drawn together to produce a finished product or task. Activities may be carried out in parallel to each other, in which case they are independent, or they may be dependent and have to be carried out in series. In this type of planning the activities are usually long term and measured in days, weeks or months. This occurs in the jobbing type production noted in Chapter 5 where the task might be the building of a submarine, spacecraft or offshore oil platform. These aids are also applied to projects such as the construction of a bridge, power station, house, the organisation of a trade exhibition and conference or the installation of a new machine or layout in a factory. The basic principles are illustrated here by manual methods for a simple project, however, for larger or more complex projects computer-based programs will normally be used.

We will assume that a robotic installation is being planned. The installation will comprise a conveyor line that will deliver boxed items to a pick up point. An industrial robot arm will pick up the boxes and stack them on a wooden pallet that, when full, will be removed by a fork lift truck. The robot and pick up point at the end of the conveyor are to be surrounded by a safety cage. The full installation is to be controlled by a PC type computer. We can therefore now identify a number of activities that must take place for the installation to be completed. The conveyor, robot and computer must be ordered and delivered. It is assumed that the safety cage is already on site awaiting installation. The work area must be marked out and a foundation prepared for the robot. These items must all be installed, mains electricity connected and the computer interfaced to the robot, conveyor and safety cage.

Having defined the project we can now list the activities and allocate an estimated duration to each. We can also specify what activities need to take place before others can begin. For example, the foundation preparation cannot take place until- the work area has been marked out, the robot cannot be installed until it is delivered and the foundation prepared and the safety cage cannot be erected until the installation of the robot and conveyor is complete. Mains electricity cannot be connected until the conveyor, robot and cage are installed. The computer can be installed any time after its delivery but cannot be interfaced to the rest of the equipment until the mains electricity has been connected. These activities and their expected durations can now be listed as shown in Figure 22.3.

The simplest technique to use to represent this data is the basic Gantt chart. This is shown in Figure 22.4. The major project activities are listed down the left-hand side of the chart and the days represented by columns; calendar dates would probably be used in a real world application. Completed activities are shown as broad bars, activities not yet started or incomplete are shown as thick lines. The thick vertical line joining the two pointers is the present time indicator, the situation at the end of day 9 is shown in the chart. The main advantage of the Gantt chart is immediately apparent, that is, it is easy to obtain a quick impression of the state of the individual project activities at the present time. It can be seen quickly that activities A, B, C and E have all been completed on schedule, that activity G has been completed ahead of schedule and that activities D

Activity		Duration (days)	Immediate predecessors
A	Mark out work area	2	Start
B	Prepare foundation	5	**A**
C	Delivery of conveyor	8	Start
D	Delivery of robot	10	Start
E	Delivery of computer	7	Start
F	Install robot	3	**B, D**
G	Install conveyor	3	**C**
H	Install computer	1	**E**
I	Install safety cage	3	**F, G**
J	Connect mains supply	4	**I**
K	Interface all equipment	5	**H, J**

Figure 22.3 Example project list of activities and durations.

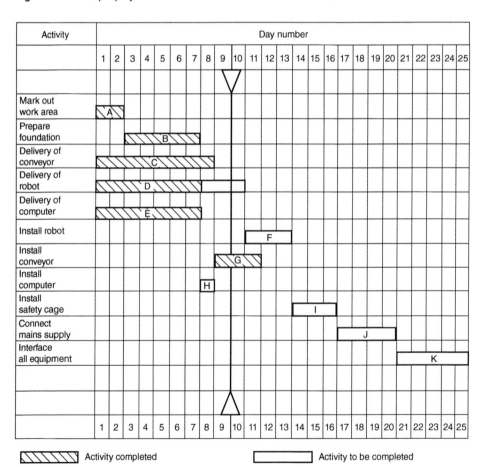

Figure 22.4 Project Gantt chart.

and H are behind schedule. It also shows that the project activities are all expected to be completed by day 25.

Although the Gantt chart is quick and easy to read regarding the progress of the project, it does have some disadvantages for use as a planning aid and for obtaining more detailed progress information. For example, it does not make it obvious that although activity H is behind schedule by two days it will not interfere with the overall completion time of the project, whereas the two day delay on element D will probably also delay the project completion by two days. As well as not giving a clear indication of the interdependence of the various activities, it also does not allow the planner to see what activities must be completed on schedule to ensure that the project is completed as planned. Neither does it allow the planner to see how much an activity may be delayed without influencing the start of other activities; this difference between the earliest and latest activity start time is termed 'float'. A knowledge of activity interdependence and float is essential to the planner, as it can allow scarce resources to he applied where and when they will be most effective. This is particularly important where a large number of items and work elements are involved: in a real project these may number thousands. Network planning techniques have been developed to provide this necessary information, the most popular of them being CPA and the Program Evaluation Review Technique (PERT) both developed in the USA in the 1950s. We will briefly examine the Critical Path Analysis (CPA) method here.

The first four steps in the CPA technique have already been carried out, that is, the project has been defined, the activities of which the project will be comprised have been listed, precedence relationships have been established and estimates of the activity durations have been made. The next stage is to construct a network of nodes and arrows. The convention we will use involves representing an activity by an arrow and representing an event, that is, the beginning or end of an activity, by a node. Thus the arrows represent a period of time whereas the nodes represent a point in time. Figure 22.5 shows the network for our project. As can be seen, it provides all the information we need regarding precedence relationships and start and end times for each activity; it is constructed as follows:

Using the previously compiled list of activities and precedence statements the start node is identified; in this case it may be assumed to be the point at which final approval is given for immediate commencement of the project. From this node arrows are drawn for each activity that can begin immediately the project starts. Each node is shown as a circle. The top half of the circle shows the node number; these are in multiples of 10 to allow the addition of extra nodes should this be found necessary. The bottom half of the node circle is split into two; the reason for this is explained later. For example, in Figure 14.6, node 10 is the start point for activities 'A', 'C', 'D' and 'E', and node 40 is the finish point for activity 'C'. Node 40 is also the start of activity 'G' and node 60 its finish; the logic of the network is such that activity 'G' cannot start until activity 'C' is finished. If we now look at node 60 we see that two arrows enter this representing activities 'F' and 'G'; this shows us that activity 'I' cannot commence until both activities 'F' and 'G' are completed. Using the arrows and nodes in this way, the network can be constructed. Further refinements are often necessary, such as the insertion of 'dummy' activities to allow precedence relationships to be represented that would otherwise be difficult to show.

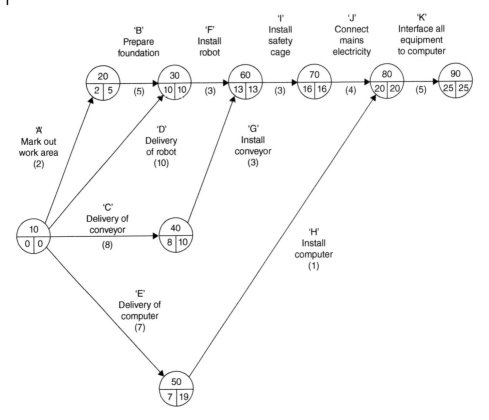

Figure 22.5 The critical path analysis for the project.

The next stage is to carry out a forward pass through the network to determine the earliest event time for each node. For example, node 10 marking the start of the project is assumed to be day zero, although a calendar date would probably be inserted here in practice. Activity 'A' takes two days, thus the earliest time for node 20 is after two days; the number 2 is therefore entered in the bottom left of the node circle. Node 30, however, represents the completion of both activities 'B' and 'D'. This means that it cannot exist until both of these activities have been completed. We must therefore insert into the bottom left space the maximum event time elapsed at that point. Since activity 'B' takes five days, a total of seven days will have elapsed by the time 'B' is completed.

However, since activity 'D' takes 10 days it is this value that must be inserted into the node as the earliest event time. This process is carried out throughout the network until the earliest completion date for the project can be entered into the final node. Examination of Figure 22.5 shows that this earliest completion date is 25 days after the project start.

Next a backward pass is carried out to determine the latest event time for each node. This will show the planner the latest each activity can be completed yet still maintain the earliest completion date. It follows logically that this process should also highlight those activities that must be completed according to schedule. The backward pass is carried out in a similar manner to the forward one, except that this time the activity durations are progressively subtracted and the results entered into the bottom right-hand sector of the node circle. The difference in value between the right- and left-hand numbers represents the 'float' time at that point. For example, the computer may arrive any time between day 7 and day 19 without affecting the earliest completion date of the project – that is, activity 'E', the delivery of the computer, has a float of 12 days.

The backward pass will also show up those activities with zero float. These activities meet at nodes where the earliest and latest node event times are identical. The network path created by these activities is termed the Critical Path and all activities on this path must be completed on time if the project is to finish on schedule. The critical path for our project passes from node 10, through nodes 30, 60, 70 and 80, to node 90.

Using this information the planner will now create a schedule for the project. Usually, each event will be scheduled to start at the earliest possible time to provide maximum float; this may be modified if two parallel activities demand the same resources, for example, specialised labour. To monitor the progress of the project, the planner may also construct a work schedule chart. This is a variation on the Gantt chart and is illustrated in Figure 22.6. The situation at the end of day 12 is shown and all work to date has been completed to plan. However, since the critical path is shown along the top of the chart, the planner or project manager will immediately recognise a serious delay should one occur on these activities. The float available on the other activities is also shown as broken lines; therefore, as long as the block showing the activity duration does not move beyond the broken lines the project may still be completed according to schedule.

22.5   Process Planning

In order to make a component by any production process it is necessary first to create a plan of how the component will be made. The selection of which process to use has already been covered in Chapter 17; here, we consider the means used to plan the detail of manufacture. Some processes require a smaller number of operations than others to make a component. For example, the process planning for making a component by injection moulding is simple, for example, inject the plastic into the mould, remove component from mould when cool, trim excess material and place component on pallet. However, the sequence of operations is often more complex and this is particularly apparent when producing components by machining. As well as planning the manufacture of individual components, the assembly of these components into a finished product must also be planned; this requires the use of slightly different planning techniques. Process planning also allows labour and capacity requirements to be studied and an accurate estimate to be made of the time taken to produce the component or complete product. These aspects are considered more fully in Chapter 24.

Various types of charts may be constructed as aids to process planning. 'Assembly charts' are used to show the sequence of assembly and relationships between manufactured and bought out components and sub-assemblies. They are particularly useful when complex products comprising many components and sub-assemblies are being produced.

More detailed descriptions are then produced by creating 'operations charts' that now show the individual manufacturing and inspection operations, processes and equipment required. Further refinement is obtained by using 'flow process charts' that include

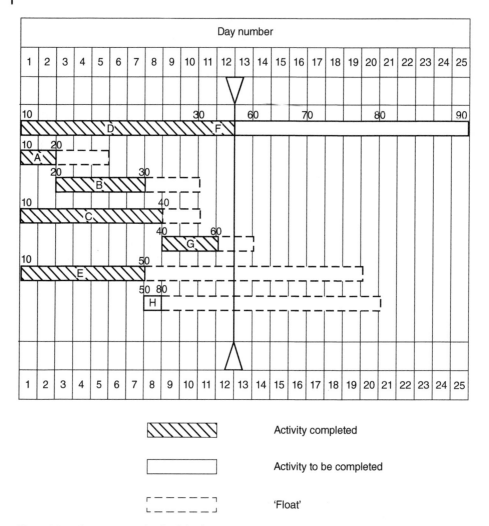

Figure 22.6 The project work schedule chart.

additional information on transportation and storage. An indication of the type of charts being discussed can be seen in Figure 24.4, where a process chart for a product is shown.

Production of components by the machining processes often provides the most challenging aspects of this type of planning; despite the relative inefficiency of the process it remains a common problem for the planner. Computer aided process planning, CAPP, is used in industry, but due to the large number of variables, complexity and general need for experiential decisions, some amount of human input is usually required. The type of machines and the skill of the labour involved have a great influence on the length and detail of the process plan. For example, the use of numerically controlled machine tools reduces the amount of handling of the workpiece and the number of times the tooling has to be changed or adjusted, inspection operations are reduced and, apart from relatively simple work, holding devices jigs and fixtures are not required. Skilled craftsmen, such as toolmakers, do not require such detailed planning instruction as unskilled workers. In fact, an experienced toolmaker will be able to make a component or product by simply working from the engineering drawings, whereas an unskilled production worker will require detailed instructions on what material and machines to use, and the sequence of operations to be followed. Naturally, if a computer controlled process is

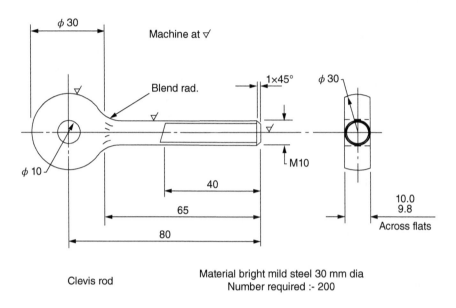

Clevis rod

Material bright mild steel 30 mm dia
Number required :- 200

Operation number	Description	Machine
10	Feed 70mm length through collet & tighten to clamp	Lathe 1
20	Rough turn ϕ 10mm shank, turn 1x45° chamfer, face end	Lathe 1
30	Finish turn ϕ 10mm shank screwcut m10 thread	Lathe 1
40	Unclamp collet, withdraw material to allow parting off to length	Lathe 1
50	Reclamp collet, part off to length + 1 mm	Lathe 1
60	Clamp shank in collet, using form tool turn 30 mm dia sphere	Lathe 2
70	Inspect	Bench
80	Clamp in fixture and mill both flats	Mill
90	Remove ragged edges with file	Bench
100	Place in jig & drill 10mm dia hole & deburr	Drill
110	Inspect	Bench

Figure 22.7 An operation layout sheet for a clevis rod.

being used such as a numerically controlled machine, the required information is held in the machine program and the machine operator simply has to load the material and unload the finished component.

To conclude this section a simplified example of an 'operation layout sheet' is shown in Figure 22.7 for a clevis rod (the clevis assembly is shown in Figure 24.4). For more complex machined components there are general guidelines that should be followed when preparing the operation layout. For example, at least one datum surface should be established as soon as machining commences. This datum will be used as a reference surface for all subsequent operations. To maximise accuracy, as many surfaces as possible should be created at the same setting; that is, without unclamping the workpiece in the machine. The use of surfaces as secondary references other than the original datum should be avoided as long as possible. Precision operations or those producing a smooth finished surface should be created last in order to reduce the possibility of damage. Finally, inspection operations should be included at appropriate intervals to minimise scrap and rework.

Review Questions

1   List 10 criteria a company might use to decide where to locate a new factory and discuss the implications of each.

2   Describe how a company might use a decision matrix to assist with its location decision.

3   Briefly describe six types of plant layout.

4   What benefits should a company gain by careful attention to its factory layout?

5   Compare the advantages and disadvantages of functional and product layouts.

6   What do you understand by the term. 'workpiece classification/coding' and what benefits might a company gain by carefully applying its principles?

7   Explain the term 'Group Technology' (GT) and the type of plant layout associated with it.

8   Apart from material flow, list and discuss some of the things that should be given careful consideration when making decisions on plant layout.

9   What is a 'Gantt Chart', how is it used and what are its limitations for project planning?

10   Identify and list the steps to be followed in carrying out a CPA of a project.

11   What is meant by the terms 'activity', 'node', 'activity interdependence', 'float' and 'critical path' in CPA?

12 Consider the following project information (St = 'start')

Activity	A	B	C	D	E	F	G	H	I	J	K	L
Duration	4	6	4	2	8	6	12	4	4	12	4	4
Predecessor	St	A	St	St	B, C	E	D	G	D	I	J	F, H, K

Construct the network for this, identify the float for each activity and show the critical path.

13 Describe the function of the 'work schedule chart' and explain how it is an improvement on the basic Gantt chart.

14 What is the purpose of 'process planning' and how might the type of machinery and labour in use in a factory influence the process planning operation?

15 What is the purpose of an 'operation layout sheet' and what sort of guidelines should be followed by the planner as he or she constructs it?

23

Production Control

23.1   Introduction

Having considered the techniques used for planning a production system, we now need to examine how this production can be controlled. Strictly speaking, the problems and techniques discussed here include planning as well as control; however, we use the term 'production control' to emphasise that it is the ongoing operation we are considering rather than 'one off' planning activities. This chapter concerns itself not only with the control of the flow of work and material through a factory but also with aspects of planning that are necessary for overall control on an ongoing basis. Other chapters will deal with the control of quality, costs and labour.

It should be noted that the content of this chapter, and indeed much of this section of the book, is part of the domain of 'Production and Operations Management' (POM). The principles and techniques of POM can be applied to almost any organised human activity, for example, the planning and control of a hospital, fast food chain, air terminal or manufacturing plant.

Efficient management of any operation demands a constant supply of up-to-date information; the means by which this is obtained is often termed a 'Management Information System' or MIS. Within any company today computers permeate the fabric of the organisation. It is through the use of these computers that MISs have gained the ability to become proficient at providing topical information when needed. This allows management to monitor and control the current manufacturing operation and also to make informed decisions concerning the future. The core of the MIS is the 'data base'. This acts as a common store for all the data required for the efficient operation of the company. Ideally, all the data, textual and graphic, should be computer accessible and stored in a logical form to facilitate access. The importance of the common data base is that changes made by one department are instantly recognisable by others, for example, a component design change by engineering can be immediately acted upon by purchasing and production.

This chapter begins by looking at the basic elements of production and material control. However, the concepts and techniques used broaden in scope until by the end of the chapter the implications of their use can be seen to include the whole structure and organisation of the company.

Essential Manufacturing, First Edition. Gordon Mair.
© 2019 John Wiley & Sons Ltd. Published 2019 by John Wiley & Sons Ltd.

23.2   Elements of Production Control

Here, we are primarily concerned with the ongoing monitoring and control of production within the manufacturing facility.

Systems need to be in place that allow answers to questions such as those that follow. When should the material for product A be issued and how much should be used? Where is job B at this moment and is it where it should be? Why is job C in the machining cell when it should be at assembly? How many units of product D can be made by next Friday? What jobs are not running to schedule and why? The tasks of production control are, therefore, to schedule production based on sales orders or forecasts, to determine work priorities, to control stock levels, successfully to achieve production programmes, that is, on time delivery and to obtain optimum production equipment utilisation.

The achievement of these tasks involves finding some way to determine what job should be done, how many items should be made, when the work should be done and where it should be done. In a large company this production control problem is an extremely complex one. Consider a product comprising hundreds or even thousands of parts being made in the one factory. The task of making sure each piece of material or part arrives at the correct process or assembly point at the correct time is notoriously difficult. Attempts to schedule the work must be made but it must also be realised that in all probability the schedule will be disrupted to some extent. Disruptions occur due to random events such as failure of a supplier to deliver material on time, machine breakdowns, labour problems or a host of other problems almost impossible to anticipate. It is for this reason that no one materials management or production control system is absolutely foolproof. Also the type of production, that is, process, mass, batch or jobbing, will determine the most suitable type of system to adopt.

Whatever system is used, before the scheduling of the work within the factory can commence it is necessary to know the sales demands which must be satisfied. This information may be contained in a document called the 'Master Production Schedule' (MPS). This MPS provides the input to our system; it shows the quantities of the end product that are required and when they are to be made. It is derived from placed customer orders or forecast sales.

Also essential is a full 'Bill of Materials' (BOM). The BOM is a document, or computer file, which contains information on all the materials, components and sub-assemblies required to produce a product. This information may also be amplified to show how the product is manufactured, the quantity of each component and subassembly required for each product unit and the actual manufacturing facilities to be used to produce the components and assemblies. The BOM is produced from information supplied by the design and manufacturing engineering departments. As a simple example the basic BOM for a toy car is shown in Figure 23.1; as can be seen, this also communicates the product structure.

Other prerequisites for a successful system are operation layouts for each part, a thorough knowledge of the work in progress (WIP) within the factory and a record of all material, components and finished products held in stock.

All items paid for by the company that are destined to become part of the finished product are classed as 'inventory'. Thus we may have inventory, (i) held in stores waiting to enter the production system, (ii) somewhere in the system at various stages of completion – this is the WIP inventory and (iii) finished products or spares awaiting

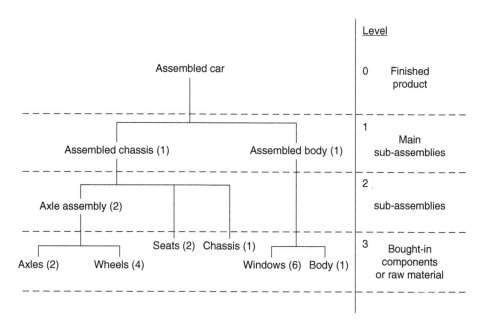

Figure 23.1 Bill of Material structure for a toy car.

delivery to customers. The aim of all company personnel, and specifically the materials or production control manager, is to reduce inventory levels to the minimum necessary for smooth production. Inventory costs the company money, which will not be recovered until the stock is transformed into a sold product. This is why the 'inventory turnover ratio' is so important. The ratio is obtained by dividing the cost of goods sold during a year by the average cost of the inventory held during that period. The actual ratio depends on the type of manufacturing business. For example, a ratio of 10 would be excellent for a company engaged in the manufacture of high value added products like machine tools and a ratio of 100 would make a producer of consumer durables, such as televisions or cars, top of the league of world-class manufacturers.

The sequence of events is therefore as follows. The sales department informs the manufacturing facility of specific customer orders plus forecasted orders that may be statistically derived. From this, an MPS is compiled. This MPS is used as authorisation to allow production of BOMs and subsequently the creation of production schedules. Information derived from the BOMs will also be used by the purchasing department to order the necessary raw material and bought out components. These detailed production schedules should take into account the capacity and availability of machines within the facility and the probable labour requirements. Areas should be identified where machines will be in high demand and where it might be necessary to find extra labour capacity by running overtime, extra shifts, hiring additional workers or contracting out work to other factories. The production control department will also determine priorities, that is, the sequence in which jobs should be done, and in the case of conflicting demands will make a decision on which job should be done first.

Depending on the type of system being used a 'works order' or its equivalent will be issued. This contains detailed information on what has to be done and how it is to be done; the information to produce this will be found in the full BOM. The individual documents contained in the works order provide instructions, allow progress records to be logged and inspection reports to be recorded and will allow cost reports to be compiled. In some cases, for example, car manufacture, inspection records and other information can be held electronically in a microchip 'tag' that accompanies the car through the manufacturing process. The tag can be 'written' to and 'read' using a radio transmitter and receiver.

Progressing work through a factory is often necessary. Ideally, especially with computerised production control systems, 'progress chasing' should be a thing of the past. Factories producing large numbers of complex products composed of many machined components, each of which has to go through a series of many different operations in many different departments, have always employed people as 'progress chasers'. This job is not completely obsolete since no system ever operates perfectly; this individual will make sure that the appropriate jobs are being worked on and not simply those that will allow the production department to produce good productivity statistics. This 'chaser', sometimes called an expediter, will also be sensitive to the implications of machine breakdowns, scrapped work and the late arrival of raw materials; he or she will be sufficiently knowledgeable on sales requirements and production schedules to make appropriate decisions as to what alternative actions should be taken. The progress chaser can therefore carry out on-the-spot investigations and should have the ability to anticipate rather than 'fire fight' production control problems. Continuously up-dated progress reports can be provided by the use of coloured Gantt-type wall charts, textual printouts from the computer system or graphic displays on a computer terminal screen.

Assuming a large factory with a wide variety of products and associated components, there will be a number of conflicting criteria that could be used to make priority decisions. One common criterion needs to be chosen to enable a consistent workable scheduling system to be established. For example, it has been mentioned that WIP inventory costs money; a valid criterion on which to make priority decisions might therefore appear to be the minimisation of WIP. Alternatively, since machines cost money and indeed continue to cost money whether working productively or standing idle, then maximum machine utilisation could be a possible criterion to use. This ensures that the equipment is seen to be 'earning its keep'. Taking this idea further, maximum labour utilisation would also seem to be a criterion. However, it is making what the customer wants that keeps a company alive and competitive: that is to say the basis on which the work schedules are created should be simply on-time delivery of the finished product to the customer. This implies that if a machine cannot be used to produce a component for which there is a known sale then there is no point in utilising it on unnecessary work, no matter how much the machine initially cost.

As well as the scheduling activity and progress monitoring, the production control system, as part of the factory MIS, also allows collection of other useful data. For example, quality records are obtained, logged and analysed; labour reporting data such as time spent on particular jobs or time spent on reworking bad parts can be gathered and the maintenance of stock records on raw materials, WIP and finished components, sub-assemblies and products can be achieved. Having considered the elements that comprise the production control function and the problems associated with it, there now follows a brief discussion on some current concepts and techniques associated with the subject.

23.3   Material Requirements Planning

This technique, in common with other contemporary methods, fully utilises the power of computers to manipulate large amounts of data that would be impractical to handle manually. It was originally developed in the USA during the 1960s to overcome the problems associated with the traditional methods of inventory and production control then in use. It has been much refined and developed since then and is used in a variety of forms all over the world. We will first consider the basic material requirements planning (MRP) technique, then see how the capacity of the facility can be taken into account by incorporating Capacity Requirements Planning or CRP. The concept is then further developed by examining Manufacturing Resource Planning or MRP II followed by some comments on Enterprise Resource Planning (ERP).

MRP is a technique for handling the planning and control of the inventories required to satisfy the dependent demand. In this case the demand for materials and components is dependent on the number of products to be produced; this number is derived from the MPS. Thus the input to an MRP system will be the MPS based on customer demands, BOMs for the product and a record of existing inventory (see Figure 23.2). The output from the system will be a report that provides: purchasing with a schedule for buying raw material and components, materials control with a schedule for controlling the inventories and manufacturing with a schedule for actually making the parts.

As a simple illustration of the benefits of fine control of inventory, look at Figure 23.3. In Figure 23.3a we see a chart showing the finished product inventory level for a product that is sold at a constant rate. The product is made in batches, production of a batch begins at point X and continues to point Y. Due to sales the inventory level decreases at a constant rate represented by the gradient between point Y and point Z. At Z the production of a new batch is started. Figure 23.3b shows the inventory levels for the purchased materials and parts used to manufacture the product, obtained when using a simple reordering system. In Figure 23.3b it can be seen that the stock of raw material and bought in parts steadily decreases from point X when production begins, to point Y when production ends. It then remains low until a new order of stock arrives at the

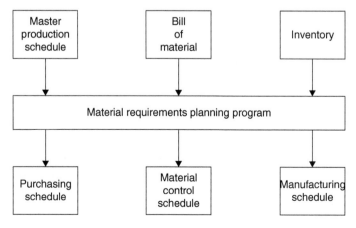

Figure 23.2 Basic MRP system.

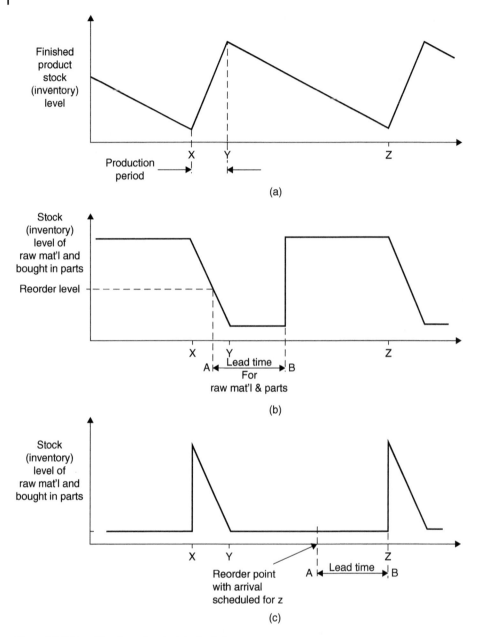

Figure 23.3 Benefits of good material scheduling.

factory for the next manufacturing cycle. The order for the raw material and parts is placed at point A in Figure 23.3b and they are actually delivered to the factory at point B. The time between A and B is called the 'lead time' and although this is undesirable there will always be some elapsed time between ordering and receiving goods.

Figure 23.3b shows that between times B and Z the inventory of raw materials and purchased components within the factory remains constantly high until manufacturing begins again. This inventory and the storage space it occupies costs the company money; it therefore follows that costs could be reduced if the time between inventory delivery and inventory use could be reduced. This approaches a minimum when the lead time for the inventory elapses just as production begins; this is shown in Figure 23.3c. Of course the absolute minimum inventory cost will occur if each piece of raw material or purchased part arrives in the factory just as it is actually needed. Theoretically, this would produce a zero storage cost and this is what is attempted with 'just in time' (JIT) production, discussed later. Returning to Figure 23.3c, the achievement of the pattern shown may be relatively easy if only one or two different pieces of material or parts are involved. But in a large factory with a large variety of products starting and finishing their production runs at different points in time, and each comprised of hundreds or thousands of discrete components, the achievement of anything approaching Figure 23.3c would be impossible without precise information and the computerised handling of this information, as is found in MRP systems.

The basic MRP-type system considers only the control of inventory and scheduling of work. However, to ensure that the decisions made are valid there must be some account taken of the capacity of the manufacturing facility to ensure that the schedules can actually be achieved. This check can be carried out at two levels. Initially a preliminary MPS is made, followed by a rough estimate of whether the capacity available is adequate to achieve the schedule. This is called *rough cut capacity planning* and it basically uses as its input the resources available and equates this with the resources required; existing inventory is not included at this stage. If the capacity is not adequate and alternatives such as contracting out work are not possible, then the preliminary MPS is adjusted accordingly. Using the attainable MPS the MRP programme can be used to obtain a schedule. This schedule is then used as an input to the detailed capacity analysis stage; this is CRP. This provides much more detail than the rough cut capacity planning stage. Inputs here are individual machine and work centre capabilities, labour hours available, existing production commitments, existing inventories and standard times. This information may be accessed from the previously mentioned data base or, if a comprehensive one does not exist, from records and reports from all the relevant departments. The steps in the overall process are shown in simplified form in Figure 23.4; this is sometimes called Closed Loop MRP since the outputs from the capacity planning stages are looped back and used as feedback to the MPSs.

The three functions of MRP, the control of inventory levels, the specification of priorities and the determination of detailed capacity requirements, are all immensely useful and assist the company in maximising its efficiency. However, as mentioned earlier, many large companies have a variety of computer systems in use, for example, for financial, labour reporting, quality reporting and design. It has also been shown that an MRP system has the ability to make use of the data produced by these systems. Thus, if we take the closed loop MRP system and expand it to incorporate the information from all other information generators and users within the factory, we obtain a very powerful tool for actually running the whole factory operation; this tool is MRP II.

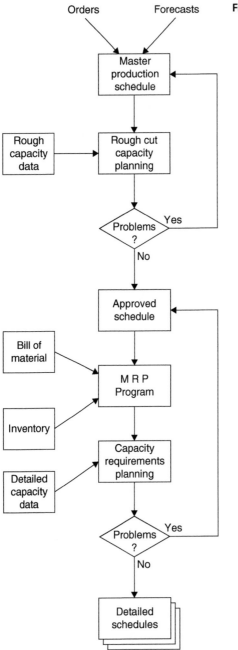

Figure 23.4 Closed loop MRP.

23.4 Manufacturing Resource Planning

Oliver Wight, one of the designers of MRP, coined the term 'Manufacturing Resource Planning' or MRP II. The term imitated MRP since MRP II is really based on MRP with CRP plus input from other computer systems within the organisation. MRP II broadens the scope of MRP to allow financial and production planning to be carried out at a strategic level by being able to use simulation to answer 'what if' type questions. Various possible scenarios can be created to analyse alternative business strategies, the effect of changes to the MPS and consideration of the implications of machine breakdowns and material shortages.

MRP II when fully implemented constitutes an MIS, as described earlier. It allows efficient management of such diverse aspects as finance and personnel, consumables and tooling used on the shop floor and the effects of design changes. It requires a reliable data base from which to draw its data. The old computer adage of 'garbage in – garbage out' is obviously applicable here. The integrity of all the data used by the system must be extremely high since even a small error might have large repercussions – just like the 'butterfly wing' effect in chaos theory where the aerodynamic effects of the flap of a butterfly wing in, say, Australia, can lead to the creation of a hurricane over Florida. This implies a regular and ongoing sampling audit for verification of inventory accuracy to ensure that up-to-date information is being used.

The system should be flexible and easily modified and expanded to cope with changing business conditions at strategic and tactical levels. Modularity of the system is common to allow various modules to be introduced; this allows the system to be built up progressively. Examples of modules might be MRP, CRP, payroll, sales analysis, accounts payable and receivable, financial analysis and forecasting. The system should be able to operate satisfactorily as each module is added; this allows the company to introduce full MRP II in a gradual step-by-step manner. This enables the users of the system to be trained and it also spreads the system cost. Full systems are very expensive since as well as the hardware and software costs there is the cost of tailoring the software to suit the individual company; however, the expense is usually justified on the basis of estimates of reduced inventory costs and increased profit. The time to install a full MRP II system, may be as long as 1–2 years depending on the size of the company and the complexity of the operation.

Finally, adequate training of all personnel is important since once the formal system is installed no informal systems or 'shortage lists' should be necessary; use of these would only corrupt the integrity of the formal system. To facilitate this the system must be 'user friendly' and interactive, that is, it should be able to be interrogated easily and provide up-to-date information in a clearly presented manner.

23.5   Enterprise Resource Planning

Just as MRP II grew out of MRP so ERP has grown out of MRP II. It includes all the functions of MRP and MRP II and broadens the scope of the MIS to include real-time updating of all databases, ensuring they are integrated and compatible. It goes beyond the manufacturing aspects of an organisation to include all aspects of the business. As companies of all types have their business functions able to make more use of the internet and cloud facilities for integrating e-business elements so the whole concept expands to include such areas as customer relations, project management, supply chain management and strategic decision making.

As organisations make use of real-time data harvested from a wide variety of sources such as the internet, customer feedback, information of political and economic significance, material cost trends, style trends and anticipated market changes, so the planning and control will become more responsive and efficient. It can be expected that terminology such as MRP. MRP II, ERP and ERP II will continue to evolve and change to suit.

23.6 Recognising Constraints

Within a manufacturing facility it becomes rapidly obvious that not all elements of the system can operate at the same rate. For example, if this is not properly considered on a production line then machine and labour utilisation can be poor and WIP or inventory can increase to the detriment of cash flow (CF). As a means of addressing this problem in its widest sense in the 1980s Eliyahu Goldratt developed a Theory of Constraints (TOCs) based on earlier management science research. Although used for production scheduling the broad philosophy of TOC is valid for other aspects of manufacturing such as costing, investing and measuring performance. A major difference between TOC and MRP is that it states that since it is the constraints within the manufacturing system that determine what can be produced and when, then the MPS should not be based on the schedule for the end products, but rather the schedule for the constraining resources. It assumes that valid production schedules are essential for any manufacturing system to be successful. It further defines 'success' as how well a company achieves its primary goal, that is, to make money. All other goals are really sub-goals that help the company towards success. The financial criteria-chosen to quantify the goal are, Net Profit (NP), which is an absolute measure how much money is made, Return on Investment (ROI), which is a relative measure of how much money has been made in relation to the money invested and CF that is really the life blood of the company. These three measures need to be maximised simultaneously.

As noted in Chapter 5, when discussing Management by Objectives, goals and objectives have to be translated into concepts that can be easily recognised by those responsible for achieving them. Thus in manufacturing management the efficiency with which the production scheduling system contributes towards achieving NP, return on investment (ROI) and CF can be measured by looking at the following. *Throughput,* the rate at which money is generated by the system through sales; it should be noted that a finished product cannot increase throughput until the day it is sold and paid for. *Inventory,* this is all the money the system has invested in purchasing things that it intends to sell; thus inventory value is the purchase price of materials only. *Operating expenses,* all the money that the system spends in order to turn 'inventory' into 'throughput'; this means that the cost of carrying inventory is part of the operating expense. Thus the goal of the manufacturing manager is now to simultaneously increase throughput and decrease inventory and operating expense. By doing this NP, ROI and CF will all be simultaneously increased and money is made.

Crucial to the TOC concept therefore is the identification of the critical constraints of marketing, capacity, batch size and time. Again, these should all be considered simultaneously. The marketing constraint relates to what should be produced and when, capacity relates to the capacity of the production process or machine, batch size relates to the number of components to be processed at each operation and time relates to the time consumed in carrying out the work and when it can be done. TOC's principal concern is with the identification of production resources whose capacity is less than or equal to market demands; these are *bottlenecks*. For example, assume a batch of material to be processed has to pass sequentially through three machines A, B and C. Further assume that these machines have the capacity to produce 200, 100 and 125 parts per hour, respectively. If the production schedule demands 100 parts per hour then operation B is the bottleneck. Machine B is also the only machine that can be 100% utilised unless work for other jobs that do not require machine B can be scheduled through A and C. It is these bottleneck operations that TOC says must be identified and used as the basis for creating the production schedule.

This is similar to the well known 'line balancing' problem often found on assembly lines. Assume there are four operations on an assembly line a, b, c and d. Each of the operations takes 2, 2, 4 and 2 minutes, respectively. The total time for assembling the product is 10 minutes. As this is being done by four consecutive operations it would be ideal if one finished assembly was being produced every 2.5 minutes. However, this is not the case as operation 'c' takes 4 minutes, therefore the maximum output will be one assembly every 4 minutes. Also operation 'd' will be idle for 2 minutes while waiting for the output of operation 'c'. Additionally, an even bigger problem would occur if operations 'a' and 'b' continued to work at full capacity as work would simply build up in front of operation 'c' costing the company money in terms of unprocessed inventory and storage space. The solution to this line balancing problem is to split operation 'c' into two operations of 2 minutes each by adding another machine or assembly operator. The cost benefits resulting from an increase in efficiency and reduction of WIP will far outweigh the cost of the extra operation.

23.7 Just in Time Manufacture

'Just in Time' manufacture is a phrase used to describe the concept of having zero inventory, that is, having material arrive in the factory JIT to make the product. Basically the JIT philosophy is to achieve zero inventory, lead times, set up times, and breakdowns, and to be able to handle batch sizes of one economically. It is therefore used as a material and production control system and as a productivity improvement system. Best operation is obtained where the product variety is relatively limited and most of the parts are repetitively manufactured, for example, a car assembly plant. It is also beneficial to have suppliers of parts in close physical proximity to the customer plant. The system has as its goal the production of parts or purchase of material in just enough quantities to produce one finished product at any one time. To obtain this JIT style of manufacture a number of enabling concepts and techniques need to be used. The ones noted here concern the factory type, the factory layout type, operation set up time, schedule smoothing and the Kanban production control system.

23.7.1 The Focused Factory Concept

Throughout the first half of the twentieth century the big manufacturing companies had large factories employing many thousands of people in the one plant. These massive manufacturing complexes would usually have many different types of production existing in close proximity, for example, mass, batch and 'jobbing'. Since the work produced by these areas was interrelated and the skills and management techniques required by each were very different from each other, control of the factory was very difficult.

Today the advantages of smaller factories employing fewer people and specialising in one type of activity are recognised. The range of labour skills required is reduced and management can utilise standard methods of control throughout the plant. Also this idea conforms with the modular design philosophy mentioned in Chapter 4, with factories producing common modules that can be used by a range of different customers. These factories are termed 'focused factories' since they focus on one type of product and production system, thus gaining the advantages that accompany specialisation and standardisation. Hard disc drives for computers are a good example of this; one factory may produce hard disc drive modules that are bought by a number of different customers for use in the assembly of their own company's computers. Other advantages are that the span of control and the number of hierarchical levels in the management structure are decreased, production and material control becomes simpler and the productivity of the factory is maximised. A focused factory also generally provides the right conditions for the low variety and repetitive manufacture necessary for JIT.

23.7.2   Line and Cell Layout

Within the factory a smooth flow of material is absolutely necessary. Thus, of the layouts described in Chapter 22, either a line layout for mass production or a cellular layout formed from using group technology for batch production should be used where practical. The functional type of layout is not efficient as it produces too much WIP, material handling routes that are too long and difficulties with control and so on.

23.7.3   Set Up Times

It has already been mentioned that JIT manufacture should be able to be used economically for batches of one. This would imply that as this batch quantity is even approached there will be frequent changes in the set up of machine tools. These changeover times from one type of product to another must be minimised by approaches such as the use of reprogrammable equipment, quick change tooling, analysis of the operation by method study techniques, storing all required tooling close to the machine or work area and use of conveyors to move tooling as well as workpiece material.

23.7.4   Smoothing of the Master Production Schedule

Although the MPS may be constructed for a forward period of up to three months, an attempt should be made to smooth this out so that an equal quantity of parts is required each day. This minimises expediting and reduces delays to produce a smooth flow of parts through the factory.

23.7.5   Kanban

The proper coordination of production operations is the foundation upon which the whole JIT concept rests. In conventional systems it is very difficult to schedule the production of work for a factory manufacturing a product that consists of thousands of discrete components. If the production consists of a sequential series of activities then it is very easy for parts to be made at one operation yet not be able to move on to the next in series because of other part shortages, machine breakdowns or temporary lack of capacity, thus creating a rapid build up of costly WIP inventory. This scheduling and production uses 'inventory pushing'. In JIT the opposite concept is adopted, that is, parts are not made at one operation until they are requested by the following operation. This 'inventory pulling' method reduces WIP and when applied to bought in components and material removes the need for large in plant stores. The technique is also suited to the repetitive manufacture of a limited variety of products and to a smooth unchanging production schedule.

The method developed in Japan to achieve this inventory pulling is called *Kanban*, the word itself simply meaning 'card' since the system utilises cards to control the flow of work. It operates as follows. First, a daily production rate is established and frozen for a period of, say, one month. Although the contributing departments and vendors are informed of this production rate they are not given a detailed production schedule; this is given only to the final assembly department. Components are held in containers in the raw material and finished goods stores of each manufacturing unit. Within each container is a card (Kanban).

The method of operation is shown in simplified form in Figure 23.5. The assembly department initiates the work flow of the system by requesting parts from a supplier work cell or outside vendor; in Figure 23.5 we consider Work Cell No. 2 as the supplier. These parts are held in a container in the supplier's finished goods store and are transferred to the assembly's raw material store. The number of components in the container should be as small as practically possible, say constituting an hour or less of production time. At this point in time the assembly raw material store will be full and the finished goods store for Cell No. 2 will be empty.

At the point when the container holding the parts is transferred the Kanban card from it is returned to Cell No. 2; this 'production' Kanban constitutes the order for the cell to start producing a new batch of parts for its finished goods store. This means that it must withdraw material from its own raw material store. When it does this it removes a card called a 'conveyance' Kanban from the container. It sends this back to the finished goods store of Work Cell No. 1 and this acts as authorisation for it to replenish Work Cell No. 2's raw material store.

When the material is removed from Work Cell No. 1's finished goods store the production Kanban is removed from the container and sent to Work Cell No. 1 to allow it to start production on a further set of parts. The process continues in this way, with the production and conveyance Kanbans reciprocating back and forth between their respective stores and work centres. No work may be carried out by any work centre unless it receives a production Kanban. If it does not have one the workers in that centre can be employed cleaning machines, sweeping floors or engaging in Quality Circles – anything but producing unnecessary WIP inventory.

In concluding this chapter, it may be noticed that considerable space has been devoted to material and production control. This is because of the importance of the activity within the manufacturing system. In fact, in modern factories, the percentage of total manufacturing costs attributable to materials and inventories is increasing and the relative labour cost decreasing. This is due to the greater use of automation and better, more efficient, organisation of labour. The problem of maximising the efficiency of labour is

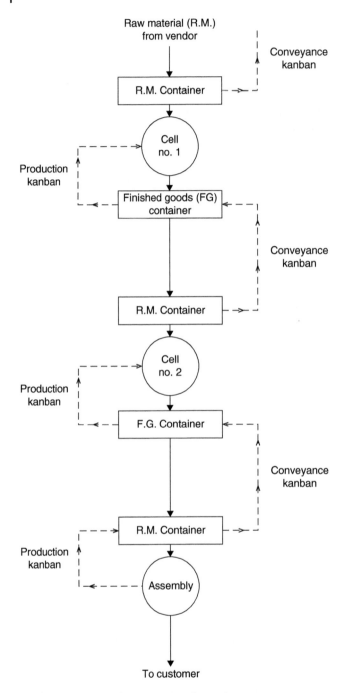

Figure 23.5 JIT manufacturing using the Kanban system.

included in Chapter 24 on Work Study and the use of automation has already been considered in Chapters 18–21.

Review Questions

1 Why is the integrity of the data base so important within a computerised management information system (MIS)?

2 What type of questions regarding the ongoing manufacture of products should a production control department be able to answer?

3 'Manufacturing schedules may become disrupted for various reasons': list some of the most common disruptive events.

4 What do you understand by the terms, Master Production Schedule (MPS) and Bill of Materials (BOM)?

5 The Inventory Turnover Ratio is one indicator as to the health of a company; how is this ratio obtained and why is it useful?

6 What is the purpose of a 'works order'?

7 Progress chasers are still necessary in many factories; what is their function?

8 What is the prime criterion a company should use when creating a work schedule? What other criteria might additionally be used?

9 What are the basic inputs to and outputs from an MRP system?

10 What financial benefits might accrue from careful consideration of raw material reorder points?

11 Closed loop MRP systems are necessary to produce practical schedules; why is this so, and how do the closed loop systems operate?

12 How does MRP II differ from closed loop MRP?

13 What is the essential difference in the basis used for constructing the work schedule between MRP and TOC?

14 When using TOC, what are the sub-goals used by manufacturing to ensure they are achieving simultaneous maximisation of net profit, return on investment (ROI) and cash flow?

15 What is the essential philosophy of JIT?

16 How does JIT differ from MRP in the way the flow of inventory is controlled?

17 Under what type of production does JIT manufacture work best?

18 Explain the 'focused factory' concept.

19 Why are set up times of particular importance in a JIT type environment?

20 Explain the operation of the Kanban system.

21 Why is the importance of tight material and production control increasing?

24

Work Study

24.1 Introduction

Although modern manufacturing systems are not as labour intensive as they were some years ago, the efficient use of the human resources of energy, skill and intelligence remains very important. This applies not only to manufacturing but to all other industries where labour is employed. The techniques used to ensure efficiency and measure human work are the subject of this chapter. They are generally considered under the term 'Work Study'; this is a rational discipline in that the techniques ensure a systematic investigation of any situation examined. The two main components of Work Study are Method Study and Work Measurement.

Work Study is generally regarded as the basic tool necessary for increasing human productivity. The concept has probably been around informally since work began, though the scientific methods really saw the greatest period of improvement around the beginning of the twentieth century. These major developments are widely conceded to be mostly attributable to four people, Frederick Winslow Taylor, Frank and Lillian Gilbreth and Charles Bedeaux.

F.W. Taylor was concerned principally with the time factor in work. He realised that the overall times for jobs were of little value as standards of performance and that times for 'elements' of jobs were more appropriate if methods were to be examined. Frank and Lillian Gilbreth applied themselves to the methods by which jobs were done. Frank Gilbreth was responsible for defining the 17 fundamental movements by which all manual work could be described. Together, the Gilbreths developed the 'principles of motion economy' by which optimum work methods could be developed. Charles Bedeaux was responsible for introducing the concept of rating. This is used to determine how actual observed times differ from the times that should be required. Bedeaux attempted to construct an objective system of Time Study by which work methods could be compared and on which incentive schemes could be based. In Bedeaux's system a common unit was used to describe work on any particular job. This time unit also included rest and relaxation allowances.

Work Study or Motion and Time Study (MTS) remains primarily concerned with discovering the best ways of doing jobs and with establishing standards based upon such methods. It also considers the 'human element' and that individuals differ in performance potential. Factors such as sex, age, health, physical size, strength, aptitude,

Essential Manufacturing, First Edition. Gordon Mair.
© 2019 John Wiley & Sons Ltd. Published 2019 by John Wiley & Sons Ltd.

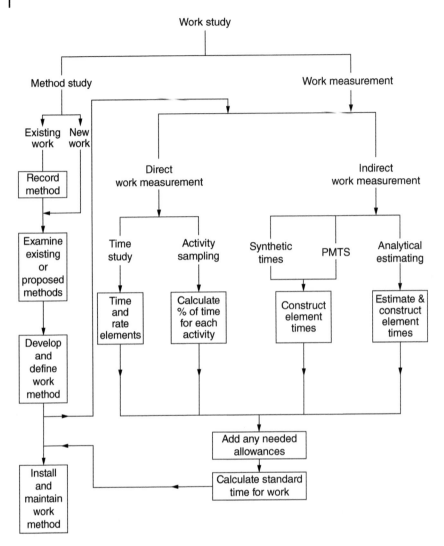

Figure 24.1 The fabric of Work Study.

training attitudes, response to motivation and other psychological factors have a direct bearing on output. Workers also dislike being treated as machines, therefore careful consideration must always be given by the Work Study practitioner to worker contact and the interrelationship of both to management.

The relationship between Method Study and Work Measurement, and their essential components, is shown in Figure 24.1. Although apparently shown in parallel, it is important to note that Work Measurement should not be carried out unless a proper Method Study has already been undertaken. If this is not the case, then it is quite possible that erroneous times will be applied to jobs, thus making all subsequent costing and estimating exercises worthless.

24.2   Method Study

Method study involves recording and analysing all factors relating to the manner in which a piece of work is carried out. The work may either be existing or proposed. The aim of the study is to make the job easier, maximise efficiency and minimise costs.

Method Study therefore provides a thorough, step by step approach to achieve the goal of the 'best' way of carrying out a work task. The procedure to be followed is widely recognised as that now shown and, although these steps apply to a job already in existence, the general concept also applies when initiating a completely new job.

- *Select* the work to be studied.
- *Record* the existing work method and all relevant data.
- *Examine* the record obtained in the previous step.
- *Develop* the most efficient or optimum method of work.
- *Install* this method and adopt as standard practice.
- *Maintain* this practice.

These six steps are now considered in more detail.

24.2.1   Select

To ensure maximum benefits from the effort put into the Method Study exercise, it is important to select appropriate situations. For example, there would be little benefit in conducting an extensive Work Study investigation in a situation where there was very little manual work and production was dependent mostly on the design of machines. Similarly, if the manual work was temporary or for any reason expected to be short lived, it may be impractical to carry out a full Work Study. However, situations where the duration of the work is short but repetitive, for example, maintenance work, will be suitable for Work Study. The economic results of the study, whether they are increases in output, reductions in scrap, improved safety, reductions in training time or better use of equipment or labour, should always outweigh the cost of the investigation. This maximum cost benefit should not only be sought after by looking at direct costs, but should also be looked for in the indirect cost areas. The human factor should always be considered; thus not only areas where there is apparent wasted time should be studied, but also areas where there is high labour turnover and absenteeism.

24.2.2   Record

There are many recording techniques available for the modern Work Study practitioner. Factors that influence the choice of technique are the rapidity with which the record has to be taken and the amount of detail desired. When first investigating a situation, one of the broad rapid techniques might be used, but as the necessity for examining the detail of the job becomes more important a more time consuming technique may be necessary. Tools such as video cameras are ideal for making permanent records of work for future analysis. Integral digital time displays assist the practitioner to determine durations of

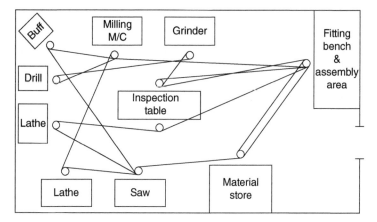

Figure 24.2 Templates, pins and string used to plot the movements of a toolmaker in a small machine shop.

work elements. High speed and time lapse photography can also be used. To change the visual record of a job into a form suitable for further analysis, techniques using charts and tables are used. Examples of some industrial problems and their relevant recording techniques are now briefly described.

- The techniques associated with plant layout, such as flow and string diagrams, templates, three-dimensional models and travel charts. Some of these are illustrated in Figure 24.2. They are suitable where large scale movements and areas are being examined without requiring detail to be analysed.
- When the activities of one or more men and/or machines are being examined, then multiple activity charts, activity analysis, or operator and machine charts can be used (see Figure 24.3).
- Process charts. There are many variations of these, for example, the outline process chart, the flow process chart and the two-handed or operator process chart. These are the most common of the Method Study recording techniques, the sequence of events being represented by a series of symbols that are basically the same for each type of chart. Figure 24.4 shows a clevis assembly chart that uses the clevis rod from Figure 22.7.
- Where long or irregular work cycle times occur, and where studies of groups of workers are required, then memo-motion photography or work sampling may be employed. In this technique, movements are recorded using a video camera set to take pictures at longer than normal intervals. It is therefore a method of sampling, as activities are observed at regular intervals rather than continuously. When the recording is played back at normal viewing speed, patterns of movement unnoticed in real time often become apparent. Passenger movements in airports and rail stations, the movement of ships in rivers and the flow of customers in large shops or post offices are typical applications of this technique.
- Cyclegraphs are recordings of the paths followed by moving objects obtained by attaching light sources to the objects and photographing the scene using a time exposure camera. The effect is observed by taking a time exposure at night of a city

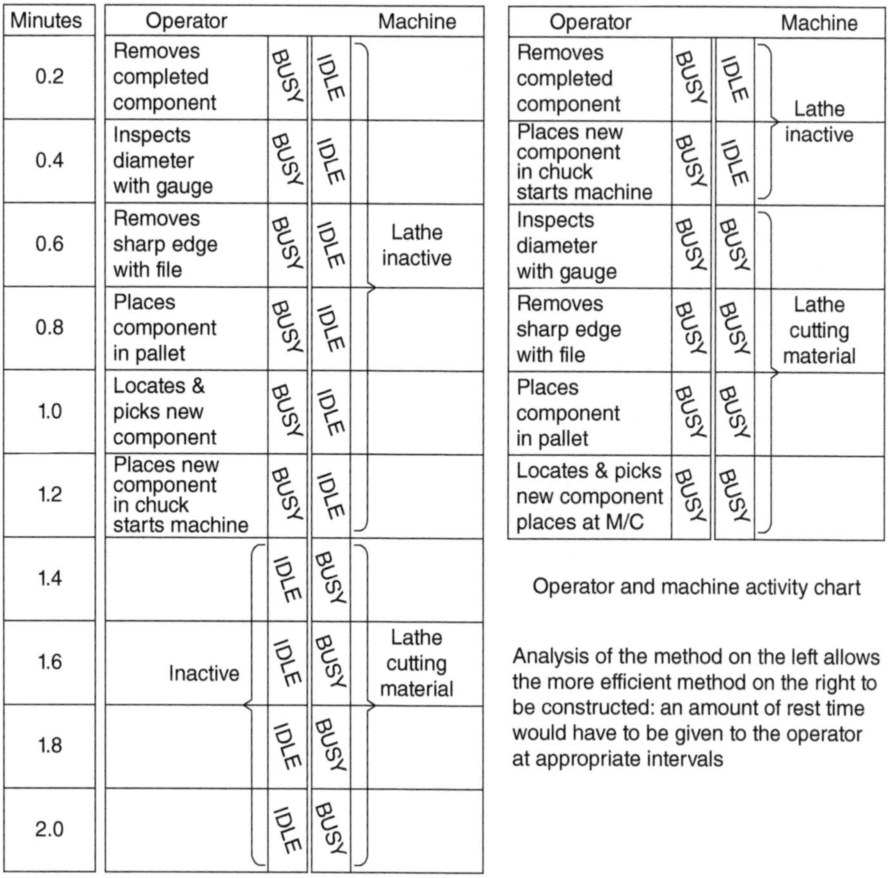

Minutes	Operator			Machine
0.2	Removes completed component	BUSY	IDLE	
0.4	Inspects diameter with gauge	BUSY	IDLE	
0.6	Removes sharp edge with file	BUSY	IDLE	Lathe inactive
0.8	Places component in pallet	BUSY	IDLE	
1.0	Locates & picks new component	BUSY	IDLE	
1.2	Places new component in chuck starts machine	BUSY	IDLE	
1.4		IDLE	BUSY	
1.6	Inactive	IDLE	BUSY	Lathe cutting material
1.8		IDLE	BUSY	
2.0		IDLE	BUSY	

Operator			Machine
Removes completed component	BUSY	IDLE	
Places new component in chuck starts machine	BUSY	IDLE	Lathe inactive
Inspects diameter with gauge	BUSY	BUSY	
Removes sharp edge with file	BUSY	BUSY	Lathe cutting material
Places component in pallet	BUSY	BUSY	
Locates & picks new component places at M/C	BUSY	BUSY	

Operator and machine activity chart

Analysis of the method on the left allows the more efficient method on the right to be constructed: an amount of rest time would have to be given to the operator at appropriate intervals

Figure 24.3 Operator and machine activity chart.

traffic junction. The paths taken by the vehicles using the junction are observed as continuous paths of light on the final print. This effect is utilised by attaching lights to operators' wrists and observing the paths travelled during a particular piece of work, for example, folding a shirt in the garment trade. By rapidly pulsing the lights on and off for specific periods the resulting traces will show up as broken lines. These are known as chronocyclegraphs and by counting the pulses the length of time for each movement can be found.

- When the work movements to be recorded are so complex or fast that the other techniques are inadequate, then SIMO (Simultaneous MOtion cycle) charts are often used (see Figure 24.5). To facilitate the detailed study required, the fundamental movements which together constitute all types of manual work are classified and identified by the use of 'Therbligs' (the name 'Gilbreth' reversed, see Figure 24.6). The use of these 'Therbligs' in conjunction with SIMO Charts is referred to as 'Micromotion Study'. By using high speed video techniques, complex and intricate finger and hand movements can be filmed and observed in slow motion or frame by frame, and from these observations the SIMO chart can be drawn.

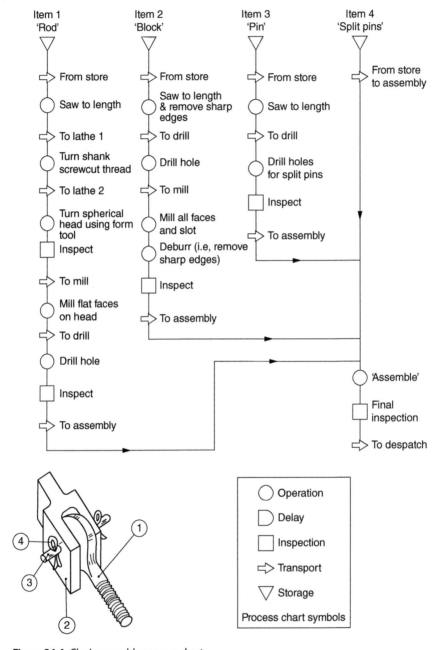

Figure 24.4 Clevis assembly process chart.

Left hand	Elapsed time shown on video display (secs)	Right hand
Operation: Clevis assembly		Operator:
Date:		Study taken by:
Reach for block		Reach for rod
Grasp block		Grasp rod
Place block in assembly fixture	5	Idle
Release block		Place rod in slot
Reach for pin		
Grasp pin	10	Hold rod in position in slot
Insert pin through holes in block & rod (push fit)		
Release pin	15	Release rod
Reach for split pin		Reach for split pin
Grasp split pin		Grasp split pin
Insert split pin	20	Insert split pin
Release split pin		Release split pin
		Reach for tool
Orientate pin & support assembly while inserting & opening split pins	25	Grasp tool
		Use tool to open both split pins
Place assembly in transport box	30	Place tool in tool holder
Release assembly		Release tool

Figure 24.5 SIMO (simultaneous motion chart) for the assembly operation in Figure 24.4.

24.2.3 Examine

In this stage, the information recorded is constructively analysed by systematically questioning each activity observed. Assuming that here we are concerned with a manufacturing task, then the recorded activities will fall naturally into two main categories: (i) those in which the material or workpiece is being worked upon, moved or examined and (ii) those in which it is not touched, as it is either in storage or at a standstill owing to a delay. The activities in category (i) may be further subdivided into three groups: (a) 'Make Ready' activities, (b) 'Do' operations and (c) 'Put Away' activities. Thus while

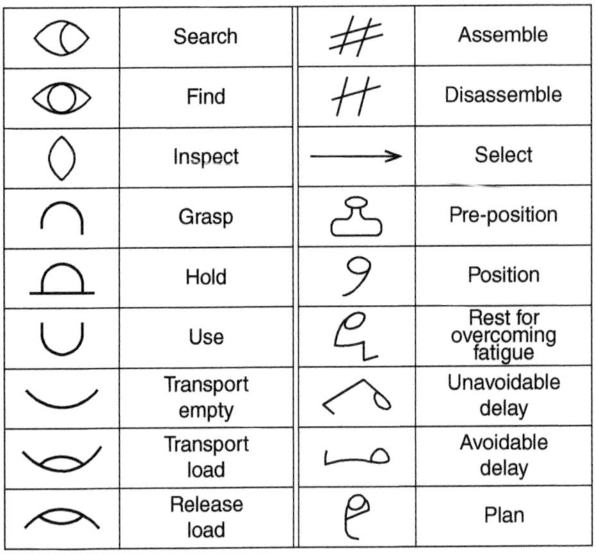

(eye with arrow)	Search	#	Assemble
(eye)	Find	H	Disassemble
( )	Inspect	→	Select
∩	Grasp	(pre-position symbol)	Pre-position
⌂	Hold	9	Position
∪	Use	(rest symbol)	Rest for overcoming fatigue
⌣	Transport empty	(unavoidable delay symbol)	Unavoidable delay
(transport load symbol)	Transport load	(avoidable delay symbol)	Avoidable delay
(release load symbol)	Release load	ℓ	Plan

Figure 24.6 Therbligs.

'Make Ready' and 'Put Away' activities can be represented by 'transport' and 'inspection' symbols, 'Do' operations can be represented only by 'operation' symbols. The object of the exercise must be to achieve as high a proportion of 'Do' operations as possible, since these are the only ones that carry the product forward in its progress from raw material to completed product. These are 'productive' activities and 'add value' to the product; all others, however necessary, may be considered as 'non-productive'.

There is a well-established questioning sequence which examines the purpose, place, sequence, person, and means of the activities. This has the aim of eliminating, combining, rearranging or simplifying them. The questioning technique can be set out as follows.

- *WHAT* is done?
- *WHY* is it done? What else might be done? What should be done?
- *WHERE* is it done? Why is it done there? Where else might it be done? Where should it be done?
- *WHEN* is it done? Why is it done then? When might it be done? When should it be done?
- *WHO* does it? Why does that person do it? Who else might do it? Who should do it?
- *HOW* is it done? Why is it done that way? How else might it be done? How should it be done?

These questions are the basis of a successful Method Study and should be asked, in sequence, every time a study is undertaken.

24.2.4 Develop

Just as in the 'Examine' stage, there is a methodology for implementing the 'Develop' stage. It consists of four steps: eliminate, combine, sequence and simplify. It is applied to each separate activity in the job, that is, each meaningful group of work elements.

The total elimination of unnecessary actions is obviously the first and most important step towards an improved work method. These redundant activities can arise due to changes that have occurred in the product or bad work practices that have gradually been adopted. Once the possibilities of elimination have been exhausted, the combination of actions must be considered, for example, parting off a component in a lathe and facing the following component simultaneously. Next, the opportunities for changing the sequence of actions must be examined, with a view to subsequent further elimination and combination. Finally, if additional improvements are still required then the more costly simplification of the activity may be necessary. This can be done by reducing the number of operations and optimising delays, storage and transportation.

To facilitate simplification, the 'Principles of Motion Economy' can be used. These are rules, originally formulated by the Gilbreths, that help ensure that a job will be carried out with the minimum of effort and the maximum achievement. Essentially, they state that work should be designed so that movements are minimised, simultaneous and symmetrical (e.g. two hands working in unison at an assembly task), rhythmical, habitual and continuous.

Once the steps of elimination, combination, sequencing and simplification have been completed, some additional experimental and practical adjustments will probably have to be made. When the practitioner is finally satisfied with the method the last two steps can be implemented.

24.2.5   Install

Due to the human element the steps of installation and maintenance of the method can often be the most difficult. Active support of the practitioner by management and the workers involved is essential. The practitioner should be able to explain clearly and simply what he or she is trying to do, and he or she must have the ability to win the trust of all those affected by the study.

Installation can be divided into five stages:

1) Gaining acceptance of the change by the departmental supervision.
2) Gaining approval of the change by works and general management.
3) Gaining acceptance of the change by the workers and their representatives.
4) Retraining the workers to operate the new methods.
5) Maintaining close contact with the progress of the job until satisfied that it is running as intended.

24.2.6   Maintain

To ensure the new method is maintained it should be monitored by the Work Study department. This is because human nature is such that a drift away from the method will probably occur if there is no check. Many disputes over time standards arise because the method being followed is not the one for which the time was specified: foreign elements have crept in. If the method is properly maintained, this cannot happen. If it is found that an improvement can be made in the method, then this should be incorporated officially, a new specification drawn up and new time standards set.

Now that the most appropriate method for doing the work has been established, it is necessary to establish the times required to complete the work elements.

24.3 Work Measurement

Work Measurement is the application of one or more of a number of techniques used to determine the time required to carry out a specific piece of work. The principal work measurement techniques are as follows.

First, there are the direct methods of Time Study and Activity Sampling. In these techniques, the Work Study practitioner directly observes the work being measured.

Second, there are the indirect techniques which can be implemented without the physical presence of the work. These are: Synthesis, Predetermined Motion Time Systems (PMTS), Estimating, Analytical Estimating and Comparative Estimating.

The measurement of work is not just desirable but is absolutely essential for both the planning and the control of production. Without work measurement data, it would not be possible to determine what output can be achieved from existing facilities, quote delivery dates and costs, measure the efficiency with which the existing equipment or labour is being used, operate incentive schemes, use standard costs for budget control or determine what extra equipment and labour will be required to achieve a certain output. It may be said that where Method Study is the principal technique for reducing work by eliminating unnecessary movement, work measurement is concerned with investigating, reducing and subsequently eliminating ineffective time. It should be recognised that Method Study and Work Measurement are complementary to each other, one aiding the other to attain the desired organisational goal.

24.3.1 Basic Procedure

As with Method Study so also with Work Measurement a systematic approach is necessary. The steps of this are as follows:

- *Select* the work to be measured.
- *Define* the methods to be used. Break the job down into elements. Critically examine the recorded data and the detailed breakdown. This will ensure that the most effective method is being used and that unproductive elements are separated from those that are productive.
- *Measure* the quantity of work involved in each element, in terms of time, using the appropriate direct or indirect work measurement technique.
- *Obtain* the total work content plus any allowances.
- *Establish* the Standard Time for the operation, which will include the time allowances to cover relaxation, personal needs, contingencies and so on. It must be ensured that the series of activities and methods for which the time has been compiled, is defined. Finally, the time will be issued as standard for the activities and methods specified.

24.3.2 The Standard Unit of Work

The concept of using a unit for measuring work is founded on the notion that the human work content of many different types of job can be expressed quantitatively in terms of a common unit. Some terms used by Work Study practitioners in connection with this concept are given next:

Standard Performance. This is the rate of output at which qualified workers will naturally achieve without overexertion as an average over the working day or shift, provided they adhere to the specified method and provided they are motivated to apply themselves to their work.

Standard Unit of Work. This is composed of both work and relaxation, the proportion of each varying with the nature of the job. In the British scale, 60 of these units would be created in one hour of unrestricted work at standard performance (standard performance is explained in the Section 2.3.3.1 'rating').

Work Content. This comprises the basic time and relaxation allowance plus any other allowance for additional work.

Standard Time. This is the total time that should be required to complete a job at the Standard Performance. This will be composed of the total Work Content plus allowances, plus any other time that may have to be added for delays and other unoccupied time.

Thus, because the unit of work represents a specific amount of work plus allowances, it has no absolute time value. However, since it must be given some relevant dimension, a figure of 60 units of work per hour at Standard Performance may be adopted. These units are called 'standard minutes', so that if one unit of work is carried out at standard performance then it will be completed in one minute of time.

24.3.3 Time Study

Time Study is normally required when there has been a change in the nature of an existing job, or a new job has been started: remember that Time Study should not be attempted until a proper Method Study has been carried out. Time Study is a work measurement technique used for recording the times taken and the rates of working, under specified conditions. It can then be used to establish the Standard Time for the job.

Time scales used when conducting the study may be seconds or decimal fractions of a minute, the latter being the more common and precise. Other time scales, for example, decimal hours, can be used but only in special circumstances.

Traditional Time Study equipment includes a stopwatch, a Time Study board and forms. Video cameras may be used in which the precise timing of the activity is displayed to enable accurate separation of work elements and establishment of their times.

24.3.3.1 Rating
The real object of a work measurement and Time Study is to determine not how long it actually takes to perform a job, but how long it should take. It is therefore necessary to compare the actual rate of working of the operator with a standard rate of working, so that the observed times can be converted into basic times, that is, the time required to carry out an element of work at Standard Performance. This performance rating is the comparison of an actual rate of working against a defined concept of a standard rate of working. This standard rate corresponds to the rate that workers would naturally adopt when working, assuming that they know and follow the specified method, and that they are motivated. On the British Standard performance scale, standard rating is equal to 100. Thus if a worker was rated at 50 he would be working at half the expected standard rate and the time observed would be halved by the practitioner and used as the basis for further calculations. Some practical examples help to create an idea of a standard rate. For example, British Standard 100 is considered to be equivalent to a walking

speed of a constant four miles per hour, or to dealing a pack of cards into four hands in 22.5 seconds. We therefore have a scale with the standard rate at 100 and complete inactivity at 0. The practitioner is trained to be able to recognise the conditions of the standard rate of working and to assess to the nearest five points the degree to which a worker's observed speed and effectiveness varies from the 100 standard. The procedure in which the practitioner carries out this assessment, while recording the observed time, is known as rating. The rating is used to find the basic time for a job or an element as shown next:

Basic Time = (Observed Time) × (Rating/Standard Rating), for example, if an operator is rated as being very fast and is allocated a rate of 125, and the observed time was 0.2 minutes, then the Basic Time = $0.2 \times (125/100) = 0.25$ minutes.

24.3.3.2 Procedure

The procedure for implementing a Time Study is shown in the following.

1) First everyone that will be involved in the study should be notified; they should be told the reasons for the study and how it will be conducted.
2) Collect and record all relevant information. The conditions under which the work being studied is being done must be recorded; this will provide useful information for the future when estimating times for similar jobs. The record will also be useful if disputes arise at a future date concerning the time established for the job; for example, the method of the tooling may have changed from that in the initial study.
3) The work being studied is divided by the practitioner into elements. This facilitates the analysis of the work.
4) The elements are timed using a stopwatch or from the digital clock on a video recording. This will be done for a number of work cycles.
5) The *Basic Time* for the work is now calculated, using the gathered data and with consideration given to the rate of working of the operator.

24.3.3.3 Allowances

After the Basic Time has been calculated there is one further step required before a fair Standard Time can be allocated to the job. This involves the allocation of allowances to compensate for the operator fatigue, delays and interruptions that are unavoidable in every work situation. These allowances are composed of the following elements: (i) A Relaxation Allowance to allow the operator to attend to personal needs and to recover physically and mentally from exertion; this may add 10–30% onto the Basic Time, depending on the nature of the work. (ii) Unoccupied Time, which occurs when the worker is unavoidably prevented from doing productive work; this should obviously be minimised if it cannot be eliminated. (iii) A Contingency Allowance to cater for occasional interruptions, adjustments to equipment and so on.

It is now possible to devise the Standard Time for the work that, as mentioned previously, is the total time in which a job should be completed at standard performance; it is shown schematically in Figure 24.7.

The foregoing sections have shown how it is possible to use Method Study to determine the most efficient way of carrying out a job, and then to use the Work Measurement technique of Time Study to establish the Work Content and Standard Time.

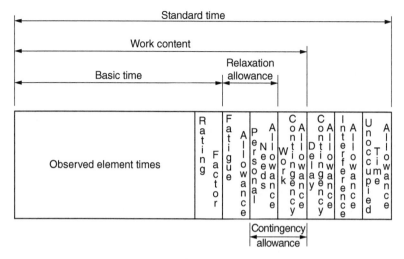

Figure 24.7 The build-up of the 'Standard Time' for a work activity.

There are other techniques used in Work Measurement. Another direct measurement technique is Activity Sampling. Here the work is not measured using a timepiece; it is sampled. Activity Sampling is used in situations where it is desired to measure the work of large amounts of people, say in an office. Over a period of time a large number of individual observations are made. At each observation a record is made of what is happening at that moment and the percentage of the observations recorded for a particular activity is taken as representing the actual time during which that activity occurs.

Indirect Work Study techniques include Estimating and PMTS. Estimating is a technique used for determining the time required to carry out a job by using knowledge and experience of similar work done previously. PMTS is a technique that uses times previously established for basic human motions. This synthetic method allows times for jobs to be built up even before the work has started. There are many varieties, but one of the most popular is Methods Time Measurement, which has been developed at three levels. MTM 1 is the most accurate and detailed; it will also take the longest time to implement. MTM 2 is used where the detail of MTM 1 would economically prevent its use. MTM 3 is intended to be used in work situations where at the expense of some accuracy times are required as soon as possible.

24.4 Work Study As a Service to Management

To keep any system in control, for example, an aircraft in flight, an industrial robot or a factory, feedback on the state of that system is required to the system controller. In an industrial organisation the controller is management. Feedback from the shop floor is therefore necessary in the form of quantitative and objective information that can be used to monitor the effectiveness of the labour force. Work Study is the means whereby this information can be produced. The information can also be used by the management for forward planning, for example, to estimate how many man hours will be required to satisfy future customer orders and to estimate future costs.

24.4.1 Labour Control and Reporting

The information coming from the shop floor to the management must be sufficiently detailed to identify specific areas of poor performance, unproductive work or excessive cost. How this is done is best understood by following the process through progressively.

Throughout the day the shop floor workers will spend their time either directly on work for which there is a Standard Time, on diverted work such as repairs or on waiting for material or equipment. Only the direct work is productive and adds value to the product. The operators will record, by writing on report sheets or keying into a computer, how they have spent their time each day. This record will show what products and operations were worked on, how many were produced and how much time was spent on direct and diverted work. It should be noted at this stage that although terminology varies between countries and even factories, the concepts remain the same. Indirect workers such as labourers will also submit reports detailing how they have occupied their time during the day.

From these reports the data processing department will be able to compare the times the operator spent on specific operations against their previously devised standard times. It will then be possible to provide management with information on the performance of the direct operators, the performance of a department excluding the indirect workers and the overall performance that will include all the workers on the shop floor. A breakdown of how the unproductive time was spent will also be supplied; this will enable problem areas to be pinpointed.

The performance figures will probably be supplied in a similar manner to the following:

OPERATOR PERFORMANCE = (Total Standard Times produced/
Total time spent directly producing work for which there is a
Standard Time) × 100

DEPARTMENTAL PERFORMANCE = (Total Standard Times produced/
Total attendance time minus time spent on indirect work) × 100

OVERALL PERFORMANCE = (Total Standard Times produced/
Total attendance time) × 100

A short example illustrates the use of these indices.
During a five-day week a department reports the following information.

Total Standard Time produced	450 h
Time spent on direct work	500 h
Time spent on indirect work	80 h
Number of operators	20 h
Number of hours available per day	7.5

Calculate the overall performance, the departmental performance, and the operator performance.

Total Attendance Time $= 7.5 \times 5 \times 20 = 750$ hours.

Overall Performance	$(450/750) \times 100$	60%
Departmental Performance	$(450/(750-80)) \times 100$	67%
Operator Performance	$(450/500) \times 100$	90%

A detailed report on performance is therefore provided to the management and supervisors. This would normally be supplied to shop floor supervision, for example, foremen, on a daily basis on the day following that on which the work took place; this will enable prompt corrective action to be taken. The supervisor will receive a full report on each individual operator and a composite report for the performance and time utilisation for his department as a whole. Reports will probably be summarised on a weekly basis for line management and factory management will probably receive weekly or monthly reports as appropriate. Action should be taken immediately on the reports, for example, low operator performance needs prompt investigation, as also do instances of high rework or other diverted time. Communicating productivity data to the shop floor is also important. Charts showing various performances for each department can be erected in prominent locations. Charts of quality, for example, reject rates and costs of scrap, are also useful.

24.4.2 Payment by Results (PBR)

In this type of payment system the workforce is usually paid a basic rate plus an amount proportional to performance. This may be on an individual or a group basis. It is apparent that the type of performance indices shown in Section 24.4.1 provide useful data on which to base such systems.

24.4.3 Planning

Since Work study provides a data base of information on how long specific operations should take, the total time required to build products and the methods and equipment required to produce specific quantities of a product, it can be used to predict future manpower and capacity requirements based on anticipated product sales volumes. The total standard times involved in a product can simply be multiplied by the anticipated weekly production target to find the total standard times required per week. This can then be related to the overall performance the factory or departments are achieving to arrive at a total time required for the anticipated production volumes. Knowing the total man hours available per week, this total time can then be used to determine the required manpower.

Review Questions

1   What is 'Work Study' and what is its purpose?

2 What is the principal aim of Method Study?

3 Explain the six basic steps that must be taken to complete a successful Method Study.

4 What types of work are particularly suitable for Method Study?

5 Describe three techniques used in Method Study for recording how work is done.

6 What type of questions should be asked when examining the record of how work is done?

7 Discuss the four steps necessary for developing a new work method.

8 What personal characteristics should the person responsible for implementing a new work method possess?

9 What is the principal aim of work measurement?

10 Work measurement can be carried out directly or indirectly. Discuss the important differences between these two methods and name two techniques associated with each.

11 Discuss the important uses of the information obtained from work measurement.

12 Describe the basic procedure for carrying out a work measurement exercise.

13 Explain the concept of the 'standard unit of work'.

14 What is the purpose of 'rating' in a Time Study exercise?

15 List the steps to be followed in conducting a Time Study exercise.

16 What is meant by the 'Standard Time' for a job? State what elements contribute to its value.

17 What is PMTS and under what conditions might it be used in preference to Time Study?

18 Discuss how the information obtained from Work Study can be used to analyse the performance of workers on the factory floor.

19 Discuss how Work Study can provide information useful for planning future operations.

25

Manufacturing Economics

25.1 Introduction

25.1.1 Accounting in Manufacturing

The financial aspects of manufacturing cover such a broad range of activities that to explain them in one chapter is impossible. What is therefore attempted here is an outline of some of the financial features involved in running a manufacturing organisation. Most of what is discussed comes under the classification of 'Management Accounting', a term widely used to describe those accounting procedures used as an aid to management.

Within a manufacturing company the finance department utilises accounting information for three basic purposes. The first is *decision making*; here, information on probable costs and increases or decreases in profits associated with alternative courses of action are analysed. Second, information is provided to assist with efficient *control* of the manufacturing system; included here will be the preparation of budgets and the subsequent analysis of costs incurred against those planned. Finally, the information is used for *reporting and recording*; this involves the compilation and updating of financial records and reports, not only for management, but also for shareholders who will be interested in profit and loss accounts and so on, and for the government who will need the information for tax purposes. We will limit our concern here to the decision making and control activities.

25.1.2 The Composition of Manufacturing Costs

Figure 25.1 shows the approximate distribution of costs within a manufacturing organisation, and how they contribute to the selling price of the finished product. The ratios of each cost will vary considerably between companies. For example, a company producing complex products incorporating state of the art technology might have a higher proportion of the selling price composed of research, development and engineering costs. A company trying to sell in a highly competitive consumer goods market might spend a higher proportion on sales, advertising and marketing. Within the manufacturing costs themselves it can be seen why much attention is given to the control of inventory: parts and materials can constitute around 50% of the total. This again will vary depending on the type of production; for example, compare the mass production of steel bolts to the one off construction of a communications satellite. In the former, material costs would

Essential Manufacturing, First Edition. Gordon Mair.

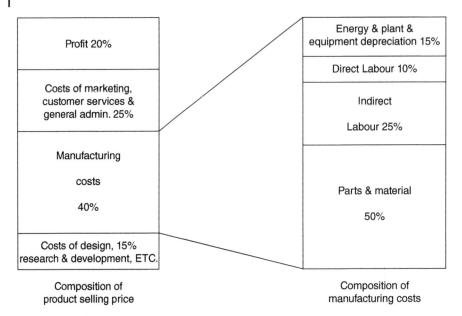

Figure 25.1 Distribution of costs for a manufacturing company.

be high and labour costs low; in the latter, the ratio is reversed. Two types of labour are shown in Figure 25.1. Direct Labour is composed of the cost of those who work directly on the product, for example, machine operators and component assemblers. Indirect labour is composed of the cost of all those whose work indirectly contributes to the production of an individual product; for example, labourers, material handlers and store personnel.

From what has been said it is apparent that *manufacturing costs* are essentially composed of the three main elements with which this book is concerned; that is, *manpower, machines* and *materials*. These costs represent resources that must be used *effectively* and *efficiently* if the company is to survive. The effectiveness of their use is easily determined by observing results, that is, was the product produced at the right quality, at the right time and at the right cost? However, the efficiency with which the resources were used involves some calculation and the resulting figure is termed *productivity*.

25.1.3 Productivity

The efficiency of any system is represented by the ratio of the value of outputs from the system to the value of inputs to the system. For example, the efficiency of a machine is measured by the value of the energy output to the value of the energy input; this can never be greater than unity – otherwise 'perpetual motion' would be possible. In the case of our manufacturing system we do not use units of energy but units of money, in whatever currency is appropriate. Thus the efficiency or productivity of our system is represented by the ratio:

Overall Productivity = Total output in monetary terms/

Total input in monetary terms

This measure of efficiency differs from that of the previously mentioned machine in that the overall productivity must be greater than unity. That is, in the manufacturing system money is being used to create even more money, while in the engine energy is being 'lost' due to friction, heat and so on in the process of producing work. The output from the manufacturing system is the market value of the goods produced; the input is the total cost of the manufacturing operation. The overall productivity ratio is the true indicator of the company's efficiency, but it is difficult to calculate in absolute terms. In practice, it is often split into three different efficiency measures by looking at how the three major manufacturing resources are utilised. Thus, we get Labour Productivity, Capital Productivity and Material Productivity. These ratios are usually used for comparison purposes, for example, a company might compare its labour productivity from one year to the next, or make a comparison with similar companies in other countries to determine its own competitiveness.

Labour productivity is the one most often quoted by the media. This is the ratio of the value of goods produced to the cost of the labour used to produce the goods. The value and costs would be taken over a specific period of time, for example, per hour, day or year. Capital productivity compares the value of goods produced with the cost of the machines, equipment and buildings used to produce the goods. Material productivity compares the value of goods produced with the costs of the materials and energy required to produce the goods. It can be seen, therefore, that if a company invests in a great deal of automated equipment it will require less labour, and therefore its labour productivity will increase. However, due to its investment in capital equipment its capital productivity may have decreased. Also if the company discards old, inefficient equipment and purchases new machines it might find its capital productivity decreased in the short term, but its material productivity increased due to less scrap being produced and more efficient use of energy, for example, electricity, gas and so on. It is for these reasons that although the individual ratios are good for comparison purposes it is their net result, that is, the overall company productivity, that is of prime importance.

Now that we have considered the way in which costs can be used to indicate the health of a manufacturing company, let us take a different perspective and consider how the company might use cost information for the purposes mentioned earlier; that is, decision making and control.

25.2   Costs for Decision Making

25.2.1   Risk

Before embarking on any project involving the outlay of money it should be acknowledged that there will be some amount of risk present. This risk can either be ignored, which is extremely unwise, or analysed in depth mathematically, which is extremely difficult. Here, we are simply acknowledging its presence. The 'riskiness' of a project is determined by the possible variation in the future returns from the investment. For example, it is generally riskier to invest money in helping a new manufacturing company start up than it is to deposit the money in a reputable bank. The investor will therefore look for a much higher potential return on the investment if it is put into the manufacturing company.

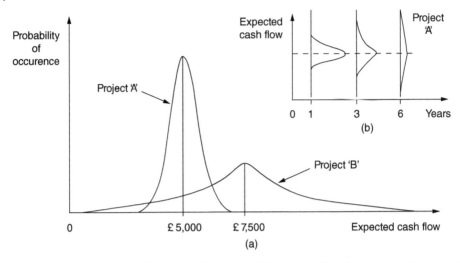

Figure 25.2 'Risk' represented by probability curves. (a) The probability of occurrence of expected cash flow for two projects, A and B. There is a high probability A will achieve the £5000 cash flow expected, whereas project B could achieve a higher cash flow but is far 'riskier'. (b) How 'riskiness' of projects increases with time.

It is worth noting that in times of recession the risk is greatly increased, thus making it more unlikely that the 'capitalist' with money to use will look to industry as a profitable place to invest. This is because most people are 'risk averters' rather than 'risk seekers'. This creates a downward economic spiral difficult to change since, as we noted in Chapter 1, manufacturing industry is the main creator of wealth and therefore a major aid for pulling out of recession. Only a few risk seeking entrepreneurs will go for the industrial investment; the potential rewards are high, but so also is the risk.

A distinction can be made between 'risk' and 'uncertainty'. Uncertainty occurs where there is insufficient evidence to evaluate the riskiness of a situation. Risk, however, occurs when enough information is known about a situation to allow a probability of achieving certain results from the investment to he estimated. Naturally the further into the future one looks the more uncertainty there will be and so the riskiness will also increase. This will be evidenced by a broader range of possible outcomes from the investment. This concept is shown in Figure 25.2 in the form of probability distributions. Figure 25.2 shows that the broader the distribution the greater the standard deviation and hence the greater the risk. A tight distribution signifies low risk.

Risk analysis is not normally carried out by the management accountant, or an engineering manager trying to appraise a potential capital investment, but the concept of the presence of risk and the possibility of including it rationally in decision making should be remembered.

25.2.2 The 'Make or Buy' Decision

A manufactured product is made up of various parts and sub-assemblies; these may either be bought in as completed items or made in the factory. The decision on whether to make or buy the product will be determined by a number of factors, for example, a comparison between the cost of making the item using existing equipment and labour compared with buying it. If the company has the capacity, the proper equipment and appropriately skilled labour, then it will probably be better to make it on site. However, if the plant is already working to capacity a comparison of costs should be made including the costs of possibly renting or purchasing more space, buying more equipment and hiring and training new labour. In this case, it may be found that it is cheaper to purchase, say sub-assemblies for a PC computer, from a local sub-contractor or from a country abroad where there are cheaper labour rates. The supplier should be reliable and fully committed to company quality standards, and more than one source per item is often necessary to avoid suppliers developing monopolies. The economics of the 'focused factory' mentioned earlier in the book will also be important here, that is, control of all aspects of manufacture including costs is easier in smaller factories with a focused activity.

25.2.3 Fixed and Variable Costs

To enable costs to be used for decision making it is helpful to classify them under the terms 'fixed' and 'variable' costs. The conditions under which the classification is made should be understood since changing conditions can alter the nature of the costs; for example, a cost which is 'fixed' in the short term may become 'variable' in the long term. A fixed cost is one that is unlikely to change with production volume: for example, depreciation on a machine, the cost of special equipment, or the cost of shop floor supervision. A variable cost is one that is likely to change with production volume: for example, the cost of materials to make the product and consumables such as welding rods or cutting tools. In the past, direct labour would have been classed as a variable cost since wages would have been related to output; however, with guaranteed basic wages and the increasing use of automation these are really now fixed costs. The only time when direct labour becomes a variable cost is when additional overtime or shift payments have to be made to produce increased output. Although not always the case, variable costs are often taken as being directly proportional to production volume so that they increase linearly as output increases (see Figure 25.3).

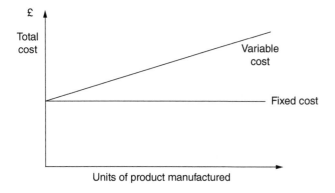

Figure 25.3 Fixed and variable costs.

25.2.4 Direct, Indirect and Overhead Costs

Direct costs are the costs of the company that can be directly attributed to the manufacture of a specific product; they comprise direct material and direct labour costs. Direct material is all the material purchased and used in the finished product, including any scrap incurred. This cost is the cost of material required per unit multiplied by the number of units produced. Direct labour is all the labour used to work directly on the product. This cost is the direct worker's wage rate multiplied by the time spent working on the product. This cost therefore comprises both variable and fixed cost. Indirect costs are those costs that can be only indirectly attributed to the manufacture of a specific product. They are also called overhead costs.

Overhead costs include rent and rates, the cost of heating, lighting, office staff, general labourers, store personnel, computer systems, telephones, fire insurance, management, maintenance, depreciation on machines and other equipment and so on. Although most are fixed costs in that they exist irrespective of the quantity of product manufactured, there are also some that vary depending on volume produced.

Although subject to certain limitations, the simplest and clearest method of illustrating the relationship between cost, volume and profit is the break-even chart. A typical break even chart is shown in Figure 25.4. It demonstrates that for a particular project, with the associated fixed and variable costs shown, the company must obtain sales to allow *X%* of capacity output to be achieved before a profit can be made. Charts like these can be constructed to assist the company in deciding the level of investment it should make in manufacturing facilities in order to satisfy perceived market needs.

The break-even chart can also be used to decide which project or process to adopt for the manufacture of an anticipated annual component volume. Fixed and variable costs for each alternative should be determined and a chart similar to that shown in Figure 25.5 constructed. This will indicate which one to choose for a specific production volume.

25.2.5 Marginal and Differential Costing

This type of costing is suited to decision making, it is also known as 'direct' costing. At any given production volume, the marginal cost is the amount by which the total

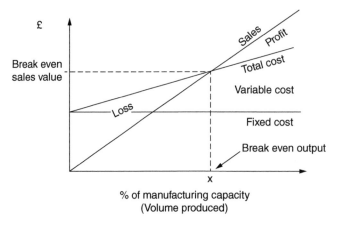

Figure 25.4 Cost-volume-profit break-even chart.

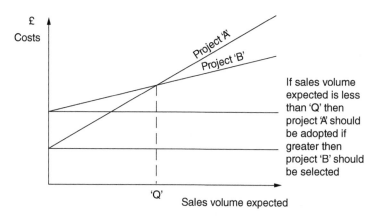

If sales volume expected is less than 'Q' then project 'A' should be adopted if greater then project 'B' should be selected

Figure 25.5 Break-even chart for project selection.

costs are changed if the production volume is increased or decreased by one unit. Fixed costs are therefore excluded from the calculation. It is basically a simple technique and relates costs to production volume rather than time. It is therefore useful for providing information for decisions related to the volume and type of production to be adopted.

The marginal cost of one product, at a specific level of output, is shown in Table 25.1. It is shown as the sum of the variable costs per unit; note that the currency may be any that is appropriate. If the company produces a number of products, then the difference between their marginal costs and their selling prices is called their contribution towards fixed costs and profit. Assuming the company has ample capacity, then the effect of additional orders for each product on the company profit is shown in Table 25.2. This information can be used by the company to decide which products it should concentrate on should resources be scarce, that is, which products provide the highest contribution.

The concept of using differences in costs for decision making is also used in differential costing. Differential costs allow projects, or any alternative courses of action, to be ranked in order of desirability. Differential costing does not concern itself with costs that have already been incurred or costs that are common to all courses of action, but only with future differential costs. For example, supposing a decision has to be made between

Table 25.1 Marginal cost of a project (bold terms are for emphasis).

Cost source	Cost per unit
Direct Material	5
Direct Labour	1
Direct Expense	2
Prime Cost	**8**
Variable Overheads:	
Manufacturing	1
Administration etc.	1
Marginal Cost	**10**

Note: Level of Output = 1000 of Product A

Table 25.2 Effect of increased production on a company's profit.

Project	Selling price per unit	Marginal cost per unit	'Contribution' per unit	Additional units	Additional profit
A	15	10	5	500	2500
B	12	8	4	800	3200
C	16	11	5	400	2000
D	14	12	2	600	1200
					8900

producing components by process A in department B, or process X in department Y. Assuming that the processes and departments are well established, and other factors such as process capability and capacity are equal, then the decision should simply be based on the future differential costs for each process. This would include the anticipated costs of materials, labour, process energy and consumable costs. Any other costs should be ignored for the purposes of decision making.

25.3 Investment Appraisal

25.3.1 The Payback Period

Assume a company has the opportunity to invest in either a single project that will generate a cash inflow or select one project from a number of alternatives. One of the simplest and quickest ways of gauging the project's worth or comparing alternative projects is to use the payback method. In this method of appraisal, the time required to recoup the cost of the initial investment is calculated. This involves estimating the cash inflows resulting from the project for each year of the project's life. These cash inflows could be increases in cash flows or simply estimates of net cost savings as a result of the investment. When the cumulative sum of these inflows equals the original investment, then the payback period is said to have elapsed. The calculation is shown in Table 25.3 for three alternative projects. As can be seen, it does show quickly that the investment can be recouped from either project A or B in three years, but it does not take into account the fact that project B gives returns much earlier than project A: also, project C is potentially very profitable in later years but using the payback method this project would be discarded in favour of projects with a shorter payback period. The method is therefore only a rough guide and really only suitable where high risk makes future anticipated cash flows uncertain.

25.3.2 Decision Criteria

A major problem of the payback method is that it provides no indication or comparison of the profitability of projects. Since the company is likely to be in business to make a profit this must be its prime objective. Now the company's funds that it uses for investment have a cost associated with them, for example, if it borrows money it will have

Table 25.3 Problem with the 'payback' method of investment appraisal.

Year		Project A	Project B	Project C
0	**Investment**	**60 000**	**60 000**	**60 000**
1	Cash flow	10 000	40 000	10 000
2	Cash flow	20 000	15 000	10 000
3	Cash flow	30 000	5 000	20 000
4	Cash flow	0	0	20 000
5	Cash flow	0	0	20 000
6	Cash flow	0	0	20 000
	Payback Period	3 y	3 y	4 y

Note: bold type shows the investment year for comparison.

to pay interest on the loan. If the company was to fund itself by loans that cost 15% and made profits of only 10%, then it would soon be out of business! This means that the company should not invest in projects that produce less than a 15% return on investment. In this case, the company therefore has a marginal investment rate of 15% and any investment should produce a return at least as high as this. Thus an investment appraisal method must be found that takes into account this marginal investment rate, sometimes called the minimum profitability criterion. This will ensure that the company does not enter into investments or projects that are unprofitable.

A full investment appraisal calculation should therefore take into account: (i) the marginal investment rate, (ii) the amount of capital to be tied up in the project, (iii) the net cash flows resulting from the investment and (iv) the timing of these cash flows. It should use this information to produce a net present value (NPV) for the project. If produced for a number of projects, their NPVs may then be compared to determine the one that is the most attractive. Calculating NPVs involves using the Discounted Cash Flow (DCF) technique. *Note that figures given here for interest rates and any government grants and tax incentives are hypothetical as actual figures depend on the economic circumstances prevailing in the country where the investment is taking place. Also note that the currency used in the examples is the pound sterling, however, other currencies such as dollars or euros can be used just as easily.*

25.3.3 Discounted Cash Flow (DCF)

In DCF all inflows and outflows of cash are related to a base year, usually called 'year 0'. Cash flows in years after the base year are transformed into their present value by 'discounting' them. This is done because it is recognised that one unit of money today is worth more than the same unit one year hence. This can be explained by assuming that if we have £10 today; we can either spend it on goods or services worth £10 or invest it in, say, a bank account. If it is invested the money will, in one year's time, be worth £10 plus the interest earned, for example, at 10% it would be worth £11. Conversely, £11 income one year in the future is worth only £10 today. In fact, if offered the choice we would accept the £10 now as there is some risk involved in obtaining the £11 in the future. For the purposes of investment appraisal calculations, the rate at which anticipated future cash flows will be discounted will be the company's marginal rate of return. This rate will be determined by current interest rates and the company's business risk and financial structure. The actual discounting calculation is quite simple since tables of 'discount factors' for different rates are readily available. The present value of the future cash flows is found by multiplying them by the appropriate discount factor from the tables.

The net present value or NPV for the project is found by taking the present value of expected cash inflows and deducting the present value of expected cash outflows. For example, if the present value of expected returns from the project is £15 000, and the net cost of the project is £10 000, then the NPV will be £5000.

As an example, assume that a company is considering the purchase of a machine tool in year 0 at a total capital cost of £140 000. Because of the cost of spares and other costs directly associated with the project the company will require an additional £10 000 working capital that will become available to the company again at the end of the project's life. There are no cash inflows expected to be received from the project until the end of year 1 when they should be £40 000. In each following year the cash inflows should be £70 000, £60 000, £40 000 and £30 000. These cash inflows will probably have been estimated by using the previously mentioned marginal costing techniques. The tax system at the time will allow the machine to be depreciated at 25% of the reducing balance each year; this means that the government recognises that, due to use and increasing obsolescence, the value of the company's machine is decreasing every year. The amount of this depreciation is called the Capital Allowance and the government permits this to be used as a tax allowance, thus reducing the amount of tax to be paid in any year. For the duration of the project we will assume the corporation tax on profits to be 35%. The company is aided in its financing of the project by the availability of £20 000 in grants towards the purchase of the machine tool. The marginal investment rate of the company is 16%. The question the company must now ask itself is: what is the NPV of this project?

Figure 25.6 shows the calculation made to determine this. NPV. First the net cash investment made in year zero is calculated; this is the sum of the costs to the company less any incentives, grants and so on, available. Column *a* of the calculation shows the year in which the cash flows take place; column *b* shows the increase in cash inflows expected by the end of each year; column *c* shows the tax to be paid on the previous year's profits; column *d* shows the capital allowances or depreciation allowed on the machine; column *e* shows the tax saved because of these allowances (it is assumed the company is paying tax from profits on other investments allowing tax to be saved in year 1); column *f* shows the net cash flow at the end of that year; column *g* shows the discount factors found from tables used to discount the future cash flows to a year 0 value and column *h* shows the resulting present value or PV of the cash flows. When the project is finished the machinery and other equipment used will still retain some second-hand value. This contributes to what is called the 'residual value' of the project. This expected residual value is discounted at the appropriate rate and added to the total of the present value column; this provides the grand total of the NPV of the investment. The NPV is now calculated as described earlier. In this example the NPV of the project is £26 816. It should be noted that although this type of calculation seems very precise

Net Cash Investment in year 0:

New machine tool	140 000
Additional working capital required	10 000
Total	150 000
Subtract value of grant	20 000
	130 000

a	b	c	d	e	f	g	h
Year	Cash inflows	Tax (35%)	Capital Allowances	Tax saved	Cash flow	Discount factor	Present value (PV)
1	40 000		35 000	12 250	52 250	0.862	45 040
2	70 000	14 000	26 250	9 188	65 188	0.743	48 434
3	60 000	24 500	19 688	6 890	42 390	0.641	27 136
4	40 000	21 000	14 766	5 168	24 168	0.552	13 340
5	30 000	14 000	11 174	3 876	19 876	0.476	9 460
6		10 500			(10 500)	0.410	(4 306)
						Total	139 104

Residual values:

Machine tool	32 200
Working capital	10 000
Total	43 000	multiply by Year 6 Discount factor of 0.410 = 17 712

Grand Total   156 816

Net Present Value = 156 816 – 130 000 = 26 816

Figure 25.6 Investment appraisal using DCF.

it is in fact based on estimates. It should therefore be used more as an indication of the quality of the investment rather than a precise quantitative analysis.

If the NPV of the project is greater than zero then the project will be profitable, and a worthwhile investment. If a number of alternative projects are evaluated in this way then their NPVs can be compared and the most attractive selected.

25.4 Cost Analysis and Control

25.4.1 Absorption Costing

In order to determine the cost of producing a product we need to take into account not just the direct costs but also the overhead costs. That is, the selling price of the product must be such that the overhead costs are recouped, otherwise the company will be unprofitable. The problem is to find a realistic way to apportion the costs to the product. This must be done carefully since serious problems have resulted in the past with accountants allocating costs to departments and eventually to products that make them artificially expensive.

In absorption costing, cost centres are identified within the factory. These centres are seen to be composed of cost elements, that is, material cost, labour cost and expenses. These elements may themselves be composed of direct and indirect costs. The direct costs are incurred within the cost centre and are therefore easily allocated. However, the indirect costs are incurred both within the centre and apportioned from the overhead costs incurred by the manufacturing operation in general, for example, the costs of maintenance, rent and rates, administration, finance and power. The most important aspect of absorption costing is the selection of appropriate criteria on which to base the apportionment of costs. An example of how overhead cost might be allocated, that is, 'absorbed' by a cost centre, is shown in Figure 25.7. This indicates how an hourly overhead rate can be obtained for a department. Should the number of operating hours

The 'Chunky Chip' and 'Crunchy Crisp' Factory

Cost Item	£	Classification
Potatoes	140 000	Direct material
Operator wages	15 000	Direct labour
Maintenance wages	16 000	Indirect labour
Rent and rates	100 000	Indirect expense
Electricity	200 000	Indirect expense

Notes:
Floor area of chip making machine department = $15m^2$
Floor area of crisp making machine department = $35m^2$
Breakdowns on chip machine = 300 hours
Breakdowns on crisp machine = 500 hours
Chip machine power rating = 100 kW
Crisp machine power rating = 300 kW
Production time for chip machine = 36 000 hours
Production time for crisp machine = 36 000 hours

Overhead analysis:

Expense	Basis	Total	Chip Department	Crisp Department
Maintenance wages	Breakdown time	16 000	6 000	10 000
Rent and rates	Floor area	100 000	30 000	70 000
Electricity	Machine power rating	200 000	50 000	150 000
		Totals	86 000	230 000

Overhead rate for chip machine department = 86 000 / 36 000 = £ 2.4/h

Overhead rate for crisp machine department = 230 000 / 36 000 = £ 6.4/h

Figure 25.7 Allocation of costs to cost centres (chip and crisp departments) using absorption costing.

change the overhead rate would have to be recognised as being flexible and be adjusted accordingly.

25.4.2 Historical and Standard Costs

In a manufacturing environment costs may be either historical or standard. In an historical costing system the costs are analysed after operations have occurred whereas in a standard system the costs are estimated or calculated before the costs have been incurred. Both systems are usually employed together, that is, standard costs are prepared, then after the event they are compared with the actual historical costs incurred. The difference between the actual and the standard cost is called the variance, which may be 'favourable' or 'adverse'; these variances can be analysed to 'home in' on problem areas. It should be noted that these variances need to be made available to managers almost immediately. It is no use trying to control an operation by using information that is provided perhaps a week or even a month after the event.

25.5 Conclusion

This chapter has highlighted some of the financial aspects of the running of a manufacturing operation. The subject is extremely wide and only a sample of the most relevant aspects has been presented, for example, the different ways in which a company might finance its operation and the construction of annual reports and statements to shareholders have not been included.

This does not imply that accounting techniques and practices are unimportant to the manufacturing engineer or manager. In fact, manufacturing engineers and managers should be thoroughly familiar with the techniques and practices to ensure that the company allocates costs realistically and makes sound decisions based on real costs, and not on those that have been artificially derived.

Decisions concerning the manufacturing operation should not be made for the expediency of producing short term cost savings or creating cost reports that simply look good. The long term outlook should always be adopted to ensure growth and ongoing profitability. Poor costing practices can lead to work being unnecessarily sent out to sub-contractors because their labour rates seem lower, when in fact most of the company's costs could be in overheads that may increase if sub-contracting is adopted due to the extra administration involved. However, the company must identify its core competencies – those areas of its business in which it excels and that are the most profitable. Obviously, if outsourcing has been thoroughly analysed and has been clearly proven to be more profitable then a well-structured supply chain and a tier structure, as mentioned in Chapter 3, will be the correct course of action. Thus while costs must be measured, controlled and used with finesse, it should be remembered that a manufacturing company is in business to make money by making products.

Review Questions

1 For what three basic purposes does the financial department in a manufacturing organisation use accounting information?

2 Discuss the distribution of costs within a manufacturing company.

3 What is 'productivity' and how might it be measured?

4 Why is it important to consider risk when making an investment decision?

5 What factors should be considered when making the 'make or buy' decision?

6 What are 'fixed' and 'variable' costs?

7 What are 'direct' and 'indirect' costs?

8 What is a 'break-even chart' and how might it be used?

9 Why is marginal costing a useful accounting technique for manufacturing companies?

10 What is the 'payback' method of investment appraisal, when might it be used and what are its deficiencies?

11 What is a company's 'marginal investment rate'?

12 What factors should be included in a full investment appraisal calculation?

13 In DCF, why are future cash flows discounted back to year 0?

14 What is the significance of the net present value (NPV) of a project?

15 What is 'depreciation'?

16 What is 'absorption costing' and what is its purpose?

17 What are 'historical' and 'standard' costs, and how are they used?

18 Present a reasoned argument stating why a manufacturing manager should be familiar with management accounting practices.

Part VI

Maintaining Manufacturing Quality

26

Quality Defined – Quality Management and Assurance

26.1   Defining Quality

What is 'quality'? We often use the word *quality* to imply that something is the 'best' of its kind, for example, we may say a six-star hotel provides quality accommodation, but this is a rather vague use of the term. Consider that as consumers we expect that whatever we spend our money on should be of 'good quality'. This means that the producer of the goods, or supplier of the services, has to find a more specific definition of 'quality' against which the standard of the product can be measured. Now, since we cannot all afford a six-star hotel, yet we say we expect our purchases to be of good quality, we really mean that we perceive a product or service to be of good quality if it provides what we believe to be good value for money. Thus, we see that quality defines how well the product conforms to our expectations. In turn, our expectations are based on how much we paid for the item and the specification supplied by the provider. We, the customer, have the responsibility of deciding how much money we are willing to pay and the supplier has the responsibility of ensuring that whatever is supplied conforms to specification. We can therefore define quality as how well a supplied product or service conforms to the customers' expectations and the suppliers' specifications. Implied in this definition is the principle that the customers' expectations are related to price and the suppliers' specifications related to cost to produce.

This concept suggests that there is a certain minimum level of quality acceptable; this is usually embodied in consumer protection regulations approved by government. Typically, any goods that are sold must be 'fit for the purpose intended'. Many countries have product liability laws that hold the designer and manufacturer of goods responsible for their safe design and construction.

The purpose of this chapter is to examine how a manufacturing company can ensure that it produces good quality products, that is, products that conform to specification, are fit to use for the purpose intended by the consumer and are cost effective.

26.2   Quality Management

The quality environment in a manufacturing company is shown in Figure 26.1. An awareness of quality should permeate the whole structure of the manufacturing organisation. Senior management have the responsibility of taking the lead in establishing a climate that fosters a dedication to quality.

Essential Manufacturing, First Edition. Gordon Mair.
© 2019 John Wiley & Sons Ltd. Published 2019 by John Wiley & Sons Ltd.

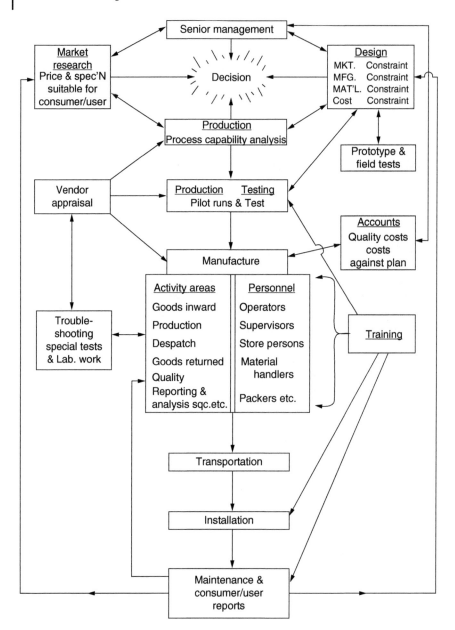

Figure 26.1 The Quality Environment for a manufacturing company.

An important aspect is that, to ensure quality standards are being met, some sort of quantitative or qualitative targets have to be set. These need to be measurable so that actual performance can be compared against that planned. Market research people must ensure they make quality judgements regarding the price and specification criteria acceptable to the consumer; feedback on this will be available only when consumer reports on the product begin to return from the market. Product designers must create a quality design based on conformance to marketing, manufacturing, material and cost constraints. Suppliers – we will call these vendors – and sub-contractors selected by the company must be fully aware of the company's expectations regarding conformance to stated specifications. This applies whether the vendor is supplying components that must conform to dimensional specifications, material that must conform to metallurgical specifications or a service such as cleaning, catering or transport, which will have their own specifications determined after consultation between the vendor and the company.

Figure 26.1 also makes it apparent that within the manufacturing operation itself physical work areas can be considered for allocation of quality targets to be achieved, these being monitored, reported and analysed by techniques such as statistical quality control (SQC), which is explained later. However, to ensure that the personnel within these areas are capable of achieving the targets, appropriate training must be given to operators and supervisors and so on. The company also needs to consider the quality of transport used to deliver the product to the customer. This means that specifications regarding the handling and storage of the goods have to be created and adherence to these ensured. With some products, the installation may also be part of the supplier's liability; again specifications and customers' expectations need consideration. The customer closes the loop by feeding back information directly through complaints or returned questionnaires or indirectly by increasing or decreasing purchases.

26.3 Organisation for Quality

Throughout the company the responsibility for quality belongs to everyone. However, some sort of organised structure is usually necessary to ensure that 'quality awareness' is maintained and appropriate systems, techniques, methods and technologies are in place for the monitoring and maintaining of quality. A possible organisational structure such as that shown in Figure 26.2 indicates the functions necessary in a large manufacturing company. Three major functions are identified, that is, the Quality Assurance Function, the Quality Control Function and the Test Equipment Function. Therefore, although a modern company may devolve these responsibilities in a different manner than is shown here, for example, inspection may be devolved to the operators producing the goods or may be fully automated, this figure is useful for showing the necessary tasks to be done to ensure that quality products are made.

26.3.1 Quality Assurance

There is a difference in areas of concern between Quality Assurance and quality control. Quality Assurance is a broad function that includes the design and implementation of systems, standards, procedures and documentation throughout the organisation's activities to ensure the awareness and achievement of quality standards. Quality control is a more specific function, concerned with the operation of methods and procedures for measuring, recording and maintaining the quality levels established by Quality Assurance.

To implement a quality assurance programme there must be a full commitment to quality throughout the company and it must be seen to be led by senior management.

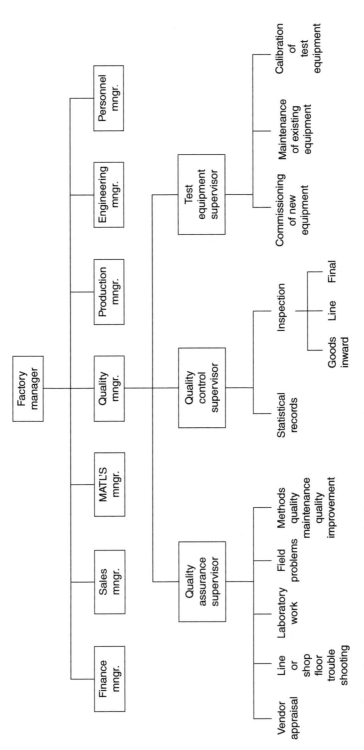

Figure 26.2 A quality organisation structure indicating some necessary functions.

The importance of teamwork is essential here. A quality improvement team may be gathered under the Quality Assurance Manager's direction. This team will be composed of representatives from various functions within the organisation, for example, design, purchasing, sales and production.

One major responsibility of the Quality Assurance Function may be the compilation of a 'quality manual'. This will contain all relevant information required by personnel to ensure that they are conforming to the quality procedures and standards acceptable to the company. As implied by our original definition, these procedures and standards will also be closely related to the needs of the customer. In fact, where the customer is a major client, such as the military, a large aerospace company or the nuclear power industry, then it is the client who often specifies the procedures and standards to be followed. The client will also wish to carry out on-site inspections of the manufacturer's premises to verify this conformance.

International standards have been developed to indicate which companies have in place quality practices that are of a specific grade. For example, the International Standards Office ISO 9000 provides a family of standards that lays down for manufacturers and suppliers exactly what is required of a quality system. By adopting these standards, and subsequently gaining the approval of the awarding body for conformance, the company is meant to be providing proof of its commitment to total quality and its ability to achieve it. Within this family the ISO 9001 standard provides the company with the skeleton on which to build its own quality management system encompassing all aspects of the company's activities from design through manufacture to the final customer care, this is the part of the standard for which companies can be certified. It should be noted, however, that although gaining certification shows that the company has the necessary systems in place, there remains the ongoing requirement of the company to adhere to these systems. There are also other quality standards in use worldwide; some are broad, such as ISO 9000, and some are more specific, just one example from the USA is the American Society of Mechanical Engineers ASME NQA – 1 Quality Assurance Program Requirements for Nuclear Facilities.

As well as assisting in establishing company quality policy and creating a quality programme and associated documentation the Quality Assurance Manager will also be responsible for other activities. For example, 'vendor appraisal' is an important activity involving close communication with the suppliers of raw materials, components and services. The vendor must be made fully aware of what is expected by the company in terms of conformance to specification, delivery time, cost and so on. It will also be necessary to verify periodically that the vendor has the capability to produce the goods or service to the level expected. Thus a requirement is often that the vendor should be approved to ISO 9001. Other activities might involve 'trouble shooting' quality problems on the factory floor, for example, suggesting alternative processes or components for short term solutions to urgent problems. The analysis of quality problems experienced by the customer is another necessary activity; this again will require a teamwork approach involving manufacturing and design engineers. An additional analysis type activity will be that of investigating 'quality cost'; this will be done in close cooperation with the company's financial personnel.

26.3.2 Quality Control

This function has two main areas of concern; these are inspection of the product and its elements, and the gathering, recording and analysing of statistical quality records.

In the past, there could exist a long line of command structure used to implement this activity. The quality control manager would have reporting to him various inspection supervisors responsible for different parts of the manufacturing process. Each of these supervisors would have control of a team of inspectors who would report on the quality of goods being processed in their area. For example, a team would be responsible for checking vendor's material arriving into the factory; this would be called 'goods inward' inspection. Each department or work cell on the shop floor would have its own inspection team, and an inspection team would be employed at the final assembly, packing and dispatch areas. These teams would require thorough knowledge of the product and production processes to allow them to make intelligent decisions on the acceptability or otherwise of what was being inspected.

However, today the armies of inspectors and a tiered quality control hierarchy are often no longer required. The use of automated inspection is one reason for this. Manufacturing equipment often incorporates automatic inspection devices such as sensors or machine vision systems that can check the product quality in process before the component leaves the machine or between operations. This allows 100% inspection and the potential for zero defects. Another feature of many factories is that where manual operations are still required, the responsibility for inspecting the product quality is devolved to the operators themselves. In this case the operators are also responsible for the rectification of their own work, thus giving them a greater incentive to produce good parts first time. Finally, the just-in-time manufacturing concept (JIT) means that when material arrives at a work centre it is assumed to be already inspected and ready for use. This means that the goods inward inspection function is much reduced and a greater responsibility put onto the vendor to ensure that his product arrives on time and already quality approved for immediate use. This principle has developed to such an extent that, although absolute perfection is not attainable, the number of defects supplied should now be measured in parts per million rather than parts per thousand.

The second activity in this function is that of SQC. Although it was mentioned that automated equipment can be used to achieve 100% inspection, this is not widespread or possible in every situation. There is therefore still a need on many occasions for statistical sampling of incoming goods to the factory, work in process and finished products. Also, even where automation is used, data can be gathered by electronic means and analysed by computer to monitor quality and rapidly identify any worrying trends that may eventually lead to bad products. This function of gathering, recording and analysing the information produced by the inspection function remains important. The results of the analyses should be communicated rapidly to all relevant areas of the factory to enable any necessary action(s) to be taken. It is worth noting that in fully automated systems this communication may be virtually instantaneous with the records and analyses being able to be viewed by interrogating the networked computer system. If required this data can also be viewed contemporaneously by other company locations throughout the world. Owing to its importance we will consider SQC a little more fully in Chapter 27.

26.3.3 Test Equipment

The third function within the quality organisation is that of the acquisition and maintenance of all the quality 'hardware' needed for inspection and testing throughout the manufacturing process. The actual equipment involved will be dependent on the type of product being produced and the manufacturing processes used. For example, simple handheld gauges may be required to inspect a manual machining operation, a machine vision system may be used for inspecting component positions on a PCB (printed circuit board) or a complex environmental chamber might be used for testing the performance of a product under various humidity and temperature conditions. All inspection and test equipment has to be regularly and systematically calibrated and checked for wear and drift and appropriate adjustments made, again this is covered more fully in Chapter 27.

26.4 The Cost of Quality

26.4.1 The Importance of Quality Costs

It has been mentioned elsewhere in this book that the primary goal of a manufacturing company is to make money. This means that to express ideas and events in the lowest common denominator, monetary terms have to be used. Therefore, we try to find some way of identifying the costs of quality, just as we identify costs associated with raw material or labour. There is of course an ethical limit to this. For example, no-one should attempt to compare the cost of an added safety feature with the probable costs of lawsuits for injuries resulting from the absence of that feature. Safety is a quality cost that has to be accepted; some manufacturers will spend more on this than others and the cost will ultimately be reflected in the selling price of the finished product.

The determination of quality costs will provide the company with additional information for control purposes and allow the identification of opportunities for reducing these costs. This is important since the cost of quality is increasing due to the growth in products that require high reliability and longer working lives, for example, in the nuclear, chemical, aerospace and oil industries. Even in everyday life our reliance in industrialised countries on technology is considerable. For example, we may purchase a frozen meal from a refrigerated display, pay for it using our smartphone, drive home in a car that has a sophisticated microprocessor-based engine management system, then use a microwave cooker to heat the meal before we sit down to watch a film that may be on any one of a wide range of audio-visual display devices. This is a sequence of events that relies on millions of manufactured parts operating according to specification. We expect every element in this simple chain of events to be performed flawlessly, this means higher precision in the elements from which they are constructed and hence the possibility of high quality costs to enable this precision to be achieved. We therefore need to be able to identify, classify and manage quality costs in order to minimise them while providing the customer with the expected product performance.

26.4.2 Classifying Quality Costs

It is possible to split quality costs into prevention, appraisal and defects or failure costs. It is apparent that the prevention and appraisal costs can be both considered as the cost of quality conformance, while the cost of defects are the costs resulting from quality non-conformance. By spending intelligently on the costs of prevention and appraisal, the costs of defects should be significantly decreased and the total quality cost minimised. These costs are detailed next.

- *Costs of prevention.* These are the costs associated with preventing the appearance of poor quality work, scrap and rework. This group includes the costs of planning and implementing the Quality Assurance programme, the construction of the quality manual, development of procedures and the costs of training personnel to use the quality system and meet the quality standards. Also included are the costs associated with gathering, analysing and reporting quality data, the costs of implementing quality improvement projects and the costs of evaluating the quality of new product designs and the quality aspects associated with their introduction.
- *Costs of appraisal.* Included here are the costs of vendor appraisal, all inspection and test costs, calibration and maintenance of inspection and test equipment and the costs of products used for destructive testing and consumables such as X-ray film and dyes used for non-destructive testing.
- *Costs of defects.* These are the costs associated with failure to conform to the prescribed quality standards; they can be considered under internal and external failure costs. Internal failure costs include: the loss in labour and material in producing scrap, the cost of reworking poor work, the cost of re-inspection and retest, the cost of the resulting under-utilisation of equipment and the cost of the time involved in deciding what to do with non-conforming material and components. External failure costs include: replacing of defective products returned by the customer, investigation of complaints and associated compensation to the customer, services provided to the customer while the product is under warranty and the cost of concessions to allow downgraded products to be sold as 'seconds'.

26.4.3   Obtaining Costs of Quality

Obtaining accurate quantitative values for these costs is difficult and expensive if appropriate accounting and cost reporting systems are not in place within the organisation. However, it may be said here that the calculation can be based on one or more of the following: (i) The labour costs of personnel involved in any quality activity; (ii) the whole costs associated with a department whose sole responsibility is some aspect of quality; (iii) the cost to process one quality 'unit' multiplied by the number of units – for example, the cost of quality training for one person multiplied by the number of people trained, or the cost of handling one customer complaint multiplied by the number of complaints and (iv) the deviation cost – that is, the difference between the costs involved in producing a good product first time and one that has been reworked or made as a replacement for one that was scrap. Should appropriate cost control structures not be in place within the company it is possible to use costs based on intelligent estimates. These will suffice as a short term measure until proper procedures are established.

26.4.4   The Optimum Quality Level

There are essentially two different views of what constitutes the optimum quality level and how the different quality costs are related. These views are shown graphically in

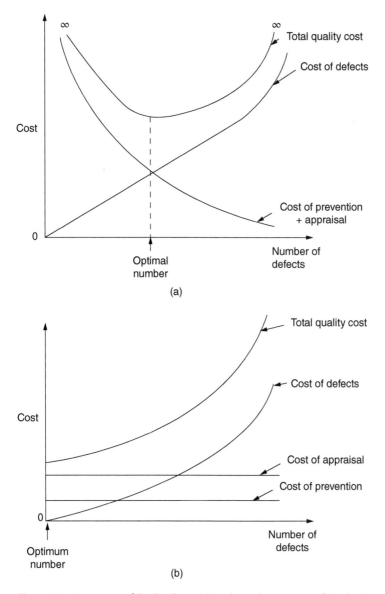

Figure 26.3 Two views of Quality Costs. (a) Traditional perception of Quality Costs. (b) Alternative perception of Quality Costs.

Figure 26.3. The horizontal axis in each graph shows the quality level, represented by the number of defective products produced as a percentage of the total production. The vertical axis indicates the quality costs involved in achieving the corresponding quality levels.

Figure 26.3a shows a traditional view. Here it is assumed that to achieve zero defects the costs of appraisal and prevention approach infinity, thus making the company uncompetitive. Conversely, as the number of defects produced increases past a certain point, due to appraisal and prevention costs being reduced, then the cost of these defects approaches infinity since the company loses all credibility in the market place and has to cease trading. Between these two extremes there is some optimum point where the costs of failure equal the costs of appraisal and prevention; this is the optimum point where the total quality cost is minimised.

Figure 26.3b shows an alternative view in which the optimum number of defects is zero. This perception comes from the belief that, once their most effective level has been found, appraisal and prevention costs should remain constant. If quality consciousness pervades the company then the awareness of every individual within the organisation ensures that they assume responsibility for the quality of their work. This results in an ongoing reduction in the number of defects. At zero defects there are no failure costs, for example, rework, scrap, customer returns and warranty repairs. Also, as defects approach zero, the company's credibility increases, thus leading to an increased market share.

In practice, the truth probably lies somewhere between both of these perceptions, that is, a zero defect situation is ideal but the appraisal and prevention costs increase as absolute zero is approached.

26.5 Conclusion

The importance of senior management in establishing a quality ethos within the company has already been noted, however, we now look at how a broad range of workers on the factory floor can contribute towards improving quality. Groups can form together at regular intervals, say weekly, to discuss quality problems related to their own sphere of influence within the organisation. This is a participatory problem solving system that also stimulates an ongoing quality awareness in those involved. The group members may apply recognised problem solving techniques. For example, Pareto analysis is a method based on the rule that 80% of effect is caused by 20% of the population. In quality terms this could mean constructing a histogram of all the different defects found from a problem process or operation. Examination of the histogram should show which defect is causing the most trouble; this should lead to identification of the problem source. A similar approach can be used very effectively for analysing quality costs. Other techniques involve the construction of 'cause and effect' diagrams that require the writing down of the problem or effect, then working backwards thus forming a tree or fishbone in which all possible causes are listed. Analysis of the diagram by systematically eliminating the possible causes should highlight the problem source. Finally, when a problem suddenly arises where none existed before, the simple question, 'What has changed?' should be asked. This may identify that a new batch of material has been used, maintenance has just been completed on a machine or a new operator has started work; changes like these often prove to be problem sources.

The concluding remarks to this chapter must therefore be that quality is everyone's responsibility and that on it, in today's intensely competitive environment, will rest the success or failure of the company.

Review Questions

1 Why is it important to define 'quality'?

2 Discuss briefly what you understand by the term 'Total Quality Management'.

3 Discuss the type of activities that would normally be classified under the heading 'Quality Assurance'.

4 Why is vendor appraisal such an essential activity for the modern manufacturing organisation?

5 Discuss the type of activities that would normally be classified under the heading 'quality control'.

6 Why should there be few full time inspectors in a modern factory?

7 Describe two methods of classifying the cost of quality in an organisation.

8 Why is the identification of quality costs of increasing importance to manufacturing organisations?

9 List the type of costs that would be included within each of the three traditional cost groups.

10 State whether you agree or disagree with the statement that 'minimum quality costs occur with zero defects'. Explain your reasoning.

27

Metrology and Statistical Quality Control

27.1 Introduction

To ensure that quality standards are being met, the actual quality achieved has to be measured. In order to do this, the quality levels must be clearly specified in such a way that measurement is possible. The process of determining whether or not a product, component, material or process conforms to specification is called *inspection*. In some situations, visual inspection can account for around 10% of the total labour cost of manufactured products, although today, particularly in high volume production, the use of machine vision systems coupled with other automated sensing and handling devices such as industrial robots is widespread. Inspection may be qualitative or quantitative. In qualitative inspection, a product is checked on an accept or reject basis by examining the *attributes*. For example, a flaw in the glass of a bottle, a crack in a casting or discoloured fruit on a conveyor would be attributes that would cause these items to be rejected. Other aspects of this would be the detection of foreign bodies such as mice or insects on a conveyor in a food processing plant, verification of a completed task in an assembly sequence or checking the degree of swarf build up on a drill bit. *Quantitative* inspection involves *measuring* the dimensional features of a product or component and this is the focus of the remainder of this section.

When manufacturing, for example, a shaft and bearing it is not good enough to say that the internal diameter of the bearing is to be exactly 25 mm and the shaft that fits into it has have an external diameter of 25 mm. This is wrong because it is impossible to make anything 'exactly' to size except by a very low chance probability and even if something was made exactly to size it would be impossible to measure that it had been achieved. Also, if the shaft was made slightly larger than 25 mm diameter or the bearing internal diameter slightly less, then the two would not be able to fit. In the early days of engineering 'making to suit' was the normal method of manufacturing this type of assembly where each component of a product was individually crafted to fit together to form the whole. Another method, called 'selective assembly' may very occasionally be used in unusual cases of high volume production. This applies where the clearance between components must be held to such a tight tolerance that the process capability is not able to produce sufficiently precise components. In this case, the components are very accurately measured and graded into various size groupings, subsequently suitable components are matched with each other. This happens with some precision shaft and

Essential Manufacturing, First Edition. Gordon Mair.
© 2019 John Wiley & Sons Ltd. Published 2019 by John Wiley & Sons Ltd.

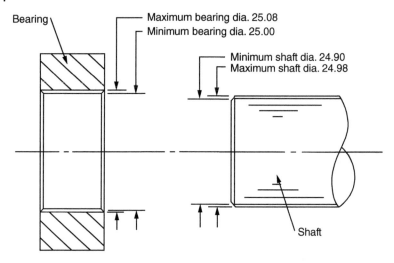

Figure 27.1 Dimensional tolerancing applied to a bearing and shaft to ensure a clearance fit.

bearing or piston and cylinder assemblies. However, the mass production systems of today demand *interchangeability* of components for efficiency in assembly and for parts replacement. For this reason, *tolerancing* must be used and although we are using a shaft and bearing as an example, the principles of tolerancing apply to all manufactured products.

A tolerance is a permissible deviation from the desired specification. Tolerancing is most frequently associated with dimensional tolerancing and geometric tolerancing (concerned with aspects such as cylindricity and parallelism) of parts, but it can also apply to any specified parameter such as weight, temperature, time, current, voltage, strength, hardness, metallurgical composition, number of blemishes in a paint finish and so on. In our example of the shaft and bearing, the designer could state that the shaft diameter should be 24.98–24.90 mm and that the bearing should be 25.00–25.08 mm internal diameter. Thus the clearance between the shaft and the bearing could vary between 0.02 and 0.18 mm, depending on the machining operation, see Figure 27.1. This is only an illustration, in practice for shafts and bearings tables of limits and fits are used to determine the tolerances to be used. These are given for different types of fits, the two most common being clearance fits and interference fits in which the shaft is a press fit in the hole.

In order to verify that tolerances have been achieved it is necessary to carry out some form of measurement and we will therefore now consider some aspects of this essential engineering process.

27.2 Metrology

As the industrial revolution progressed at the end of the eighteenth century, the need for components to be interchangeable became important. Previous to this the low volume of manufactured goods meant that craftsmen could make parts on an individual basis with dimensions hand crafted to fit an assembly. Mass production demanded that parts made in one factory could be brought to another and assembled together with components from other sources. Also, the new high pressure steam technology and mechanism complexity needed components that could fit together with greater precision. This led to the importance of reliable measurement techniques. James Watt in the UK invented a bench type micrometer around 1772 and by 1848 Jean Laurent Palmer in France invented the modern form of micrometer shown later in this chapter.

Today, precision measurement is absolutely essential with products being assembled in one country using components often manufactured in factories from all over the world. For example, a large passenger plane may have its fuselage made in one country and wings in another. When they are assembled they have to fit together without the need of further modification. A modern car, as well as being assembled from parts made in many different countries, also needs to be easily maintained and repaired by removing damaged parts and replacing them with new ones – all of these parts must be interchangeable. The measurement process which, in engineering, covers both the dimensions and geometry of parts is called *metrology*.

Metrology in general terms is the science of measurement and it is based on the fundamental units of measurement, that is, length, time, mass, temperature, electric current, luminous intensity and amount of substance. Here we are specifically concerned with the branch of 'engineering metrology' and we will focus our attention on the measurement of length based on the fundamental SI standard of the metre. Within a factory inspection is carried out during the manufacturing process to ensure that whatever is being produced conforms to the design specifications of the product. But how do we know that our measurement of the produced item is true relative to the fundamental standard metre? The answer lies in the concepts of *accuracy, uncertainty, traceability* and *calibration*.

Accuracy as a term is used in an informal manner in the following paragraphs as this is the way it is commonly used by most people and in everyday use by engineers. Formally, in relation to length, accuracy is the closeness of the agreement between the result of a measurement and a true value of the length being measured, we can therefore correctly talk about 'high' or 'low' accuracy. However, accuracy really should not be used quantitatively since the 'true' value cannot be realised perfectly. In metrology, the term *uncertainty* is preferred since it states quantitatively a range of values within which the true value will be likely to lie. *For our purposes we will continue to use the term 'accuracy' and we will use it both quantitatively and qualitatively.*

Traceability in metrology is the concept of ensuring that the accuracy of any properly calibrated measuring instrument can be traced back through increasingly accurate instruments to the ultimate known primary standard. National Metrology Institutes (NMIs) are responsible for realizing the length standard and they cooperate on ensuring the international standard is maintained. For example, the UK has the National Physical Laboratory (NPL), the USA has the National Institute of Standards and Technology (NIST), Japan has the National Institute of Advanced Industrial Science and Technology (AIST) and Germany the Physikalisch-Technische Bundesanstalt (PTB). These institutes can check the accuracy of the measuring instruments used by industry and issue a certificate to verify their accuracy. This process is called *calibration*. These calibrated instruments can then be used by industry to check the accuracy of less accurate instruments and this process continues until the calibration of instruments used by an inspector or machine operator in the factory workshop.

The metre. In order to achieve an absolute value for the metre it has been decided internationally to use the speed of light as the standard. This implies that an absolute value for time is necessary. The fundamental unit of time is the second and of all of the fundamental units this is the one that is stated with the most precision, it is based on the natural resonance frequency of the caesium atom 9 192 631 770 Hz. This is measured to an accuracy of two nanoseconds per day and although there are now clocks much more precise than this it is the caesium clock that provides the standard. Thus a sort of iterative process has taken place with the speed of light stated as 299 792 458 m s^{-1} and the metre is therefore defined as: the length of the path travelled by light in a vacuum during a time interval of 1/299 792 458 of a second, that is, 3.34 ns. *To translate this into a value that can be used as a basis for engineering metrology we can state the metre as a specific number of wavelengths of light.* In practice, the source of light is recommended to be that of a helium-neon laser (HeNe), which has a wavelength of 632.991 398 22 nm. This means that the metre can be delineated as 1 579 800.298 728 wavelengths of HeNe light in a vacuum (from 1/0.000 000 632 991 398 22).

The HeNe laser interferometer utilises the principle of light interference. The monochromatic light from the HeNe laser is in the form of a sinusoidal wave. This will have amplitude 'a' and a wavelength λ. A beam splitter divides this light into two rays that follow different path lengths before being recombined with one ray 180°, or half a wavelength, out of phase with the other. This results in the wave patterns shown in Figure 27.2. When this light falls on a surface the eye will see a light band if the amplitude of both rays combine as shown on the left side of the figure and a dark band would be seen if they are combined as shown on the right side of the figure where they interfere, thus cancelling each other out. Thus, as the surface on which the two beams are projected moves towards or away from the light source a sequence of light and dark bands will result. Each change from dark to light and back to dark again will signify a movement of half a wavelength, *that is, interference will occur (each dark band) approximately every 316.5 nm.* Lengths can then be measured to accuracies of ±1 nm by sub-dividing the fringes and counting these electronically. Note that a chart of the SI units of scale is included at the end of the book in Appendix A.

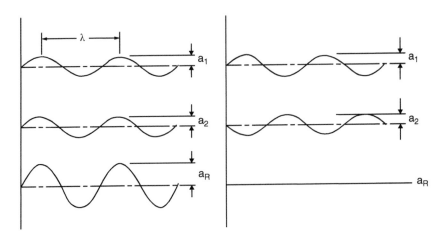

Figure 27.2 The principle of light interference of monochromatic light.

27.3   Factory and Workshop Metrology

In practice, within a factory various methods are used to check the sizes of manufactured components.

An important means of obtaining reference lengths is the use of gauge blocks, sometimes called slip gauges or Johansson blocks. These are used as the standard references within companies and are obtainable in various grades. The highest grade will have been calibrated using the HeNe laser interferometer. They are purchased as boxed sets containing a carefully devised range of blocks. The individual blocks can be 'wrung' together to create any desired length. The blocks are made from heat treated and stress relieved alloy steels or ceramic material. They are made flat and parallel depending on the grade to within the following accuracy ranges: Reference Gauge ±0.05 µm (0.000 05 mm), Calibration Gauge Blocks ±0.10 µm (0.000 10 mm), Inspection Gauge Blocks ±0.15 µm (0.000 15 mm) and Workshop Gauge Blocks ±0.25 µm (0.000 25 mm). The reference grade blocks will be the most expensive and would only be used for checking the calibration grade blocks, these in turn would be used for checking the inspection and workshop grade blocks.

Gauge blocks are known as end standards as it is the distance between the faces that is important. These are able to provide higher precision than instruments that use line standards, examples of which are hand-held micrometers, Vernier callipers and engineer's steel rules, all of which are commonly used within the factory or workshop.

Figure 27.3 shows examples of some of these instruments with their associated measurement accuracies.

Inspection methods that are used to check whether or not a part is within tolerance can be separated into *measurement, comparison and gauging.*

Measuring instruments can be either direct or indirect. Direct instruments allow a reading of size directly; for example, from a lined scale. Indirect instruments do not contain line scales and are used to transfer dimensions to a direct instrument; this increases the opportunity for error, therefore direct methods should be used wherever possible. Examples of direct measuring instruments used for measuring lengths are the steel rule, the micrometer, the coordinate measuring machine and the laser interferometer. An indirect measurement example would be using internal callipers to acquire the diameter of a bored hole, then using a micrometer to measure the distance between the two calliper points.

Comparators, or deviation type instruments, do not measure an absolute length but instead amplify and measure variations in the distance between two surfaces. For example, a dial indicator gauge as shown in Figure 27.3 translates the vertical linear movement of a spindle into a scale reading on an indicator. The means of translation may be mechanical, mechanical/optical or electrical. To describe the operation assume the height of a machined surface is to be measured. The dial indicator, the item having the reference height – this could be a gauge block assembly and the item whose height is to be measured are all placed on a flat smooth metal or granite table called a 'surface table'. The indicator spindle is allowed to come into contact with the reference surface and the reading on the indicator dial is noted or set to zero. The spindle is now lowered into contact with the item to be measured and the difference in dial reading noted. The height of the measured item will therefore be the height of the reference

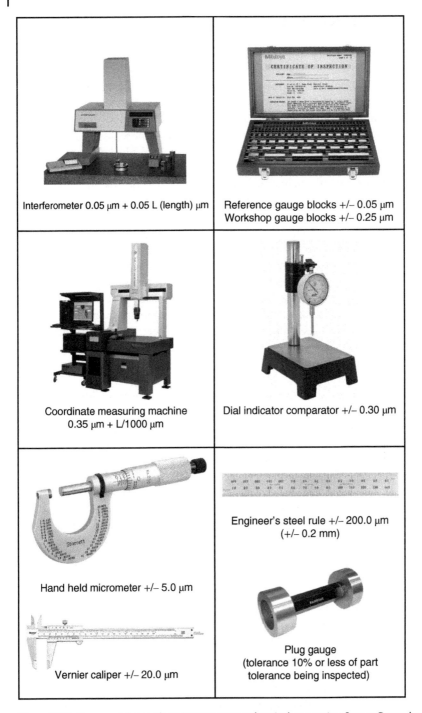

Interferometer 0.05 μm + 0.05 L (length) μm

Reference gauge blocks +/– 0.05 μm
Workshop gauge blocks +/– 0.25 μm

Coordinate measuring machine
0.35 μm + L/1000 μm

Dial indicator comparator +/– 0.30 μm

Hand held micrometer +/– 5.0 μm

Vernier caliper +/– 20.0 μm

Engineer's steel rule +/– 200.0 μm
(+/– 0.2 mm)

Plug gauge
(tolerance 10% or less of part
tolerance being inspected)

Figure 27.3 Commercial metrology instruments and typical accuracies. *Source*: Reproduced with permission of Pratt and Whitney, Mitutoyo, MSCdirect and L. S. Starrett.

± the dial indicator reading. Accuracies of ±0.0003 mm are obtained using this type of technique.

Gauges used for production purposes are the simplest, quickest and cheapest means of inspecting the size of components. They are usually in the form of GO and NOT GO gauges. For example, when checking a hole a cylindrical gauge will be used such that the GO end should be able to enter the hole while the NOT GO end of the gauge should not enter. 'Taylor's theory of gauging' states that the GO gauge checks the maximum metal condition and should check as many dimensions as possible; conversely, the NOT GO gauge checks the minimum metal condition and there should be a separate NOT GO gauge for each dimension. This will ensure that a bad component is not accepted in error. It is apparent that the tolerances to which both ends of the gauge are to be made must be much tighter, normally by a factor of 10, than those of the tolerances of the component being checked.

27.4   Surface Texture and Measurement

Until now we have focused on the measurement of length. However, another important aspect of metrology in manufacturing is the measurement of the surface texture of a component or product. In many cases it is not sufficient to simply state a surface has to be smooth or rough since the texture will strongly influence many aspects of the product. Among these are functionality, cost, aesthetics and corrosion and wear resistance. For example, the outside of a car body is expected to be aesthetically pleasing by being smooth and coloured but also corrosion resistant, the surface texture underlying the surface coating will have a strong influence on the appearance and performance of the finished product. If we consider the outside surface of the engine cylinder block it may be quite rough, this is not a problem since it is not normally seen and there is no functional need for a very specific finish. However, within the cylinder block of the car the cylinder walls will look 'shiny' but close examination of the surface texture will show that it is not perfectly smooth. If it was too smooth then lubricant would not adhere to the walls and the pistons could seize, if it was too rough then the piston could break off small particles from the cylinder walls and so reduce the life of the engine by increasing wear. In this case the specification and measurement of surface finish is therefore very important to ensure smooth and efficient operation of the engine.

Generally speaking, the more closely specified the surface finish, the more expensive the product will be. This is because the time the component needs to be worked on will usually increase as the tightness of the specification of the surface finish increases. The surface finish pattern will be influenced by the type of manufacturing process used, for example, turning, milling, grinding, extrusion, forging and casting, will all produce different textures. We are normally concerned with machined surfaces when measuring surface texture.

The short wavelength *roughness* of the completed component can be given a numerical value. This is usually termed the R_a value and is measured in micrometres. For example, an R_a value of 12.5 μm would be the type of surface finish obtained from rough turning in order to remove metal quickly, whereas a value of 0.4 μm would represent a very smooth surface and by using a lapping process on a flat surface an almost mirror like finish of 0.05 μm could be obtained. The R_a value is a measure of the amplitude of the short wavelength surface irregularities caused by the machining process. This could be due to the feed rate of the cutting tool and any material rupture that occurs during the chip forming process.

Figure 27.4 shows the surface features apparent on a machined component and Figure 27.5 shows the equipment used for their measurement. The roughness pattern is measured by drawing the stylus of a surface texture measuring machine along the surface at right angles to the direction of lay. The *lay* is the term given to the dominant pattern of the tool marks in a machined component and will be different depending on the process used such as turning, end milling or slab milling.

The *waviness* has a much longer wavelength and the roughness is superimposed on this. Waviness arises due to various factors such as inaccuracies in the machine tool structure or vibration incurred by the cutting process or generated by the machine components. In order to obtain an accurate measure of surface roughness, the waviness height should be excluded otherwise an erroneously high R_a value will be obtained. This is ensured by choosing a short enough cut-off length over which the stylus is drawn to prevent the effect of waviness being included and this is shown in Figure 27.4. A flaw or crack is also shown in the figure. In practice, this is likely to be uncommon with good quality materials. However, extremely small flaws can occur and will show up as an irregularity on the trace produced by the stylus.

As well as the specification of product dimensions and surface texture, other parameters such as sphericity, roundness, parallelism and flatness can be specified and measured. The topic is a vast one but this information provides a basic introduction to the essential nature of metrology. The next section considers the application of inspection in high volume production.

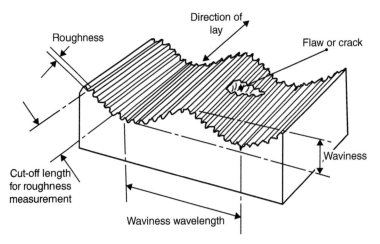

Figure 27.4 Greatly exaggerated machined surface features. This could be part of the surface of a cylindrical component turned on a lathe.

(a)

(b)

Figure 27.5 (a) A surface texture measuring machine. The stylus can be seen suspended from the column on the left, surface texture information is displayed on the screen. (b) Surface texture stylus being drawn across the machined surface of a bored hole in a casting. *Source*: (a) Reproduced with permission of Mitutoyo UK Ltd. (b) © Copyright Renishaw plc. All rights reserved. Image reproduced with the permission of Renishaw.

27.5 Statistical Quality Control (SQC)

The use of automated processes with integrated automated inspection devices providing 100% 'in process' inspection for high volume production are often found. However, whether automated or manual inspection is carried out control of the process is essential. Where high volumes of the same items are to be checked, samples of the items are taken as representative of the whole population. These samples are inspected and the results statistically analysed to determine, (i) if the population is composed of good parts and (ii) if the manufacturing process is in control or likely to begin to produce bad parts in the near future. This activity is called Statistical Quality Control (SQC). It can be used to inspect not only parts being produced by a machine but also to inspect incoming goods to a factory that have been supplied by a vendor. SQC can be implemented using manual or microprocessor-based methods. Since one of the more important elements in SQC is the construction of control charts, a brief review of the technique is given next.

SQC is based on the principle that there will always be a random variation in the work from which a nominal value is desired. For example, using the shaft described earlier, it is required to be turned to 24.94 mm diameter with a tolerance of ±0.04 mm. Provided the machine has an appropriate 'process capability' it will be found that the parts being produced will vary slightly in size but should remain within the range of 24.90–24.98 mm diameter. If the process is in control then the random variation in size will produce a normal distribution about the mean for any population or batch of parts produced. If the machine has been set properly the mean value should lie at 24.94 mm see Figure 27.6a.

The properties of the normal distribution curve are well known. It is therefore possible to use these to assist with the ongoing control of our turning operation. The properties of the curve are shown in Figure 27.6b. This shows that if we set points A and A′ at 1.96 standard deviations above and below the mean, then it will he found that 95% of all items produced will have their sizes between these two points. Also if we set points B and B′ at 3.09 standard deviations above and below the mean, then 99.8% of parts will lie between them. This principle is used to construct control charts.

We can take the normal distribution curve that we would expect to obtain from the item population produced by our process and turn it through 90°, as shown in Figure 27.6c.

If the tolerance band is now superimposed on the curve it can be seen that, in this case, the process is in control and producing 100% good parts. The actual control charts for the process are constructed by setting 'Warning Limits' and 'Action Limits' at the 1.96 and 3.09 standard deviation levels, respectively, see Figure 27.6d. This is based on the principle that there is a 1 in 40 chance of the mean of any sample withdrawn from the process being beyond the warning limit; therefore, if one such is found then subsequent checks should be made. Also if several consecutive sample means were found close to the warning limit then it could indicate the process was drifting out of control. If the process is in control there is only a 1 in 1000 chance of a sample mean being beyond the action limits; therefore if one sample mean is found to be in this region, the process should be stopped for checking.

Control charts are in practice constructed by taking samples from the process when it is known to be working satisfactorily and following a straightforward procedure using tables of figures or computer programs specially devised for the task. The 'process capability' is also taken into consideration. This is the relationship of the tolerance band – that is, the upper specification limit minus the lower specification limit – to the standard deviation of the process. This can also be expressed as an RPI or 'Relative Precision Index', which is the total tolerance divided by the average sample range produced by the process. By comparing the ratios obtained to the sample size being used it can be determined, from standard tables, whether the process is of low, medium or high relative precision. It can then be decided if adjustments have to be made to the control charts; for example, if the process is low precision, rapid action is essential if any deviation from normal operation is observed, whereas with high precision more process drift may be allowed before action is necessary.

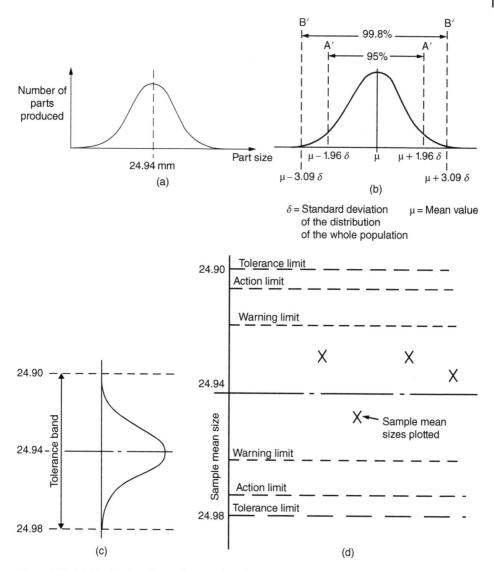

Figure 27.6 (a) Distribution of part sizes produced by turning. (b) Properties of the normal distribution curve. (c) Size distribution in relation to tolerance. (d) A control chart for the shaft turning operation.

These charts are used for monitoring 'variables' that are measurable in some quantitative manner, for example, size, weight or volume. However, they can also be constructed for monitoring 'attributes' as noted earlier in this chapter.

One of the aspects of the manufacturing process that this chapter has shown is that the tolerance, determined by the product designer, has an influence on the type of measurement system needed to check individual components. The tighter the tolerance specified the more precise and expensive the measurement method required. Also tighter tolerances imply more time required to create the product and often more time required to inspect it. All of this will increase the cost of the product and care has to be exercised to ensure optimum values are selected in order to achieve a competitively priced end product.

Review Questions

1 What fundamental physical constant is used to establish the length of the metre and why has this been used?

2 Quantitatively define the 'International Metre'.

3 What is the name given to the technique used to establish the 'International Metre' in practical terms?

4 Discuss why metrology is an essential ingredient in modern manufacturing industry.

5 Three different methods of inspection can be used in a factory for checking whether or not a part is within its specified tolerance. State these and provide a description of each.

6 Discuss what you understand by the terms accuracy, uncertainty, traceability and calibration with respect to metrology.

7 During inspection a part can be examined by using attributes or variables. Describe what you understand by each of these terms.

8 There are two different types of length standards, the type depending on the manner in which the measurement is established. State these and discuss the difference between them.

9 Describe why and how gauge blocks are used in industry. Include a description of them and the manner in which they can be utilised in order to determine the size of a component.

10 Name five instruments or tools you might find in a metrology laboratory or inspection room in a mass production factory and describe their use.

11 State three effects the surface finish can have on a machined component.

12 There are three significant features that influence the surface texture of a component. Describe each of these, comment on their significance and how they are created. You should use a sketch to illustrate your answer.

13 What is the significance of the 'sampling length' when measuring surface texture? You may use a sketch to illustrate your answer.

14 Briefly discuss what you understand by the term SQC.

15 Describe the operation of a control chart for a machining process.

Part VII

Human Factors in Manufacturing

28

Human Factors in Manufacturing

28.1   Introduction

Manufacturing is about *people* just as much as it is about technology and organisation. As well as creating products *for* people to use, manufacturing involves the use *of* people in the creation process. Some of the factors that relate to these people are considered in this chapter.

Within the manufacturing system people should be able to enjoy their work and carry it out efficiently in a pleasant, healthy and safe environment. Here, we will examine how these criteria are met by looking at the areas of job satisfaction, health and safety and ergonomics, including the work environment.

28.2   Job Satisfaction

Before people can enjoy their work and gain satisfaction from what they are doing they must feel motivated to carry out that work in the first place. In 1943 Abraham Maslow proposed a 'Theory of Human Motivation' that included a hierarchy of needs composed of five levels (see Figure 28.1). He said that needs would appear in ascending order only. For example, a physiological need would be the satisfaction of hunger and thirst, and the need for safety would be of secondary importance until these essentials of life had been acquired. In our industrialised societies with advanced economies the physiological needs are usually satisfied for most members of society. Similarly, on a general level, safety needs are satisfied except in specific cases where, for example, war, accidents or criminal acts might occur. The social needs at level 3 will be satisfied through the personal life of the individual through family, friends and often the workplace. Many people also look to work to help them satisfy the esteem and self actualisation or realisation needs suggested at levels 4 and 5. This work may be that of a parent who works full time at home raising a young family, a volunteer worker with a charity, or someone who is self-employed and so on. In this chapter, we are mostly concerned with the manufacturing employee working in an engineering environment.

In Chapter 24, Work Study was examined as a means of improving the efficiency and ease with which a job could be completed; however, additional factors should also be taken into account when designing a job for a worker. Psychological factors should be considered to ensure that the worker is able to experience fulfilment of levels 4 and 5 on the Maslow scale. It may be argued that in most cases this is unnecessary and that many

Essential Manufacturing, First Edition. Gordon Mair.
© 2019 John Wiley & Sons Ltd. Published 2019 by John Wiley & Sons Ltd.

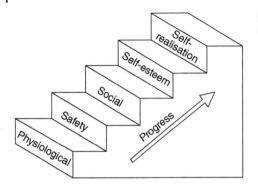

Figure 28.1 The Maslow Scale of the hierarchy of human needs.

people will be willing to do a particular job simply for the pay alone. This is all very well, but it should be remembered that it is the best performance we are looking for from a worker, not one that is mediocre or barely adequate.

Factors such as monotony of movements and boredom should be minimised. This appears contrary to the principle of specialisation and division of labour where simple repetitive tasks are used to produce high labour productivity. But, as in everything involving human beings as individuals, there is no perfect solution to any problem. There is evidence that some individuals actually prefer to have a rigidly defined, repetitive work cycle as can be found in, for example, mobile phone assembly, while others benefit from variety and flexibility in how they are allowed to tackle a job. The traditional way to assemble mass produced items such as motor cars is to use an assembly line. Here, the manual workers have to work at the same speed as associated machinery and at a rate governed by the speed of, say, a conveyor belt. In this case, each worker completes a simple operation and repeats this many hundreds or thousands of times per day. Monotony, boredom and fatigue can occur in these situations and this leads to poor quality products, absenteeism and high labour turnover rates. It has been found that by changing the job so that the individual workers can each build a larger part of the finished product and in the extreme case a complete car, then work satisfaction can improve. This leads to a greater pride in the job and better quality. Absenteeism and labour turnover also reduces and this lowers the costs involved in lost time and in training new workers. Naturally, the actual production rates will be slower as the benefits of labour specialisation are lost; this loss in production rate generally is so important that almost all mass production factories today still adhere to the production line system, though the needs of the individual will still be recognised and satisfied where possible.

Job satisfaction is therefore obtained by meeting a number of needs for the individual worker. As each combination of worker and manufacturing situation is unique, so the solution to the job satisfaction problem will be unique. In the ideal situation, individuals will be chosen to suit jobs and jobs designed to suit individuals. However, as the ideal situation seldom occurs, a 'best compromise' will probably be the only solution.

28.3   Health and Safety

As explained in the previous section people need and expect to be able to work in a healthy and safe environment. As also explained, where people are involved perfection is not possible. A human trait is that we make occasional mistakes; under the wrong circumstances these mistakes can cause accidents of varying degrees of seriousness. The mistakes may be made by an engineer at the design stage of a product, there may be a mistake made by someone preparing a procedure to be followed in operating a process plant, or someone may make a simple mistake such as pushing the wrong button or carelessly walking too near moving machinery. Each of these mistakes could cause an accident and it is estimated that hundreds of people are injured or killed every day throughout the world due to industrial accidents alone. Another aspect that must be considered is the health of the individual within the manufacturing system. Again due to human error and commercial pressure, conditions detrimental to health may appear. In this section, we will look at the provisions that should be made to make work both healthy and safe.

28.3.1 Legislation and Risk Assessment

In most countries today, laws exist to provide a minimum standard of health and safety within industry. For example, in the UK there is the Factories Act and the Health and Safety at Work (etc.) Act. There are also further Regulations and Codes of Practice approved from time to time to suit new and changing situations, for example, COSHH (Control of Substances Hazardous to Health Regulations). Copies of extracts of these publications, relevant to the particular industry concerned, must be displayed in prominent positions in work areas. This ensures that all employees have access to a general knowledge of the law as it relates to their own situation. Adherence to the law is ensured through inspection visits by the Health and Safety Executive in the UK. It is an offence to obstruct an inspector during the course of his duties and he has the power to inspect a factory at any time of the day or night. Each country has its own system of law but the broad scope and content of regulations are, by necessity, very similar. We are concerned here with manufacturing industry and the subject will be considered under the headings of health, safety and welfare.

Risk assessment by law must be carried out in all situations where people are present or likely to be affected by an activity or product, for example, teachers and pupils in education, patients and medical staff in hospitals and crew on marine craft. Risk assessment should also carried out by those designing systems such as aircraft, motor cars, buildings, research experiments, software for financial products and so on. In our case we will focus on the working environment in a manufacturing facility.

A risk assessment ensures identification and careful consideration of factors in the workplace that could cause harm. It also facilitates this by the use of a step by step procedure that is written down to ensure the analysis is logical and thorough. This also provides a record for the future should an accident occur and acts as evidence of what was considered and also provides a basis for creating and implementing improvements.

First, *hazards*, which are qualitative and *risks*, which are quantitative, must be identified.

A hazard is anything that can cause harm, such as the presence of electricity, radiation, wet floors, working from a height and so on.

A risk is the likelihood that a person could be harmed by a specific hazard. It can be allocated a value such as high, medium or low. If statistical or historical data are

Table 28.1 Categorising levels of risk severity.

Risk matrix Likelihood of harm	Severity or consequences of harm				
	Insignificant, e.g. requires basic first aid	Minor, e.g. deep bruise	Moderate, e.g. deep cut, torn ligament	Severe, e.g. fracture	Extremely severe, e.g. permanent disability, death
Rare: <1 in 1000 chance	Very low minor risk	Very low minor risk	Very low minor risk	Low moderate risk	Low moderate risk
Unlikely: 1 in 200 chance	Very low minor risk	Very low minor risk	Low moderate risk	Medium significant risk	Medium significant risk
Moderate: 1 in 50 chance	Very low minor risk	Low moderate risk	Medium significant risk	High severe risk	High severe risk
Likely: 1 in 10 chance	Low moderate risk	Low moderate risk	Medium significant risk	High severe risk	High severe risk
Probable: >1 in 3 chance	Low moderate risk	Medium significant risk	High severe risk	High severe risk	High severe risk

Severity of risk	Action to be taken
Very low, minor risk	Implement practical short term control measures
Low, moderate risk	Attention indicated
Medium, significant risk	Corrective action needed short term
High, severe risk	Discontinue operation and/or immediate action

available then a specific value in terms of a percentage can be allocated. Additionally, the *severity* of the risk should be considered. As a first step in the assessment categorising the levels of risk severity is necessary in order to decide on what action will subsequently be appropriate for reducing risk. This can be done in the form of a table as shown in Table 28.1.

The UK Health and safety Executive suggests the following five steps.

- Identify the hazards
- Decide who might be harmed and how
- Evaluate the risks and decide on precautions
- Record significant findings
- Review the assessment and update if necessary

The next step is to create an action plan using a form similar to the one shown in Table 28.2. The table would be extended downward until all of the hazards had been identified and listed.

Table 28.2 Listing of hazards, those at risk, risk level, action plan and result.

Activity: Lifting metal blank and inserting into press			Location	
Hazards List anything that could cause harm	Who is at risk List those exposed to hazard	Risk Decide on level of risk for each hazard	Control measures/ Action plan List measures to be taken to minimise risk	Revised risk after implementing control measures
Lifting metal blank: back strain; cut to hands	Press operator	Low Moderate risk	Provide lifting training. Provide safety gloves Ensure work at proper height	Very low minor risk
Inserting metal blank into press: loss of fingers or complete hand	Press operator	High Severe risk	Provide appropriate training and press guards. Ensure press will not operate unless operator's hands are clear (two hands needed to activate press)	Very low minor risk

28.3.2   Health

Within the factory the working conditions should be such that no hazard to health exists. There should be no fumes or particles in the atmosphere that could cause harm in the short or long term. If such particles do exist, then protective clothing and breathing equipment or filters should be provided and worn. For example, particles of fibreglass or asbestos are particularly dangerous and the best way of dealing with this type of hazard is to remove it at source. Substitute materials should be found, or an alternative process that does not produce these particles or effective filters at the point of particle creation should be used. Grinding operations, handling sacks of powder material such as flour and paint spraying are all tasks that involve the possibility of inhaling harmful material. A method of removing the human element in these jobs is to automate them using industrial robots and this is in fact common practice, with large productivity gains being experienced, particularly in paint spraying applications.

More general considerations include the following. There should be a high standard of cleanliness in factories and this applies not only to foodstuff and electronics manufacturing. The floors, walls and ceilings should be regularly cleaned, washed and redecorated at reasonable intervals. Refuse bins and other rubbish should be cleared daily; this improves hygiene and also reduces the risk of fire. Workers themselves should also ensure their clothing is kept clean as very serious skin disorders can occur due to prolonged contact with oily clothes. There should be no overcrowding and around $12\,\mathrm{m}^3$ per person should be regarded as a minimum, though this will obviously increase greatly if material handling is required and large machines are present. Food should also not be consumed where there is a possibility of absorbing poisonous fumes or dust.

Lighting must be adequate in every part of the factory to prevent accidents and avoid eye fatigue. Noise levels should be kept to a minimum and adequate hearing protection worn where necessary. Work areas should be well ventilated. Suitable temperatures should be maintained and the heating methods used should not in themselves be harmful. These factors will be considered again in the section on ergonomics later.

28.3.3 Safety

28.3.3.1 The Cost of Accidents

Within industry about 50% of all accidents are caused by either handling and lifting or machinery. Apart from the personal discomfort and distress experienced by the victim and their family, there is also the financial cost of accidents to be considered. These costs arise due to lost production, loss of earnings of the individual, cost of benefit paid to the worker if injury results and cost of compensation and or legal costs if legal action is taken by the victim in pursuit of a claim. The cost of production time lost may be considerable; this arises due to the lost time of the injured employee; the lost time of other employees who stop work to give assistance or are curious, sympathetic or shocked; the lost time of foremen and managers who have to investigate and report on the accident; the cost of retraining someone to do the injured person's work; the cost of repairing damage that may have resulted to equipment and so on. The loss of earnings to the employee involved in the accident may cause anxiety to himself and his family. This can be offset by compensation payments though these will probably not be immediate and cannot, of course, ever totally compensate for pain and suffering. Thus the cost to government, industry and individuals is considerable and runs into millions of pounds per year.

28.3.3.2 The Causes of Accidents

Accidents occur due to unsafe acts, unsafe conditions or both. The unsafe act might be one of the following; someone operating a machine without authority or proper training; someone disabling a safety device in order to improve their rate of work or working from an unsafe position or adopting an unsafe posture. Examples of unsafe conditions are: inadequate lighting, oil on floor, overcrowded work areas and dangerous work arrangements. These and other unsafe acts and conditions are termed 'hazards'. They occur everywhere people are present and they eventually cause accidents. Hazards are caused by people: sometimes the human error occurs very early on in a chain of events and the accident is separated widely in time from the unsafe act. An example of this would be a faulty weld on a ship's hull that becomes apparent as an accident only months later when a crack appears and propagates catastrophically. Often there is very little time between the unsafe act and the accident. An example of this would be an operator ducking under a guard rail to get at a piece of equipment and being hit by, say, the moving arm of an industrial robot located behind the rail.

28.3.3.3 The Prevention of Accidents

This can be considered under four areas: the original design of equipment and machinery used, the physical layout of the work area, the training adopted by the factory personnel and the safety devices and procedures in place.

First, equipment and machinery should be designed by qualified engineers and all materials and components to be used clearly and unambiguously specified. Quality components should be incorporated, they should have a minimum of moving parts and be solid state where appropriate. If possible, the type of components used should already have proven their reliability in other applications. The equipment and machinery should be manufactured under well supervised conditions and to specification. Records should be kept of how the work was done, who did it and the result of any inspection operations or tests carried out.

Having ensured that the equipment used has been designed and manufactured to satisfactory safety standards, the next stage is to install the equipment so that it can be operated in a safe manner.

All personnel within a factory or other industrial environment should receive basic safety training. This will inform them of unsafe practices, the need for tidiness and cleanliness and any hazards particularly associated with their work. For example, 'substances hazardous to health' have to be carefully controlled regarding their use and storage and all personnel likely to come into contact with them should be made aware of any special precautions to be taken. Those who operate machinery and other equipment must receive specialised training regarding their work. This training is essential if safe working conditions are to prevail. The supplier of the equipment being used could provide the training and should also ensure that the equipment has been installed safely. It may be necessary to compile a 'safety manual' to be read by all who are going to operate the equipment. In a particularly dangerous area a 'permit to work' might be issued to the relevant personnel and access denied to those with no need to be there.

Hazardous equipment and machinery should be guarded. This can take many forms and some of these are now noted, see also safety for reprogrammable equipment in Chapter 20.

Fencing. This may vary from a simple handrail about 1 m high, to a complete cage using 2 m high steel mesh or toughened Perspex. The caging would be used in most areas where large pieces of moving machinery are present. The space between the equipment and the fence would be adequate to allow maintenance personnel access. Gates into these work areas should be interlocked into the electrical power to the equipment. This means that if a gate is opened a switch built into the gate automatically switches the equipment off.

Light curtains. Instead of building a physical barrier it is possible to use light beams. In these systems an array of light emitting diodes (LEDs) passes light across a gap of a few metres to an array of photosensors. When someone passes through the gap, the beam is broken and the equipment is switched off.

Pressure sensitive mats. These are used either at entrances to work areas or around the hazardous machinery. Should someone step on the mat, a switch is closed, so stopping the equipment. These mats are pneumatically or electrically operated.

Guards. Guards actually on the machinery are common. These usually open to, allow the machine operator access while the machine is at a safe part of its cycle. They are also interlocked to ensure that the dangerous parts of the machine cannot move until the operator withdraws from the danger area and closes the guard.

Safety equipment. All workers should wear appropriate safety equipment. In almost all workshops it is necessary to wear protective goggles. These will prevent damage to the eye from flying particles such as may be present during grinding and cutting operations. Because of the harmful radiation given off when electric arc welding, welders must use protective face shields with tinted viewing windows. Protective helmets, or 'hard hats', should be worn where there is any overhead working taking place.

Fire precautions. There are regulations regarding the number and position of fire exits within a building. Exits should be easily accessible, easily opened from the inside and never locked. Personnel should be conversant with the procedure to be followed in case of fire. They should also be aware of the different types of fire extinguisher, what they are used for and their locations, which should be prominent. Flammable materials should be clearly marked and stored carefully; there are special regulations for explosive materials such as pressurised gas.

28.3.4 Welfare

Welfare is an aspect of health and safety in which the more general concerns of the well-being of the individual are considered. Employers are bound by law to provide adequate facilities for the welfare of their employees. For example, drinking water must be provided in ample quantity and of good quality. Clean drinking vessels or drinking fountains should be available. Clean, well maintained washing facilities should be available and supplied with soap and towels. There should be facilities for the drying and storing of clothes not worn during working hours.

A first aid box should be easily accessible and it should contain an appropriate range of materials for the type of work being undertaken in the local area. The box should be in the charge of a responsible person.

This section has shown that health and safety is everyone's responsibility. The designer, manufacturer and supplier of the equipment to be used in the factory have an obligation to ensure that it is safe to operate and that a clear instruction manual is provided. Employers have to ensure that their premises and procedures conform to the appropriate laws, regulations and codes of practice. They must ensure that adequate training and appropriate safety equipment is provided to the employee. The employee has the responsibility to make be familiar with the hazards that are likely to occur in his job, and to take advantage of any safety training, advice and equipment offered. The next section will consider ergonomic aspects that will combine methods to improve productivity with methods to make work safer and more enjoyable.

28.4 Ergonomics

Ergonomics is the study of the worker's relationship to the surrounding environment with intent to improve efficiency and work satisfaction. It can also be applied to product design to ensure ease of use of the product. The word is composed from the Greek *ergon* meaning work and 'economics'. Other terms used for ergonomics are 'human engineering' and 'human factors engineering'. An ergonomist receives and analyses input from both the engineering and the human sciences, for example, materials selection, functional design, statistics, mechanical engineering and so on, and from anatomy, physiology and psychology (see Figure 28.2).

When a worker is carrying out a job, he is operating in a 'closed loop' system. By this we mean that he or she observes a situation, analyses what is happening and what subsequently needs to happen, takes action to cause the required change to occur, observes the result of the action on the situation and so on. This is shown in Figure 28.3. This will occur in any task where a human is actively participating in a process, for example,

Figure 28.2 Engineering and human sciences contributing to ergonomics.

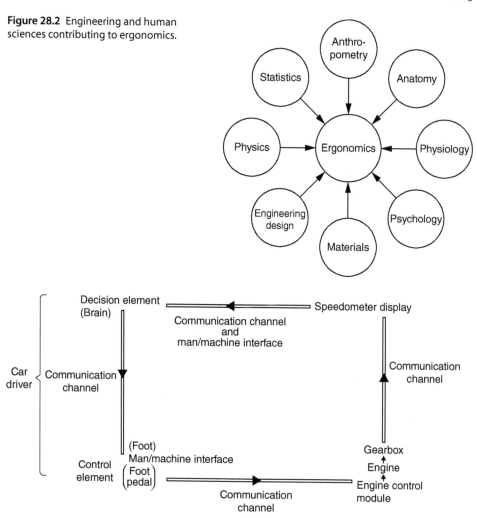

Figure 28.3 Human-machine closed loop system for car speed control.

assembling a product, controlling a chemical plant or even driving a car. All the elements involved in this control loop must perform at their optimum. Here we are concerned with the efficiency of the human operator, how well he or she can observe the situation and how well they can effect the necessary changes. Ergonomics therefore includes consideration of the design of the equipment and workplace, and the general environment surrounding the worker. These two aspects will now be considered in more detail.

28.4.1 Equipment and Workspace Design

All products have to be designed with consideration given to ergonomics. Mobile phones are examples of good ergonomic design and although they often contain excellent cameras there is still a demand for dedicated cameras for the serious amateur and professional photographer. The 'ergonomically' designed camera, for example, has its body contoured to fit comfortably into the hand, it is not too heavy, it is balanced for comfortable handling, controls are easily accessible even when looking through the viewfinder and non-slip surfaces are used where appropriate. In motor cars, gear levers are located for fast and easy shifting, the windscreen is designed for maximum visibility and the fascia panel, containing displays and controls, is easily seen through the non-slip optimum diameter steering wheel. The principles employed in the design and layout of these consumer products also apply to machine tools, aircraft cockpits and indeed anywhere humans have to interact with products. Some of these principles are now considered.

To ensure that the product design will suit the largest number of people anthropometry, that is, the measurement of people, is used. If a single item is being made, say a made to measure suit or a seat for a racing car, then it can be designed to fit exactly the person that is intending to wear or use it. However, most products must be able to accommodate a range of people, male and female, of varying heights, shapes, strengths and weights, and this is why anthropometry has to be used.

Large amounts of statistical data have been gathered over many years and these are now at the disposal of the product designer. These data are often presented in graphic and tabular form for ease of use. It is usual for a particular characteristic, say the height of the adult male population of the UK, to follow a 'normal distribution'. A normal distribution is a continuous distribution of some random variable, in this case height, with its mean, median and mode equal. The concept can be seen graphically in Figure 28.4. This symmetrical bell shaped curve represents a normal distribution. It shows that the mean height of the population is approximately 1.7 m. This means that 50% of the population are above and 50% below this height. The 'standard deviation' is a numerical factor used to indicate scatter. From the curve it can be seen that one standard deviation from

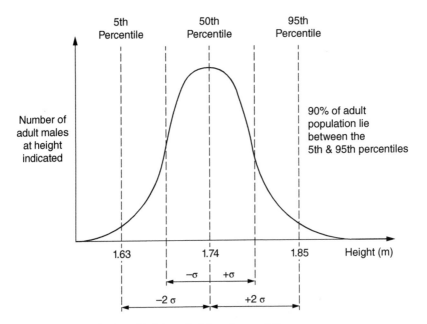

Figure 28.4 Distribution of heights of adult male population.

the mean includes about 34% of the population. Two standard deviations above and below the mean enclose about 90% of the population. Thus if a piece of equipment was designed so that people of between 1.6 and 1.9 m tall could use it, then the designer would be confident of his design suiting 90% of the adult male population of the UK.

The anthropometric data can be used to construct models of typical members of the population. These models can be used in graphic computer simulations to assist with design or two-dimensional cardboard mannikins can be made and Figure 28.5 shows examples of these models alongside the design for a lathe. Figure 28.6 shows a plan view of an operator together with suggested dimensions for a group of adult males using a seated workplace. Seats should be designed to be comfortable and provide proper back support. If the operator is working at a bench doing assembly work then components and tools should be stored at specific locations within easy reach. This will ensure that the operator can develop habitual movements without the need for the eyes to direct the hands.

As well as considering the dimensions of the equipment and workplace, the designer should consider the instrument displays and control elements that might be used. Displays should convey information to the worker in the simplest manner possible. No unnecessary information should be presented as this could create confusion.

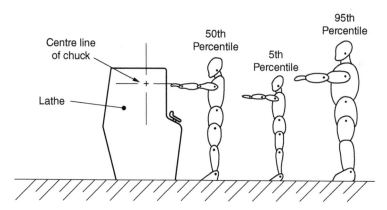

Figure 28.5 Simple 2D models to assist with lathe design.

Figure 28.6 Typical operator dimensions.

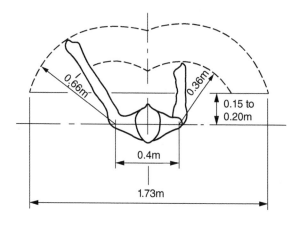

This principle is now employed in the pilot's workspace, that is, the cockpit of modern aircraft. Here computer screens display the essential information required by the pilot as the need arises. This reduces the proliferation of confusing dials and indicators that would otherwise be necessary. In the industrial situation the number of displays required is usually much less and conventional displays are the norm. These displays are of three main types, qualitative, quantitative and representational.

28.4.1.1 Qualitative Displays

These are usually visual, but auditory types such as bells, buzzers and sirens can also be used. The distinguishing feature of a qualitative display is that it conveys no numerical information. It is normally employed to convey a warning, for example, red lights to indicate ignition or oil pressure problem in a car and flashing lights on machines that are in operation. Colour coding of these displays should not be regarded as a reliable method of communication as a percentage of the population is colour blind.

28.4.1.2 Quantitative Displays

This type provides numerical information to the user in either analogue or digital form. Everyone is familiar with these two forms through their use on watches and clocks. Examples of these instruments are shown in Figure 28.7. The analogue display shows a reading on a scale analogous to the value it represents; for example, the speedometer on a car. It has the advantage that a quick glance will tell the observer an approximate value for the reading. It is also advantageous in that the rate of change of the variable being monitored is easily seen as the pointer moves over the scale. It has the disadvantage that precise readings are difficult in that if the pointer is between graduations, then an estimate of its true position has to be made. This problem can become worse if the display is read at an angle, as parallax error can occur due to the space between the pointer and the scale. The use of a reflective strip attached to the scale can remove this problem as the observer should line up the indicator with its reflection before noting the reading, see Figure 28.8.

Digital displays show the numerical information directly as a number; for example, the odometer on a car. This has the advantage that a reading to any accuracy desired can be made, provided enough numerals are on the indicator. Some digital displays are mechanical but most today are electronic, liquid crystal or LED displays. This fact

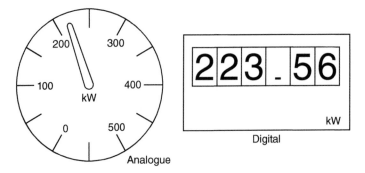

Figure 28.7 Quantitative displays.

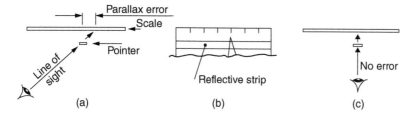

Figure 28.8 Parallax error when reading mechanical analogue displays: (a) view from above (plain), (b) front view and (c) plan.

leads to another advantage, that is, they are relatively easily integrated with electronic and electrical control systems. One disadvantage is that if the value is rapidly changing or fluctuating, then a quick glance may not be enough to determine the actual value, or direction and rate of change. In some instances when the advantages of both types of display are required then they are combined into the one instrument.

28.4.1.3   Representational Displays
A representational display provides a pictorial diagram or working model of a process. This must be kept as simple as possible to keep extraneous information to a minimum. Such displays would be found, for example, in large process plants, railway control rooms and electricity distribution centres. Computer screens can be used and simulation programs could be incorporated to show possible future events.

There is a wide variety of controls available for the designer to select for any piece of equipment. Knobs, buttons, horizontal and vertical levers, joysticks and handwheels are only a sample. Factors to be considered when selecting a means of control are the force required to be exerted, the speed with which this force is to be applied and the accuracy with which the process has to be controlled. For example, for delicate control a large diameter knob will allow fine, precise movements, but it will be unsuitable for rapid movements or the application of high forces. At the other extreme, a large vertical lever will allow the rapid application of a large force but would be most unsuitable for precision movements.

When designing integrated display and control elements it is important, for safety reasons, to adhere as far as possible to the conventions recognised in the country in which the equipment is used. Simple examples are: pushing a switch downward usually means 'on', rotating a switch clockwise also means 'on', but turning a water tap clockwise usually means 'off'. If an operator rotates a control clockwise or moves a cursor from left to right, he will expect an increase in whatever value is being adjusted. A composite of some of these control and display conventions is shown in Figure 28.9 for increasing values.

28.4.2   The Working Environment

Environmental conditions have a strong effect on the efficiency, comfort and health of the worker. We will consider them under three main headings: lighting, noise and heating and ventilation.

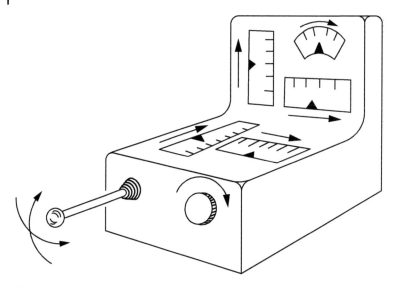

Figure 28.9 Controls and display convention.

28.4.2.1 Lighting

A lumen is the derived SI unit for luminous flux and it is a measure of the total amount of visible light emitted by a source. However, here we are more concerned with the total luminous flux falling on a work area, this is termed *luminance* and in SI units is measured in lux (in the USA foot candles are often used and one foot candle is roughly 10.76 lux). One lux is equal to 1 lumen per m^2. Irrespective of which country you may be in there will be recommended minimum lighting levels provided as guidance for workspaces. Generally, the very minimum amount of light for a continuously occupied working space can be taken as 200 lux or 200 lumens per m^2. With respect to manufacturing industry, this represents the minimum illumination that should be apparent throughout a factory for safety purposes. Much more light than this is needed to avoid eye strain when carrying out activities involving detailed tasks. In practice, levels of 1000 lux or higher are often necessary. A table showing typical lighting levels is shown in Table 28.3. These lighting levels can be measured quickly and simply by using easily purchased light meters.

Light sources should be chosen carefully. There are various types available, for example, incandescent, fluorescent, LED and so on. The range of sources available all have different characteristics regarding cost, power consumption, colour and amount of light produced.

The amount of light is only one factor. Other considerations should be the distribution of the light, for example, daylight coming in from side windows may only one illuminate side of the factory while the other is in deep shadow; supplementary artificial lighting will be required here. Glare is another problem. Light reflected from shiny work surfaces or instrument glasses can cause strain and fatigue. This can be avoided by proper design of surfaces and careful positioning of light sources. For instance the need to choose light sources carefully to avoid 'veiling reflections' on computer screens is of particular importance within the office environment.

Table 28.3 Approximate luminance requirements for interior activities.

Lux	Typical activities or locations
100–150	Occasionally visited interiors, corridors (similar to very dark day)
200–250	Simple office work, lecture rooms and non-detailed visual tasks
300–500	Libraries, normal office work, packing products, foundry work
500–750	Spray painting, engine assembly, moderately difficult visual tasks
750–1000	Craft work, colour judgement, food inspection (overcast day)
1000–1500	Electronic product assembly, detailed press tool making
1500–2000	Fine detail work, woven fabric and small assembly inspection
2000–20 000	Very detailed low contrast lengthy work (full daylight is 10 000 lux)

28.4.2.2 Noise

Noise is created by vibrations and movements that cause a rapid rise and fall in the pressure of the air surrounding the vibrating or moving source. These pressure changes are propagated through the air in the form of 'sound waves', which we hear when they impinge on our ear drum. These sounds can be pleasant, as in soft music; they can convey information, as in speech; they can be annoying, as produced by a motorbike with a poor silencer or they can be harmful, such as those experienced at close proximity to a jet engine. In the working environment noise is often described as 'unwanted sound' and it constitutes one of the most widespread and frequently encountered industrial hazards. The effect of noise can be both psychological and physiological. It can lead to a decrease in working efficiency and in some cases can present a safety risk. The most significant danger from noise is its ability to damage one of our most important senses — the sense of hearing.

We recognise noise as being composed of two elements, these are the frequency and the amplitude of the sound wave. The frequency is measured in Hertz (Hz), which is cycles per second. Low frequency sound produces low notes and high frequency produces high notes. Humans can usually hear sound between 20 and 15 000 Hz, although this range decreases with age. The amplitude of the sound is measured in Decibels (dB) and this is an indication of intensity or volume. The higher the decibel number the louder the sound. A logarithmic scale is used for measuring Decibels so that a 10 times increase in intensity is measured by 10 dB. This means that if a whisper at the threshold of hearing is 0 dB and the noise of a large machining centre is 90 dB, then the machining centre is 1000 million times as intense as the whisper.

While the ear can hear from 20 to 15 000 Hz, it is not equally sensitive to all frequencies, for example, 65 dB at 100 Hz does not seem as loud as 65 dB at 1000 Hz. Overall sound pressure level in dB does not therefore provide a good measure of 'loudness'. In order to modify objective measurements to correspond to the response of the human ear, weighting networks are used that discriminate against frequencies at which the ear is less responsive. The most commonly used network is the A-weighting and sound pressure levels measured on this basis are denoted dB(A). All international criteria related to industrial noise exposure are based on measurements of dB(A). It is thought that industrial noise first causes hearing loss to occur in the 4000 Hz region, with most annoying

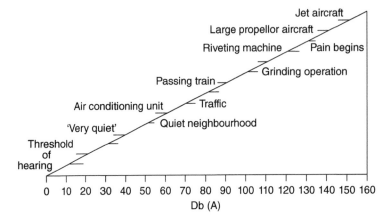

Figure 28.10 Some Db(A) values of typical sounds.

noise occurring between 1000 and 4000 Hz. Figure 28.10 shows values for typical sounds in dB(A). Equipment is available to measure noise levels in dB as well as dB(A) and to isolate and measure individual frequency components.

The risk of hearing loss due to exposure to noise depends on both the level of noise and the duration of exposure. If the noise is steady, then a direct A-weighted sound pressure level provides an adequate basis for assuming exposure. Where the noise fluctuates, as occurs in most industrial situations, the concept of the 'Equivalent Continuous Sound Level' or 'Leq' is used. This is the notional continuous level that, if experienced over the exposure period, would cause the same A-weighted energy dose to be received by the ear as that due to the actual sound. Values of Leq must be qualified by stating the exposure period to which, the level relates. For example, in the UK The Control of Noise at Work Regulations 2005 uses the equivalent continuous sound level over a standard eight hour working shift as the basis for specifying limits of exposure. In the regulations, this eight hour Leq is termed the 'Daily Personal Noise Exposure', $L_{EP,d}$. It states that the lower action level will be a daily or weekly exposure of 80 Db(A), the upper action level will be 85Db(A) and the exposure limit level will be 87Db(A).

The regulations clearly state that the employer in all cases has a statutory duty to reduce the risk of hearing damage to employees to 'the lowest level normally practicable'.

For exposures at or above the first action levels a *noise assessment* must be carried out by a competent person to identify which employees are so exposed. Where employees are likely to be exposed between the first and second action level the employer must provide hearing protection if requested by the employee. Where exposure is at or above the second action level, exposure must be reduced to the lowest level reasonably practicable other than by the provision of personal ear protectors. If exposure is still at or above the second level then the employer must provide suitable personal ear protectors and ensure that they are worn. Such areas of high exposure must be clearly identified as Ear Protection Zones.

The problem of minimising industrial noise should, however, begin at the equipment design stage. Machines and process equipment should be designed for quiet running with minimum vibration. The next level is to ensure that machinery is properly maintained, for example, bearings lubricated frequently and replaced promptly when necessary. After this comes containment of the noise within the machine's immediate vicinity. This can be carried out by using acoustic mufflers; these are soundproofed boxes that fit around the noise source. Finally, if the noise levels are still too high, personal protection will need to be worn by the workers in the local area. This takes the form of ear plugs or ear muffs. One reason that these are used only as a last resort is that they remove the ability to hear normal noises in the work area. This can be a safety hazard; for example, a warning shout from a colleague or an approaching vehicle may not be heard. Special frequency selective hearing defenders are available to reduce this problem, but they are more expensive.

28.4.2.3 Heating and Ventilation

As mentioned earlier in this chapter, harmful atmospheric contaminants, such as paint particles, dust and gases, must be extracted from the work area or protective equipment supplied to the workers. In addition, heating and ventilation should be maintained in the workplace at such a level as to ensure physical comfort for the workers.

The Factories Act requires a 'reasonable' temperature to be maintained in any working space. Where work is sedentary and there is little physical effort 'reasonable' is often interpreted as a temperature of at least 15.5°C after the first hour of occupation. The definition and assessment of 'comfort' is complex, however, and depends on variables that are both personal, such as activity and clothing and environmental. For example, if some workers are physically active in operating machines or lifting material in the same area as others are sitting at a bench carrying out assembly work, then what is too warm for the first group may be comfortable for the second.

'Thermal comfort' therefore relates to the ease with which the body's internal energy production can be balanced by its energy loss to its surroundings. This energy loss will depend on heat transfer by evaporation of moisture from the skin to the air, and by radiation and convection of heat from the body to the surrounding environment. The required environmental conditions for comfort under different work activities have been studied by many researchers. One researcher, P.O. Fanger, and his co-workers identified the following environmental variables that will influence comfort perceptions:

1) *Air temperature.* This affects the amount of heat loss by convection from the body and is a very important factor for thermal comfort assessment. A wide variety of thermometric devices are available to measure air temperature, but the mercury in glass thermometer is probably still the most convenient instrument. In cases where there is likely to be a significant difference between the air and mean radiant temperatures, a radiation shield, which may simply be a piece of aluminium foil on a frame, can be placed around the bulb of the instrument.

2) *Air speed.* Air speed affects convection as well as moisture evaporation from the body surface. The average air speed may be measured by a variety of instruments, the most commonly used at present being the hot wire anemometer. An air flow of around $0.15 \, \text{m s}^{-1}$ is suitable for most situations provided the air and mean radiant' temperatures are acceptable. Normally, at about $0.5 \, \text{m s}^{-1}$ people will complain of draughts, but at less than $0.1 \, \text{m s}^{-1}$ of staleness.

3) *Mean radiant temperature.* This controls the loss of energy from the body by radiation. The traditional instrument for measurement of mean radiant temperature is the Globe thermometer, which consists of a 150 mm diameter hollow blackened

sphere, normally of copper. The temperature sensor is usually a mercury in glass thermometer with its bulb located at the centre of the sphere. To assess the mean radiant temperature simultaneous readings of the Globe and air temperature and the air speed are required. An alternative to the Globe instrument is a blackened sphere of 100 mm diameter, which gives a reading of 'dry resultant' temperature.

4) *Relative humidity.* This affects the degree of evaporation from the surface and is therefore of much greater importance when heavy work is being carried out and the individual is sweating. It is normally uniform throughout a space, therefore, readings are required only at one position. The normal method of measurement involves using wet and dry bulb thermometers. Relative humidity is expressed as a percentage, with values between 35% and 70% considered acceptable for most situations. Below 35% static electricity build up can cause difficulties and above 70% building fabric condensation may occur.

Thermal indices are used to express thermal comfort in terms of a single number, which is often an index temperature. Examples are air temperature (a poor thermal index when used alone), Globe temperature and dry resultant temperature. Most indices of this kind are convenient to use when assessing the suitability of environments in terms of thermal comfort. Dry resultant temperature is the index that is recommended for use in the UK. This index does not take into account relative humidity, which might also have to be measured depending on the work situation. For comfortable conditions to exist, it is recommended that the dry resultant temperature should lie between 19 and 23°C.

28.5 Conclusion

This chapter has indicated some of the human factors that must be considered by the engineer or manager in a manufacturing organisation. One whole area of concern that is not investigated here is that of 'industrial relations'. Industrial relations is the study of social relationships within an industrial context and it involves 'workers', that is, non-management personnel, management and the government. The interrelationship of these three groups can be very complex and all are involved in creating the terms and conditions within which work takes place. The character of industrial relations differs widely not only between countries but also between industries and individual companies. For example, the management style in a traditional heavy engineering company is often quite different from that in a new electronics company; this will produce different worker – management relationships and hence different types of industrial relations problems. In conclusion, it can be said that while broadly following the scale of needs mentioned at the beginning of the chapter, the aspirations of individuals within an organisation will never be exactly the same. This often leads to disagreement and conflict; it is the job of a good manufacturing manager to ensure that this conflict is constructive rather than destructive. This approach can be facilitated by also ensuring that all the human factors mentioned in the chapter have been carefully considered.

Review Questions

1   Do people really need to work? Give a reasoned basis for your answer.

2   Give examples of the type of work situation that should be avoided to ensure that a factory worker is psychologically satisfied.

3   How might job dissatisfaction be manifested in an employee's behaviour and performance?

4   What is the purpose of legislation such as the 'Health and Safety at Work Act'?

5   List typical factors that should be considered to ensure the health of factory workers.

6   'All accidents are caused by someone!' Do you agree with this statement? Why?

7   Discuss where accident costs are incurred.

8   Within the factory situation, what are the three road areas that can be considered for accident prevention?

9   List three types of guarding that can be employed to improve safety when machinery is being used.

10   Who has the responsibility for safety within a factory? Discuss the functions of the personnel you describe.

11   Describe the work of an ergonomist.

12   Discuss how anthropometrics can be used to assist in the design of healthy, safe and easy to use equipment.

13   Describe the three main types of display used in industrial equipment and explain the type of application to which each is suited.

14   What is the 'control convention' in integrated display and control elements?

15   Name three aspects of lighting that should be considered to ensure worker efficiency in the factory or office environment.

16   Why is it extremely important to give attention to the noise levels in a work area?

17   Describe the two elements of which noise is composed and discuss their relevance to the employer and the employee.

18 Discuss what is meant by the 'equivalent continuous sound level' (Leq) experienced by workers.

19 At what four levels should machinery created noise problems be tackled in an industrial situation?

20 Discuss fully the factors that contribute to the thermal comfort of a worker; include in your answer comments on how they are measured.

Part VIII

Conclusion

29

Introduction

This book has provided an overview of the essentials of manufacturing industry. However, although the basic concepts have remained fairly constant; for example, forging metals for directional strength properties or minimising the time from product design to market; it is worth concluding with a comment on a few of the relatively recent concepts and technologies that are having a strong impact on the industry. It should be understood that advances in technology will continue so that what is regarded as the present will certainly be regarded as historical very soon, the comments relating to the future are entirely speculative.

Two closely related terms are widely used today to convey the current state of manufacturing, these are Industry 4.0 and The Fourth Industrial Revolution. In one respect it can be said that the only Industrial Revolution was the one that started in the eighteenth century, see Chapter 2, and everything since then has been evolutionary rather than revolutionary. However, the terms serve here as a framework for describing some relatively recent developments that are significantly influencing manufacturing industry. Originally used in Germany near the beginning of the twenty-first century the term Industry 4.0 relates to the integration of cyber-physical systems (CPSs), the Internet of Things (IoT) and the Internet of Services in order to create Smart Factories. The term fourth Industrial Revolution is in more general usage and is similar to Industry 4.0 but tends to have broader scope in that it also includes the broader social and technological implications such as employment and autonomous vehicles. What they have in common is recognition of the fact that many technologies are now providing new opportunities for wealth creation and their potential is boosted by the possibilities of integration provided by the Internet. Some of these technologies particularly related to manufacturing are now briefly considered.

29.1 Additive Manufacturing

Already covered in Chapter 13, this method of manufacturing is constantly improving in range of materials, precision, speed and product size. It is currently suitable for single or very low volume customised product manufacture but is unsuitable for large batch and mass production. It has also found many applications outside those of traditional manufacturing and prototype creation; for example, dentistry, surgery, sculpture, jewellery, clothing and footwear. Small relatively low cost domestic and hobby systems have become popular and can be found in homes, schools and other teaching

Essential Manufacturing, First Edition. Gordon Mair.
© 2019 John Wiley & Sons Ltd. Published 2019 by John Wiley & Sons Ltd.

establishments. In the future, the ability to combine increasingly different materials will allow more sophisticated products to be created using this technology. For example, combining electronic, optical and mechanical elements could allow customised and personalised mechatronic devices to be manufactured. Also, by manipulating matter at the nanometre and atomic level completely new materials and products could be created with applications in medicine, engineering, food production and electronics.

29.2   Augmented Reality (AR) and Virtual Reality (VR)

Mentioned briefly in Chapter 4, augmented and virtual reality (VR) are part of the mixed reality continuum. At one end of the continuum we have the real world environment and at the other end we have VR – a completely computer generated environment. By using an appropriate display system, such as a tablet computer or glasses equipped with a small screen or projection system, it is possible to superimpose computer generated information or images onto the real world, this is augmented reality (AR). It is also possible to superimpose real world images into the virtual environment and this is called *augmented virtuality*.

AR, although not widespread in manufacturing, is used in a number of situations. For example, it can be used in assembly training where the correct assembly sequence and torque information can be provided to an operator inserting bolts into an aircraft frame. This is done by presenting the information, which may be text or video, into the field of view of the operator via the tablet or AR glasses. The images seen by the operator can be coordinated with the real world through registration and distance calibration of the area of concern with the augmented images. AR can also be used in product design with a number of product designers able to look at a virtual model of the product in the real world space. Through the use of on board position and orientation sensors on the tablet or glasses each designer can see the virtual model from their own perspective. By the use of gloves containing similar sensors it is also possible to manipulate the virtual model in real time.

VR is usually considered as an immersive experience and can be used by product designers to examine a proposed design and its position within an artificial environment. The user of such a system will use a head mounted display that has two small video displays one of which will show the image as seen by the left eye and the other showing the image as seen by the right eye thus providing a stereoscopic image. Again the system user can manipulate objects in the virtual world through the use of gloves containing location and orientation sensors. Although not yet common in design applications, the gloves can also include small haptic actuators to provide a sense of touch for the object manipulation. The head mounted display may also have headphones to enhance the immersive experience.

In order to experience the sensation of being inside a design, for example, the interior layout of a submarine, a CAVE (Cave Automated Virtual Environment) can be used. This is a cube like room located within a larger workspace that is comprised of three or four walls, a ceiling and floor. Stereoscopic images are displayed on these surfaces, either through back projection or on flat panel displays in such a manner that the user is completely surrounded by the images in what appears as a seamless immersive environment. The user will normally wear stereo glasses that may be polarising, shutter or anaglyph depending on the display type. These systems are expensive relative to head mounted display enabled VR and are therefore only found in some Universities, research institutions and very large commercial companies with design departments.

In the future volumetric imaging systems will be of great use to product designers who will be able to view and manipulate a virtual three dimensional model without the aid of any special eyewear or hand held screen. A volumetric image would exist in real world space, it would appear solid and edges and surfaces that would be occluded in a physical model would also be occluded in the volumetric image. A number of technologies today approach this. For example, one type has a helical screen spinning at high speed in a vacuum and by using carefully controlled laser beams impinging on the surface an apparently solid image is created through persistence of vision. Another type uses an array of fast spinning light emitting diodes (LEDs) to achieve a similar effect. Other technologies are available but none can yet produce a full colour volumetric image indistinguishable from a physical model. The main problem with achieving the image is that of physics – how to create the three dimensional equivalent of pixels, that is, voxels, in a seemingly empty space.

29.3 Immersive Telepresence

Today the term 'telepresence' is commonly used to describe an advanced form of what was called 'teleconferencing' using high bandwidth telecommunication links. As well as being useful for general business meetings, telepresence is also ideal for collaborative product design. Large screen displays, which may be flat panel or back projected, show conference participants from different geographical locations in life size images. These images are arranged to give the impression of all participants being around a similar conference table. Eye contact is not normally possible as the cameras at each location are not able to be located at the same position on the screen of the eyes of the participants. Methods of attempting to overcome this can include the use of half silvered mirrors or mounting cameras at eye level behind pinholes on the displays but these are not very successful.

Immersive telepresence however has the connotation of allowing system users to feel completely immersed in a remote real world environment. This would therefore include displays that provide stereoscopic vision, binaural hearing and haptic, cutaneous and olfactory sensation. The important eye contact for personal communication as in a conference would also be possible. Also by using remote mechatronic systems with actuators responding to the system user's movements, and with appropriate sensors such as stereoscopic cameras and binaural microphones, then remote immersive tele-operated working could take place. No systems are available yet to provide all of these features in an integrated manner but by combining potential future developments in AR, VR, volumetric imaging, sensors and robotics fully immersive telepresence is a distinct possibility.

29.4 Communications Technologies and the IoT

Wireless, fibre-optic cable, satellites, standardisation of communication protocols and computer technology have all led to the ability to transmit information reliably and almost instantaneously across any distance. This is manifested most effectively in the Internet where all types of networks, such as factory wide networks, telephone networks and home networks can all be interconnected. The World Wide Web operating on the Internet provides the opportunity for everyone connected to the Internet to share personal or commercial information. The fact that this communication infrastructure is now ubiquitous throughout the world's industrialised countries has opened up the opportunity for the interconnectedness of everyone and everything with access to the Internet. Thus, in manufacturing industry, equipment within a factory will not only be part of the local factory network but will also be capable of being integrated into the global network. Individual machines may be equipped with automatic inspection systems linked to statistical quality control algorithms that can provide real time information on machine and quality performance to a parent company that may be in a different country. This of course assumes the machines have been equipped with the necessary sensors and communication interfaces.

Interconnectedness through the Internet allows the following scenarios to be possible where the manufacturer is very closely integrated into their supply chain both upstream and downstream. Consider a manufacturer of a relatively low volume consumer product that is sold in a large store. Downstream from the manufacturer the sale of the product is monitored by the store and this is registered in the store's computer system where it will be compared with the current stock levels. Using an algorithm that includes an action point when a minimum stock level is reached the system will send a message to the supplier's warehouse to provide more stock. The warehouse system with knowledge of the lead time required for replenishing its stock will then send a message to the manufacturer's factory system to produce more of the product in a specific number based on the previous and anticipated future sales. In a smart factory this can all occur autonomously with the information being transferred down through various levels until it reaches individual machines such as industrial robots and automated guided vehicles on the factory shop floor. Using this information the factory can then assess its own stock levels of raw materials and automatically request additional materials from upstream suppliers in the supply chain. In other situations the downstream supply chain can extend to individual consumers with smart refrigerators that note when the last of an item is being removed. The fridge may then send a message through the home router asking if the consumer wishes to replenish the item. If the answer is yes then it can place an order upstream with the supermarket – and so the process will then continue down to the food manufacturer. These are examples of the 'Internet of Things' and how it contributes to efficiency in the manufacturing supply chain.

29.5   Cloud Computing

Cloud Computing allows companies of any type to access information and communication technology (ICT) software, hardware and services over the Internet. This facility is a 'cloud' and is provided by a third party thus allowing companies to focus on their core businesses while the necessary ICT is provided for them. The company pays the cloud provider for these services and will access them through their own computer systems. The cloud can store and process data for companies and can gather data from many sources on a global scale. It allows companies to keep their IT infrastructure to a minimum thus streamlining their operation. As the cloud services are shared by many users security issues have to be carefully addressed to ensure commercial data is kept confidential and is not corrupted intentionally or unintentionally.

29.6   Big Data Analytics

Big data analytics refers to the activity of gathering very large, varied and timely data from multiple sources relevant to a company's areas of interest. These data are processed and analysed to enable intelligent decision making leading to increases in profits and more satisfied customers. Closely related to this is the term 'predictive analytics'. This relates to the statistical analysis of historical and new data in order to identify trends in, for example, production quality, customer requirements and costs of raw materials.

In manufacturing Big Data can be used for quality management of the production process by identifying trends in the quality of the components and products being produced. This can be done in real time based on information coming from sensors in the machines on the shop floor. This in itself is not 'big data' as automated statistical quality control techniques have being doing this for some time. However, by combining this data with other data from the company's ERP (Enterprise Resource Planning) system (see Chapter 23) correlations can be found between factors such as personnel, time of day or machine maintenance records. Just as monitoring of the data from the machines on the shop floor can be used to facilitate preventive maintenance programmes the principle can also be applied to non-factory equipment such as jet engines, locomotives and emergency portable power generating equipment. The use of wireless communication means that data from equipment anywhere in the world can be monitored and analysed.

Big data can also be used for simulating possible new manufacturing layouts and processes based on anticipated customer demand and production capacity. Production schedules can be optimised based on a combination of customer requirements, supplier capabilities, machine and labour availability and budget constraints. It can be used for analysing factory performance across a wide range of measurable parameters and correlating these to look for cause and effect relationships between them. For example, a pharmaceutical manufacturing company has used Big Data to analyse process interdependencies in order to increase successful yield values for vaccine production. A company that produced products to order used Big Data to analyse customer behaviour in order to significantly reduce manufacturing lead times. And a microprocessor manufacturing company used Big Data predictive analytics to reduce the number of tests carried out during quality assurance checks.

29.7   Conclusion

Throughout this book, I have attempted to provide the essential elements of manufacturing industry. There is much more to manufacturing both in depth and breadth and I hope that you will be able to pursue many of the facets mentioned here in more detail. While civilisation exists manufacturing industry will exist no matter the condition of the global economy. As labour rates, transport costs, material availability and political environments change throughout the world so the centres of manufacturing will move and fluctuate. New discoveries in physics and chemistry will produce new technologies for products and manufacturing processes. Consumers will demand more customised products to suit individual specifications. Automation and artificial intelligence will reduce the requirements for certain types of worker but will create demand for other skills and expertise. The need for products for both the developed and developing countries will ensure that manufacturing industry will continue as a major wealth creator and therefore as source of finance for improving the quality of life for everyone.

Appendix A

The SI prefixes and multiplication factors.

Prefix	Symbol	Multiplication factor	Common name (US)
yotta	Y	$1\,000\,000\,000\,000\,000\,000\,000\,000 = 10^{24}$	one septillion
zetta	Z	$1\,000\,000\,000\,000\,000\,000\,000 = 10^{21}$	one sextillion
exa	E	$1\,000\,000\,000\,000\,000\,000 = 10^{18}$	one quintillion
peta	P	$1\,000\,000\,000\,000\,000 = 10^{15}$	one quadrillion
tera	T	$1\,000\,000\,000\,000 = 10^{12}$	one trillion
giga	G	$1\,000\,000\,000 = 10^{9}$	one billion
mega	M	$1\,000\,000 = 10^{6}$	one million
kilo	k	$1000 = 10^{3}$	one thousand
hecto	h	$100 = 10^{2}$	one hundred
deca	da	$10 = 10^{1}$	one ten
		$10^{0} = 1$	one
deci	d	$0.1 = 10^{-1}$	one tenth
centi	c	$0.01 = 10^{-2}$	one hundredth
milli	m	$0.001 = 10^{-3}$	one thousandth
micro	µ	$0.000\,001 = 10^{-6}$	one millionth
nano	n	$0.000\,000\,001 = 10^{-9}$	one billionth
pico	p	$0.000\,000\,000\,001 = 10^{-12}$	one trillionth
femto	f	$0.000\,000\,000\,000001 = 10^{-15}$	one quadrillionth
atto	a	$0.000\,000\,000\,000\,000\,001 = 10^{-18}$	one quintillionth
zepto	z	$0.000\,000\,000\,000\,000\,000\,001 = 10^{-21}$	one sextillionth
yocto	y	$0.000\,000\,000\,000\,000\,000\,000\,001 = 10^{-24}$	one septillionth

Essential Manufacturing, First Edition. Gordon Mair.
© 2019 John Wiley & Sons Ltd. Published 2019 by John Wiley & Sons Ltd.

Index

Where a topic is referenced by more than one page then, if appropriate, bold type indicates the page(s) of greater significance. Where 'ff' appears after a page number it means there is a page or a number of pages following where the topic also appears.

Essential Manufacturing, First Edition. Gordon Mair.
© 2019 John Wiley & Sons Ltd. Published 2019 by John Wiley & Sons Ltd.